Laboratory Manual

Concepts in Biology

THIRTEENTH EDITION

Eldon D. Enger
Frederick C. Ross

Delta College

LABEXAM #1 60/60 (GRAPHS #3)

LABEXAM #2 40/40

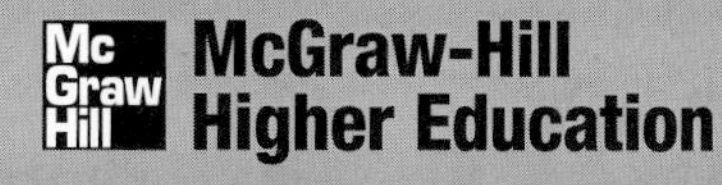

Boston Burr Ridge, IL Dubuque, IA New York San Francisco St. Louis
Bangkok Bogotá Caracas Kuala Lumpur Lisbon London Madrid Mexico City
Milan Montreal New Delhi Santiago Seoul Singapore Sydney Taipei Toronto

CONCEPTS IN BIOLOGY LABORATORY MANUAL, THIRTEENTH EDITION

Published by McGraw-Hill, a business unit of The McGraw-Hill Companies, Inc., 1221 Avenue of the Americas, New York, NY 10020.

This book is printed on recycled, acid-free paper containing 10% postconsumer waste.

2 3 4 5 6 7 8 9 0 QPD/QPD 0 9 8

ISBN 978–0–07–337792–6
MHID 0–07–337792–9

Publisher: *Janice Roerig-Blong*
Executive Editor: *Michael S. Hackett*
Director of Development: *Kristine Tibbetts*
Developmental Editor: *Debra A. Henricks*
Marketing Manager: *Tamara Maury*
Project Manager: *Joyce Watters*
Senior Production Supervisor: *Laura Fuller*
Associate Design Coordinator: *Brenda A. Rolwes*
Cover Designer: *Studio Montage, St. Louis, Missouri*
(USE) Cover Image: © *Brand X Pictures/PunchStock RF*
Senior Photo Research Coordinator: *Lori Hancock*
Compositor: *S4Carlisle*
Typeface: *10/13 Times*
Printer: *Quebecor World Dubuque, IA*

Some of the laboratory experiments included in this text may be hazardous if materials are handled improperly or if procedures are conducted incorrectly. Safety precautions are necessary when you are working with chemicals, glass test tubes, hot water baths, sharp instruments, and the like, or for any procedures that generally require caution. Your school may have set regulations regarding safety procedures that your instructor will explain to you. Should you have any problems with materials or procedures, please ask your instructor for help.

www.mhhe.com

Contents

Preface

To the Student

The laboratory gives you the opportunity to "get your hands on biology." You go beyond reading and studying and actually enter into the process of doing science. The benefit you derive from the laboratory is directly dependent on two things: first, the care with which you perform the experiments and record your observations and, second, your awareness of the relationship between your observations and the general principles under study. Because of time limitations, we cannot investigate in the slow, methodical (often tedious) way scientists do. However, you will have the opportunity to measure, experiment, observe, and discover for yourself. In general, each laboratory exercise consists of the following:

The **Safety Box** alerts you to any hazardous materials.

The **Objectives** explicitly state what you can expect to accomplish. It would be to your advantage to read through the list before coming to class. These objectives should be given special attention during the laboratory exercise. Upon conscientious completion of the exercise, you will be able to meet all the objectives for that activity. Before leaving class, check the objectives once again to see that you can meet them.

The **Introduction** consists of background information to orient you to the biological concepts being explored.

The **Preview** is a general overview of what you will be doing.

The **Procedure** provides step-by-step instructions to guide you through the laboratory exercise. Be sure to refer to this section often as you set up your experiment and collect data.

End-of-Exercise Questions help you focus on the important observations and concepts relevant to the day's investigation. These questions are a good gauge of how much was learned and understood during the exercise. The objectives and end-of-exercise questions are excellent study guides for lab quizzes.

To the Instructor

The exercises constituting this laboratory manual are intended to supplement the text *Concepts in Biology,* but they can be used with any introductory-level biology text.

A *Laboratory Resource Guide* is available on the Instructor's Edition of the text website at www.mhhe.com/enger13e. This guide aids you in preparing the laboratory experience for the students and lists needed equipment, places where this equipment can be acquired, and several other hints that may save you time in your laboratory preparation. It also contains answers to the end-of-exercise questions.

Laboratory exercises begin with a list of safety issues that your students need to be aware of. These items identify caustic chemicals, equipment that needs special care or handling, and health or safety hazards a careless student may encounter.

We suggest that students use the laboratory manual as a workbook and record data and answers to questions in the spaces provided. It seems particularly inappropriate to use a previously owned manual, because students constantly compare their answers with the answers provided by a previous user of the book.

We acknowledge the valuable help we received from the comments and constructive criticism of our colleagues and former students.

We also thank the following reviewers:

Barbara N. Beck
Rochester Community and Technical College

Laura Brand
Cossatot Community College

Geralyn M. Caplan
Owensboro Community & Technical College

Michael Crandell
Carl Sandburg College

Jason Fitzgerald
Southeastern Illinois College

Scott Johnson
Central Carolina Technical College

Carla Murray
Carl Sandburg College

Steve C. Nunez
Sauk Valley Community College

Micah W. Perkins
Owensboro Community and Technical College

Michael D. Quillen
Maysville Community and Technical College

Susan T. Rouse
Southern Wesleyan University

Jennifer Rubin
Rochester Community and Technical College

James H. Stegge
Rochester Community and Technical College

Peter J. Wilkin
Purdue University North Central

LABORATORY

Metric Measurement and the Scientific Method

1

+ Safety Box

- Exercise care using glassware. If any glass items are broken, dispose of them in the appropriate container.
- The thermometers may contain mercury. Mercury is toxic. If a thermometer is broken, notify your instructor, so that the mercury can be cleaned up and disposed of properly.

Objectives

Be able to do the following:

1. Define these terms in writing. Although some of these terms may not be used in today's lab exercise, it is still important for you to understand them.

   International System of Measurement (metric system)
   Celsius
   meter
   centimeter
   millimeter
   kilometer
   precision
   scientific method
   accuracy
   density
   theory
   liter
   milliliter
   gram
   kilogram
   milligram
   observation
   hypothesis
   scientific law

2. Measure length, volume, and weight using the metric system.
3. Measure temperature in degrees Celsius.
4. Formulate and test a viable hypothesis.

Introduction: Measurement in Science

During this exercise, you will make careful measurements using the **International System of Measurement** (SI, Le Système International d'Unités), commonly called the **metric system.** SI has been adopted as the official system of measurement by most countries and is used by scientists worldwide. In addition, the United States is increasing its use of metric units. You encounter SI units every day. For example, you may purchase soda in 1- or 2-liter bottles; you may own 7-millimeter pearls or a 35-millimeter camera; your automobile may have a 2.2-liter engine; or you may take a 325-milligram aspirin tablet for a headache.

The units and divisions of SI are given in the following tables. The standard units of measurement are as follows: Length is measured in **meters,** mass in **grams,*** volume in **liters,** and temperature in degrees **Celsius** (table 1.1). The basic units (meter, liter, gram, degree Celsius) are the starting points of the metric system. From these points, there are two directions we can go: (1) smaller than the unit indicated by using a subunit prefix and (2) larger than the unit indicated by using a superunit prefix. These prefixes and values are listed in table 1.2. If this is the first time you have studied the metric system, it may seem complicated, but SI is actually much easier to learn than the customary U.S. system because it is entirely based on the decimal plan, that is, using divisions of 10. For example,

1 decimeter = 10 centimeters = 100 millimeters (figure 1.1)

Table 1.1 Metric Units of Measure and Their Equivalents

Quantity	Metric Unit	Symbol	Equivalent
Mass	Gram	g	1 g = 0.0022 pound
Length	Meter	m	1 m = 3.280 feet
Volume	Liter	L	1 L = 1.057 quarts
Temperature	Degree Celsius	°C	1°C = 1.80°F

Table 1.2 Divisions of Metric Units

Prefix	Symbol	Value		Example
Superunits—Make Unit Larger				
Mega-	M	One million	1,000,000 = 10^6	1 megameter (Mm) = 10^6 m
Kilo-	k	One thousand	1,000 = 10^3	1 kilogram (kg) = 10^3 g
Hecto-	h	One hundred	100 = 10^2	1 hectogram (hg) = 100 g
Deka-	da	Ten	10 = 10^1	1 dekaliter (daL) = 10 L
Subunits—Make Unit Smaller				
Deci-	d	One-tenth	0.1 = 10^{-1}	1 deciliter (dL) = 0.1 L
Centi-	c	One-hundredth	0.01 = 10^{-2}	1 centimeter (cm) = 0.01 m
Milli-	m	One-thousandth	0.001 = 10^{-3}	1 milligram (mg) = 0.001 g
Micro-	µ	One-millionth	0.000001 = 10^{-6}	1 micrometer (µm) = 10^{-6} m

Compare this to the traditional U.S. weights and measures, where nearly every conversion requires multiplication or division by a different quantity:

1 yard = 3 feet = 36 inches
1 gallon = 4 quarts = 8 pints = 16 cups = 128 fluid ounces

Temperature measure is simplified with the Celsius scale because the freezing point of water registers at 0°C and the boiling point of water registers at 100°C. Compare this with 32°F and 212°F (figure 1.2).

*Technically the kilogram is the fundamental unit for mass, but we will use the gram as the fundamental unit for teaching purposes.

Figure 1.1 Comparison of metric units of length to inches.

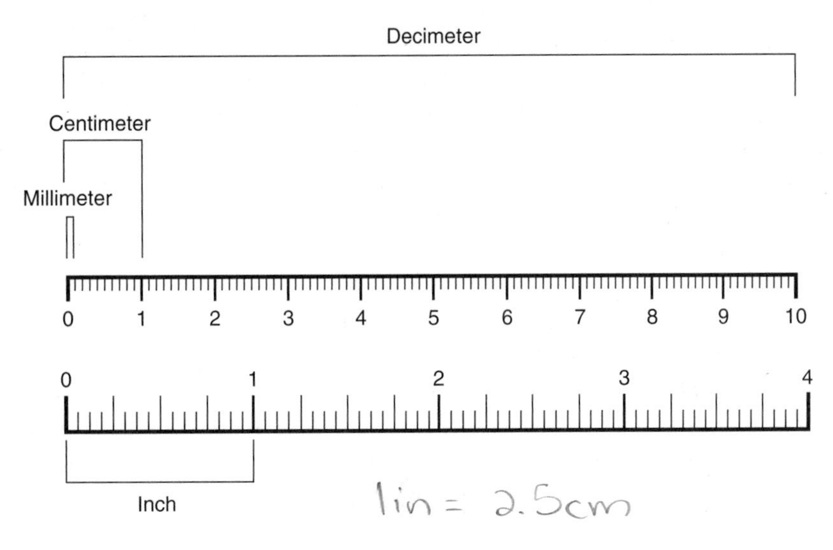

Figure 1.2 Celsius and Fahrenheit temperatures can be converted quickly by using a ruler to read from one scale across to the other. Conversion is also achieved by using the following simple formulas:

$$°F = (°C \times 1.8) + 32$$

$$°C = \frac{(°F - 32)}{1.8}$$

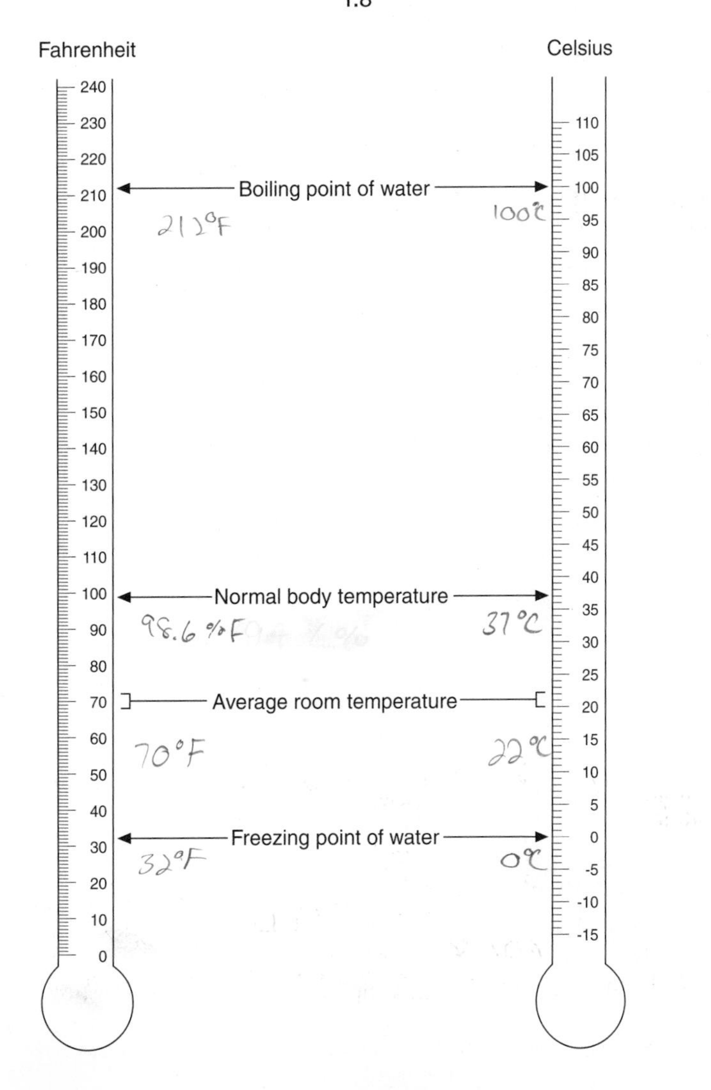

Preview

During this lab exercise, you will

1. become familiar with the metric system as it is used to measure length, weight, temperature, and volume.
2. make observations.
3. construct hypotheses.
4. make measurements.
5. test hypotheses.
6. evaluate data.

Procedure

Table 1.3 shows the most commonly used metric measurements. You will use many of these units of measure during lab activities. Work with a partner to complete the chart.

Table 1.3 Commonly Used Metric Units of Measurement

Length

Meter

Centimeter = ________ m

________ = 0.001 m

Kilometer = ________ m

Weight

Gram

Milligram = ________ g

________ = 1,000 g

Volume

Liter

Milliliter = ________ L

________ = 1,000 mL

Temperature*

° Celsius

Water freezes at ________ °C and boils at ________ °C.

Body temperature is usually about ________ °C.

Room temperature is usually between ________ °C and ________ °C.

*See figure 1.2 for temperature conversions.

Measurement

Work together to measure accurately and record the following in metric units.

1. A laboratory manual is ___27 1/2___ cm × ___21 1/2___ cm.
2. The tabletop is ___.86___ m × ___1.21___ m.
3. My height is ___1.675___ m, or ___167 1/2___ cm.
4. A ballpoint pen weighs ___5.5___ g.
5. A dollar bill weighs ___1___ g, or ___1000___ mg.
6. My biology textbook weighs ___626.8___ g.
7. My weight is ___64.5___ kg, or ___64,500___ g.

1K = 2.2lbs
220lbs
÷ 2.2 lbs/Kg
100Kg

2.2)142

The Scientific Method

The process of science involves the critical evaluation of ideas and information. Scientists have a "healthy skepticism" about information and ideas. They continually test and evaluate information for validity and reliability. Validity and reliability have to do with how consistently similar results can be produced.

The way most scientists evaluate ideas and information has come to be known as the *scientific method.* The **scientific method** is a way of gaining information (facts) about the world by forming possible solutions to questions followed by rigorous testing to determine if the proposed solutions are valid. Many scientific activities begin with an observation. An **observation** is a mental record of an event gathered with the use of our senses (smell, sight, hearing, taste, touch). The use of the senses to collect information is often aided by the use of mechanical devices, such as microscopes, sensitive microphones, film, and other tools that extend the sensory ability of humans. Scientists are like detectives. They make observations, gather facts, and attempt to reach conclusions. Often, it is very important to quantify the observations in some way. Careful measurements of time, length, volume, temperature, or other variables are necessary.

A second activity that is important to the scientific method is *hypothesis formation.* A **hypothesis** is a statement that provides a possible answer to a question or an explanation for an observation that can be tested. A hypothesis is not a fact but is meant to be tested, challenged, and refined as a result of experience. A good hypothesis must have two characteristics: (1) It must be able to account for all the available data and (2) it must be able to be tested.

Once a hypothesis has been formed, it should be tested. When it is tested, it may be supported or disproved. The hypothesis may be tested by the collection of additional data or by designing and carrying out an experiment. An experiment usually involves the construction of an artificial situation in which all variables are controlled except for the one that is being investigated. The information collected from the experiment must involve accurate and precise observations and measurements. **Accuracy** means the degree to which your measurements reflect the actual value of something. **Precision** involves being able to get the same result with repeated measurement of the same thing. An electronic balance may be able to measure to within 1/1,000th of a gram. If you weigh the same object several times and get the same result, it demonstrates a high degree of precision. However, if the instrument is miscalibrated it is not accurate. When collecting data, it is important that attention is paid to both accuracy and precision.

Once a hypothesis has been tested, it is either supported by the evidence or found to be invalid. It is important to recognize that, although a single experiment may support a hypothesis, it still may not be true. Scientists insist that many sets of data be collected about a phenomenon before they are satisfied that the hypothesis has strong support. Typically, if a hypothesis is found to be invalid, a scientist thinks about ways to revise it before discarding it completely.

After the investigator feels confident about the results of an experiment, he or she is ready to formulate a conclusion. It is important to remember that conclusions are always tentative and are subject to revision if new observations or experiments demand change. Often when scientists see a fundamental pattern in nature that has broad application and that can be used to predict the behavior of similar phenomena, it is called a theory or law. A **theory** is a widely accepted, plausible generalization about fundamental concepts in science that explain why things happen. A **scientific law** is a uniform or constant fact of nature that describes what happens in nature. *Theories and laws are considered highly likely to be true but still capable of being disproved.* They are like other scientific tools. They are used as long as they successfully predict and explain every observation they are associated with.

Another characteristic of the scientific method is that the results or observations should be repeatable by the scientist and others who use similar methods. Typically, science involves the publication of the design, results, and conclusions drawn from experiments. This allows others to test the same ideas or suggest alternative hypotheses. Most scientists would be very cautious about publishing the results of a single experiment.

Measurements and Numbers

This activity involves many kinds of measurements involving the metric system. As will all measurements, it is important to be accurate and precise. Any measurement involves the recording of a number and a unit. A number by itself has little meaning. It is a quantity but we need to know what the units are. For example, when you buy fruit at a grocery store you place it in a clear plastic bag. The clerk needs to know two things—the kind of items and the number of items—in order to charge you correctly for the items you buy. Similarly, when you measure length, mass, or volume, you need to provide information about the unit of measure you are using. Therefore, whenever you take a measurement, be sure to always indicate the units involved. The quantity expressed as *12.7 centimeters* provides much more information than just the number *12.7.*

Procedure

Work in pairs or small groups to test the various hypotheses in these exercises. Your instructor will give you directions on how to pool the data from the entire class.

Exercise 1

In this exercise, you use the scientific method to clarify the relationship of various characteristics of the human body to one another. For example, you have probably *observed* a relationship between height and weight: Tall people usually weigh more than short people. But is there a specific relationship?

We will construct a hypothesis of the relationship between height and weight and test it.

Hypothesis: *Height in centimeters divided by weight in kilograms equals 4 cm/kg.*

Test the hypothesis by determining your weight in kilograms and your height in centimeters. Remember, you want your measurements to be accurate and precise. Record your data on the following chart and publish your data, so that the entire class has access to the information.

Student	Height in cm	Weight in kg	$\frac{\text{Height}}{\text{Weight}}$ = cm/kg	
			Expected	***Observed***
Christian	167 ½	64.5	4 cm/kg	2.59 cm/kg
Tyler	183	68.1	4 cm/kg	2.68 cm/kg
			4 cm/kg	cm/kg
			4 cm/kg	cm/kg

Analysis of Results

After examining the data for the entire class, answer the following questions.

1. Is height in centimeters four times weight in kilograms? ________________
2. What is the average relationship between height and weight? ________________
3. Do the data support or disprove the hypothesis? ________________

Exercise 2

You have probably *observed* that tall people have long arms, whereas shorter people have shorter arms. Let us construct a hypothesis.

Hypothesis: *The length of a person's arm in centimeters is equal to 0.4 of his or her height in centimeters.*

Since you have already measured your height in centimeters, you can determine what we expect the length of the arm to be by multiplying your height in centimeters by 0.4. Enter this number in the following table.

1. How can this hypothesis be tested?

2. Since each individual in class will be making measurements, what are some of the factors we should consider when making the measurements, so that accurate and precise data are obtained?

3. Determine a standard method for collecting data about a person's arm length. Carefully describe the method you will use to determine the actual length of a person's arm in the following space. Include in your descriptions such things as the tools to be used, how the tools will be used, the measurement units (meters, centimeters, or millimeters) that will be used, and the accuracy desired.

Collect and record your data here and publish your data, so that everyone in class has access to them.

Student	Height in Centimeters	Expected Arm Length in Centimeters (Height in cm × 0.4)	Observed Arm Length (Measured with Meterstick)	Deviation from Expected
Chris	167.5 cm	67 cm	68.5 cm	1.5 cm
Tyler	183 cm	73.2 cm	77.5 cm	4.3 cm
	cm	cm	cm	cm
	cm	cm	cm	cm

Analysis of Results

After examining the data for the entire class, answer the following questions.

1. According to class results, what is the relationship between height and arm length?

2. Do the data support or disprove the hypothesis? ________________
3. At this point in the scientific method, what would be your next step?

4. If you are not satisfied with your hypothesis, revise it and write it in the following space.

Exercise 3

You have probably observed that hand temperature differs among people under different conditions. Some people have warm hands and some people have colder hands. Construct a hypothesis concerning hand temperature.

Hypothesis: *What factors do you think determine the temperature of hands? In your own words, state your hypothesis in the following space. (There may be several different hypotheses.)*

Our hands will have a lower temp. in the room than if we were outside which would make them warmer.

Once you have formed your hypothesis, test it by holding the bulb of a thermometer in your hand for 5 minutes. Because there will be several groups testing different hypotheses, other members of the class may ask you for your hand temperature.

Analysis of Results

Construct a table and record your results in the space provided. Use a class discussion to publish your hypothesis, the method used to test the hypothesis, the results of the test of the hypothesis, and any conclusions you reached.

	Tyler	Chris
Room	30°C	32°C
Outside	31°C	35°C

1. Did you support or refute your hypothesis? Support

Exercise 4

For our final task, we will test the validity of the concept that all objects of the same material have the same density. *Our hypothesis is: All objects of the same material will have the same density regardless of their size.* You have learned that the metric unit for volume is the liter and the metric unit for mass is the gram. The **density** of a material is its mass divided by its volume (density = mass/volume).

You may not know that units of volume are related to units of length when using the metric system. One cubic centimeter (cc or cm^3) is the same as 1 milliliter. If we know how many milliliters an object takes up, we can convert milliliters directly to cm^3 and we can describe the density of a material in terms of grams/cm^3.

Your instructor will provide four different objects made from the same material. Determine the mass and volume of each of the objects. Remember, you want your measurements to be accurate and precise.

1. To determine the mass of the object, use an electronic balance.
2. To determine the volume of the object, use a graduated cylinder.
   (When you read a graduated cylinder, you must
   —have the graduated cylinder sitting on a level surface,
   —have your eye level with the surface of the liquid in the graduated cylinder, and
   —measure to the *bottom* of the concave surface of the liquid.)
   a. Place a known amount of water in the graduated cylinder and record the volume.
   b. Place the object in the graduated cylinder without allowing any water to splash out, and record the change in volume. The change is the volume of the object.
   c. Record the data in the following chart and use the information about mass and volume to plot the points for each of the 4 objects on the graph provided.

15 mL

w/mL

Object	Mass in Grams	Volume in mL (cm^3)	Density = Mass/Volume (g/cm^3)
Aluminum	18g	22mL	.81
Brass	54g	22mL	2.45
Copper	55g	23mL	2.47

Analysis of Results

If the hypothesis is supported, the density you calculated should be the same for each of the objects and you should be able to draw a straight line connecting the 4 points on the graph. However, the density you calculated is probably not exactly the same for all four objects and the points on the graph probably do not form a straight line. These kinds of results are not unusual when measuring a variety of items, since all measuring devices have limits to their accuracy. Furthermore, those making the observations may need to make estimates. This leads to some inaccuracy. If we assume that the objects are all of the same material and should have the same density, what could have caused the variation we saw? Consider the following list of possibilities and rank them in order of most important to least important (1 as most important and 5 as least important).

—The electronic balance gave inconsistent readings.
—It was difficult to get accurate readings with the graduated cylinder.
—The units for the graduated cylinder (mL) could not be subdivided as easily as the units (g) for the electronic balance.
—Air bubbles may have been trapped on the objects in the water in the graduated cylinder.
—Wet objects were weighed, so the water added weight to the object.

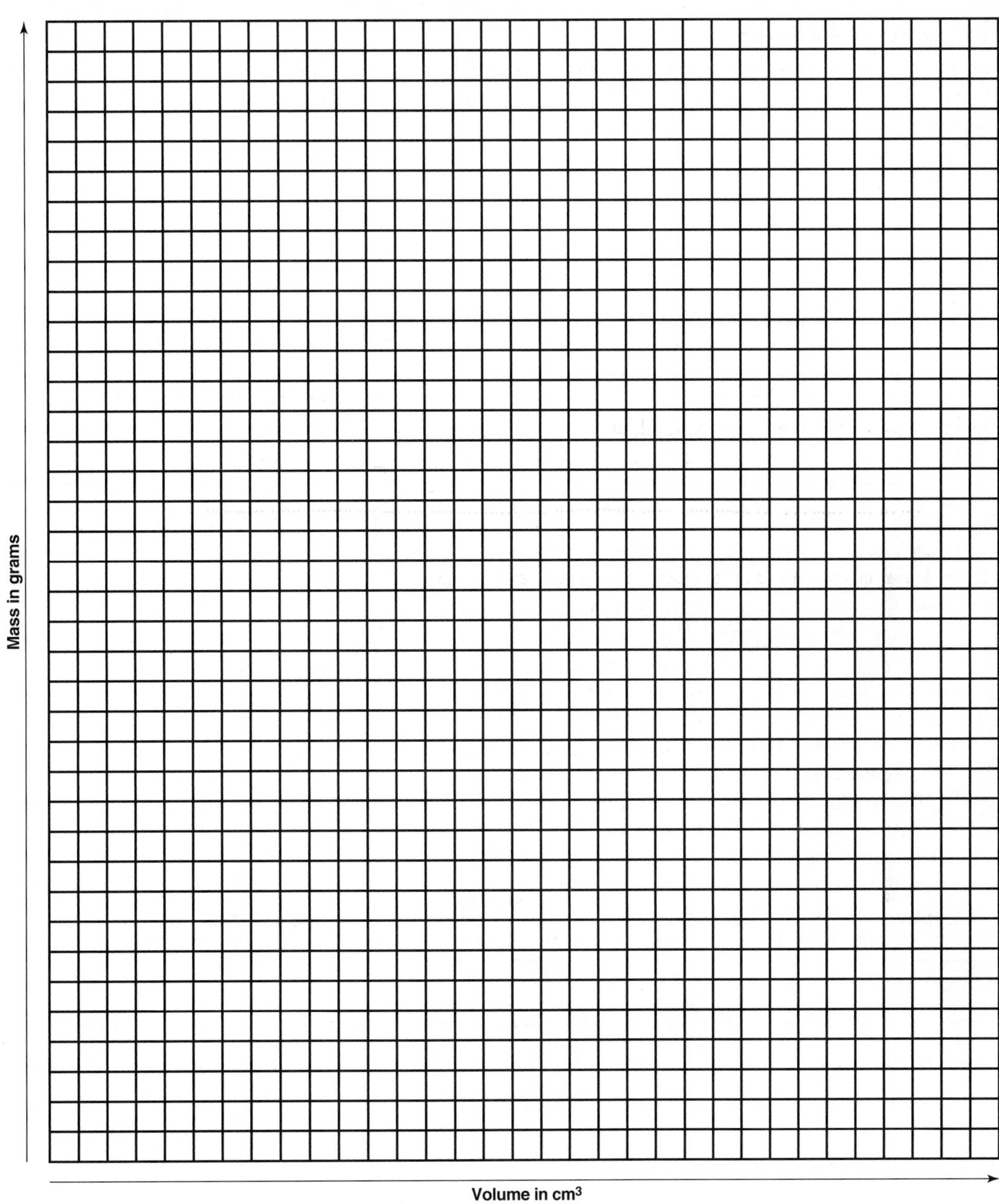
Mass in grams
Volume in cm3

1 Metric Measurement and the Scientific Method

Name ______________________________ Lab section ____________

Your instructor may collect these end-of-exercise questions. If so, please fill in your name and lab section.

End-of-Exercise Questions

1. Lead melts at 620°F. Would a furnace that heats to 200°C melt the lead? Why or why not?

2. The room down the hall has a temperature of 38°C. Would you be comfortable or uncomfortable? Defend your answer.

Since it is one degree above our body temp. which in Farenheit is 99.6 youd would be uncomfterble.

3. Albert Einstein loved ice cream. Would he rather have 1 kg or 10,000 mg of Rocky Road? Express both in grams.

1kg = 1000g
10,000 mg = 10g

Albert would rather have 10,000mg of Rocky Road

Questions 4–6 ask you to use the information you have gained about the metric system and make reasonable estimates about the quantities listed. Place a decimal point within the series of numbers so that the statements are reasonable. For example, if you were evaluating the statement, **your textbook weighs 123456 g,** the most reasonable answer would be 1234•56 g. Placing the decimal in any other position would not give a reasonable estimate.

4. Barry is 154•5 cm tall.

5. The ambulance sped by at 10000 km per hour.

6. A soft drink bottle holds 32000 mL.

7. Gravity on the Moon is one-sixth that of gravity on Earth. What would be the weight in kilograms of a 75-kg body on the moon? What would be its weight in pounds?

8. How does a hypothesis differ from a theory?

A hypothesis is an explanation waiting to be tested compared to a theory which is widely accepted

9. List the attributes of a good hypothesis.

1.) must account for all data
2.) Must be able to test.

10. How are accuracy and precision different?

LABORATORY

2 Atoms and Molecules

+ Safety Box

- Be careful not to get any of the solutions on your hands or clothing. They may be caustic.
- Disposal of waste solutions may pose environmental hazards. Carefully follow your instructor's directions and place the waste solutions in the proper containers.
- Chemicals used as reagents, such as bromthymol blue or sodium iodide, may permanently stain clothing. Use with caution.

Objectives

Be able to do the following:

1. Define these terms in writing.

atom	pH scale	mass number
molecule	ion	periodic table of the elements
proton	compound	electron
acid	ionic bond	element
neutron	base	covalent bond
hydrogen ion	orbital	nucleus
product	hydroxide ion (OH^-)	reactant
atomic number	balanced equation	energy levels

2. Determine the number of protons, neutrons, and electrons for any atom, given the information on a periodic table.
3. Draw and label a diagram of any atom with an atomic number less than 20, given the information on a periodic table.
4. Diagram a conceivable molecule using all the atoms given, and show the shape of that molecule and the proper number of bonds for each atom.
5. Describe in writing and/or by a diagram the characteristics of ionic and covalent bonds.
6. Determine if a given solution is an acid or a base by using any of the following:
   a. pHydrion paper
   b. phenolphthalein indicator solution
   c. bromthymol blue indicator solution
   d. pH meter
7. Interpret the pH number of a solution.
8. Interpret a chemical equation and point out the reactants and the products.

Introduction

The smallest units of all matter, including living matter, are known as **atoms.** There are 92 different kinds of natural atoms found in nature and several others that have been manufactured by smashing other atoms together. These quickly disintegrate. A kind of matter that consists exclusively of one kind of atom is known as an **element.** Atoms can be combined in many ways to form millions of different kinds of **molecules.** Each kind of molecule has a specific arrangement of atoms within its structure. Kinds of matter that are composed of only one kind of molecule are called **compounds.** Each kind of atom has a specific number and arrangement of parts that differs from the number and arrangement of parts in other kinds of atoms. The specific structure of an atom determines the kinds of atoms that it can bond with to form larger molecules. Each atom consists of **protons, neutrons,** and **electrons** arranged with the positively charged protons and uncharged neutrons located in a central area called the **nucleus** and with the negatively charged electrons moving around outside the nucleus in specific regions known as **orbitals.**

Figure 2.1 is a **periodic table of the elements** that arranges elements in order of increasing complexity and according to the way they react chemically. Do not attempt to memorize it. (There probably isn't much chance of that, anyway.) This table contains a lot of information. Most important, the periodic table identifies the number of protons, neutrons, and electrons in atoms of each element. Figure 2.2 interprets some of the information about a carbon atom from the periodic table. The symbol for carbon is C. The number at the top is the **atomic number,** which tells you how many protons or electrons are in the atom. The number at the bottom, the **mass number,** is the sum of the number of protons and neutrons in the atom. You can determine the number of neutrons in this atom by subtracting the atomic number 6

Figure 2.1 The periodic table of the elements.

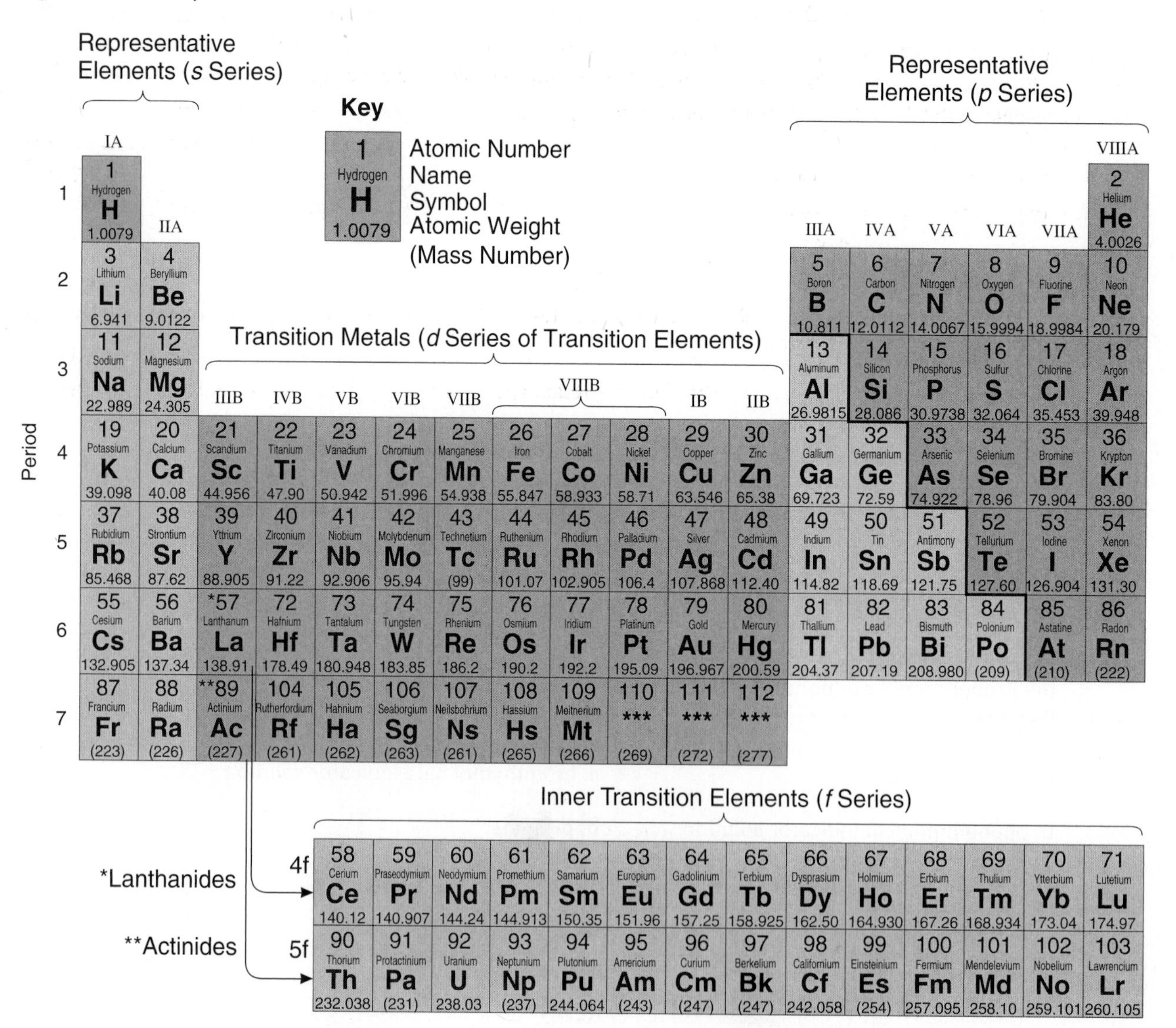

***These elements have not yet been named.

from the mass number 12 (rounded off); you will find that there are 6 neutrons in a carbon atom ($12 - 6 = 6$). To determine the number of protons, electrons, or neutrons in an atom, use the periodic table and the following equations:

1. **Atomic number = number of protons**
2. **Atomic number = number of electrons** (in a neutrally charged atom)
3. **Mass number** (rounded off) – **atomic number = number of neutrons** (in a typical atom)

Figure 2.2 Periodic table information for carbon.

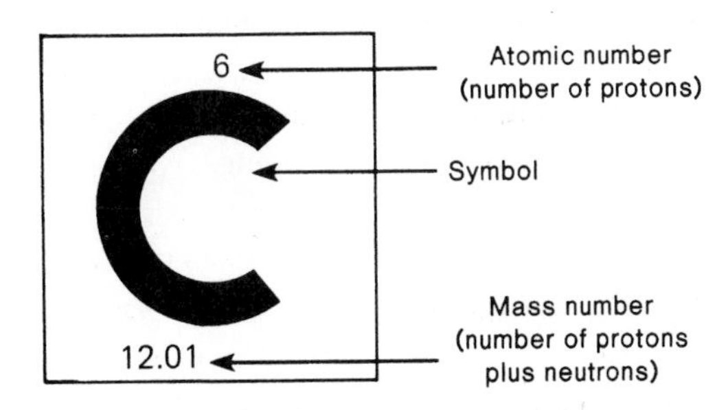

Atoms have the same number of positively charged (+) protons and negatively charged (–) electrons; therefore, the atom has no net charge. Because opposite charges are attracted to one another, the moving electrons are held near the nucleus but their movement prevents them from being pulled into the nucleus. Because the electrons repel each other and are light and move very rapidly, there is a tendency for electrons to move around the nucleus in specific regions called orbitals. The distance of electrons and their orbitals from the nucleus is determined by the amount of energy the electrons possess. Electrons with the greatest energy are found in the orbitals farthest from the nucleus. **Energy levels** (*1s, 2s, 2p,* etc.) designate differences in an electron's energy and distance from the nucleus. Two types of orbitals occur in atoms of the first 20 elements: *s* (spherical) *orbitals* and *p* (propeller) *orbitals.* All orbitals are full when they have 2 electrons. The first energy level contains just a single *s* orbital (*1s*) that may hold a maximum of 2 electrons. The second and third energy levels each contain an *s* orbital (*2s* and *3s*) and three additional *p* orbitals designated *px, py,* and *pz* (*2px, 2py, 2pz* and *3px, 3py, 3pz*). Each of these orbitals also may hold only 2 electrons but because there are four orbitals in the second energy level (*2s, 2px, 2py, 2pz*) the second energy level can hold up to 8 electrons. The general rules concerning electron distribution are

1. **Electrons always fill lower energy levels first.**
2. **The *s* orbital of any given energy level is filled before electrons occupy *p* orbitals.**
3. **One electron occupies each *p* orbital of a given energy level before a second electron is added to any *p* orbital.**

Therefore, an atom of the element fluorine that has a total of 9 electrons has 2 electrons located in the *1s* orbital, 2 located in the *2s* orbital, 2 in the *2px* orbital, 2 in the *2py* orbital, and 1 in the *2pz* orbital. This electron configuration is written as $1s^2\ 2s^2\ 2px^2\ 2py^2\ 2pz^1$ (figure 2.3).

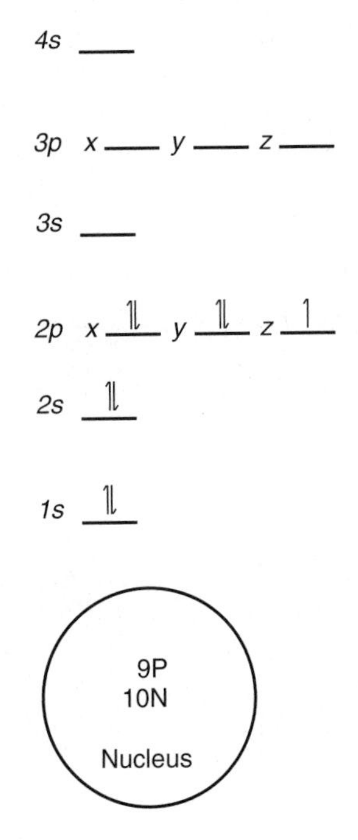

Figure 2.3 Diagram of a fluorine atom.

Preview: Part 1

This exercise is not designed to make you a chemist. To understand some of the important biological concepts, you must be aware of certain information about atoms and molecules. Work by yourself or in pairs as you proceed through this exercise.

During this lab exercise, you will

1. determine the number of protons, neutrons, and electrons in an atom.
2. diagram the parts of an atom.
3. assemble molecules that are formed as a result of ionic bonding.
4. diagram molecules that are formed as a result of covalent bonding.

Procedure

Review of Atomic and Molecular Structure

The Structure of Atoms

1. Refer to the periodic table of the elements (figure 2.1). Determine the number of protons, neutrons, and electrons in atoms of each of the following elements. The first one has been completed for you.

Element	Symbol	Protons	Neutrons	Electrons
Carbon	C	6	6	6
Hydrogen	H	1	0	1
Nitrogen	N	7	7	7
Oxygen	O	8	8	8
Sodium	Na	11	12	11
Magnesium	Mg	12	12	12
Phosphorus	P	15	16	15
Chlorine	Cl	17	18	17
Potassium	K	19	20	19
Calcium	Ca	20	20	20

2. Using figure 2.4, illustrate the correct number and position of protons, neutrons, and electrons for an atom of the element with the atomic number 7.

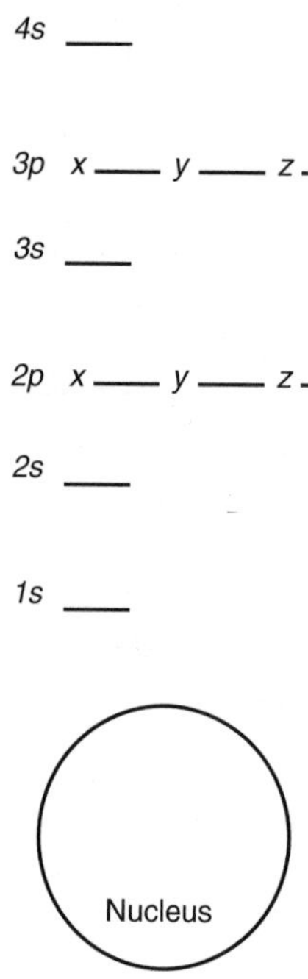

4s ___

3p x ___ y ___ z ___

3s ___

2p x ___ y ___ z ___

2s ___

1s ___

Nucleus

Figure 2.4 ____________
(Name of atom you diagrammed.)

3. Choose any other atom with an atomic number from 10 to 20 on the periodic table of the elements. Determine the name of the element and the number and position of the protons, neutrons, and electrons, and indicate this information in figure 2.5. If you have difficulty, be sure to ask your instructor for help. You may also want to sketch several other atoms for practice.

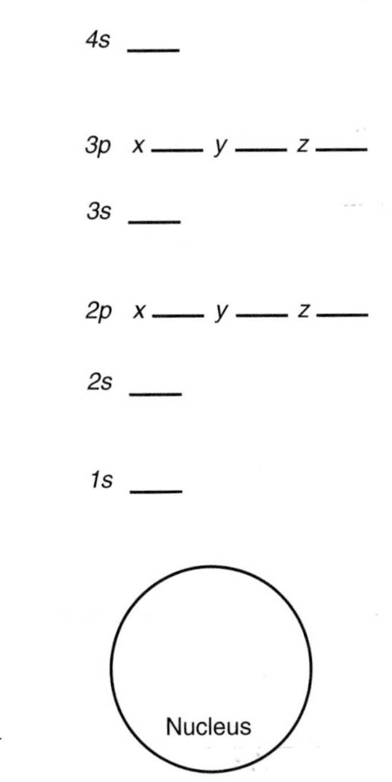

Figure 2.5 ____________________
(Name of atom you diagrammed.)

Molecular Structure

Specific atoms may be combined in certain ways to form larger units called molecules. The bonding together of atoms is a very precise process. Some kinds of atoms are very reactive and will combine with one or two other kinds of atoms. We will not try to determine why certain atoms combine into molecules but, rather, how this process happens.

Ionic Bond (Steals electrons)

Some kinds of atoms have such a strong attraction for electrons that they steal electrons from other atoms that have rather loosely held electrons. The specific structure of an atom determines whether it will form an ion. **Ions** are atoms or molecules that have gained or lost electrons and, therefore, are either negatively or positively charged. Atoms that lose electrons are positively charged (+), and atoms that gain electrons are negatively charged (–). Those ions that have the same charge (both + or both –) repel one another, whereas those with unlike charges attract one another and form an **ionic bond.**

1. Figure 2.6 on page 19 contains models of different ions. Cut them apart and assemble them to form as many different kinds of compounds (combinations of ions) as you can.
2. List at least five compounds you were able to assemble from the model pieces in the following space. (Chemists generally write the formula of ionic compounds by writing the symbol for the positive ion first and then the negative ion. They use a subscript to indicate the number of ions needed to balance the charge. For example the formula $MgCl_2$ means that there is a single magnesium ion (Mg^{++}) bonded to two chloride ions (Cl^-) to make the salt magnesium chloride.)

Directions: Refer to the periodic table of elements (figure 2.1) and determine the number of protons and electrons that constitute the following atoms, ions, and molecules. *The number of plus signs (+) indicates how many electrons have been lost and the number of minus signs (–) indicates the number of electrons gained.*

1. Protons in a sodium atom (Na) 11 Electrons in a sodium atom (Na) 11
2. Protons in a sodium ion (Na^+) 11 Electrons in a sodium ion (Na^+) 10
3. Protons in a chlorine atom (Cl) 17 Electrons in a chlorine atom (Cl) 17
4. Protons in a chloride ion (Cl^-) 17 Electrons in a chloride ion (Cl^-) 18
5. Protons in a sodium chloride molecule (NaCl) 28 Electrons in a sodium chloride molecule (NaCl) 28
6. Protons in a calcium atom (Ca) 20 Electrons in a calcium atom (Ca) 20
7. Protons in a calcium ion (Ca^{++}) 20 Electrons in a calcium ion (Ca^{++}) 18
8. Protons in an oxygen atom (O) 8 Electrons in an oxygen atom (O) 8
9. Protons in an oxide ion (O^{--}) 8 Electrons in an oxide ion (O^{--}) 10
10. Protons in a calcium oxide molecule (CaO) 28 Electrons in a calcium oxide molecule CaO 28

Covalent Bond (Share electrons)

A second kind of bond that holds atoms together to form molecules is known as a **covalent bond.** In covalent bonds, the electrons are not actually transferred from one atom to another, as in the formation of ions and ionic bonds, but are shared by two or more atoms. Each pair of electrons that is shared is the equivalent of one covalent bond. Chemists typically diagram molecules by using a line between atoms to represent a single covalent bond.

If you know how many electrons each atom is able to share, you should be able to diagram a variety of kinds of atoms. Figure 2.7 indicates that a single carbon atom (C) is sharing 4 electrons with four different hydrogen atoms (H) and that each of the four hydrogen atoms is sharing an electron with the same carbon atom.

Figure 2.7 Covalent bonding in a methane molecule.

```
     H
     |
  H—C—H
     |
     H
```

Sometimes two atoms share more than one pair of electrons, creating a double bond; for example, the carbon dioxide molecule has the following structure: O = C = O. The diagram of the carbon dioxide molecule indicates that a carbon atom is sharing 2 electrons with one oxygen atom and 2 electrons with another oxygen atom. Each oxygen atom shares 2 electrons with the same carbon atom.

Table 2.1 is a list of a few atoms and the number of electrons they usually share.

Table 2.1 **Bonding Capacity**

Name of Atom	Symbol of Element	Number of Bonds	Bonding Capacity
Carbon	C	4	–C– –C= =C= –C≡
Nitrogen	N	3	/N\ –N= N≡
Oxygen	O	2	O– O=
Hydrogen	H	1	H–

Figure 2.6 Ionic models.

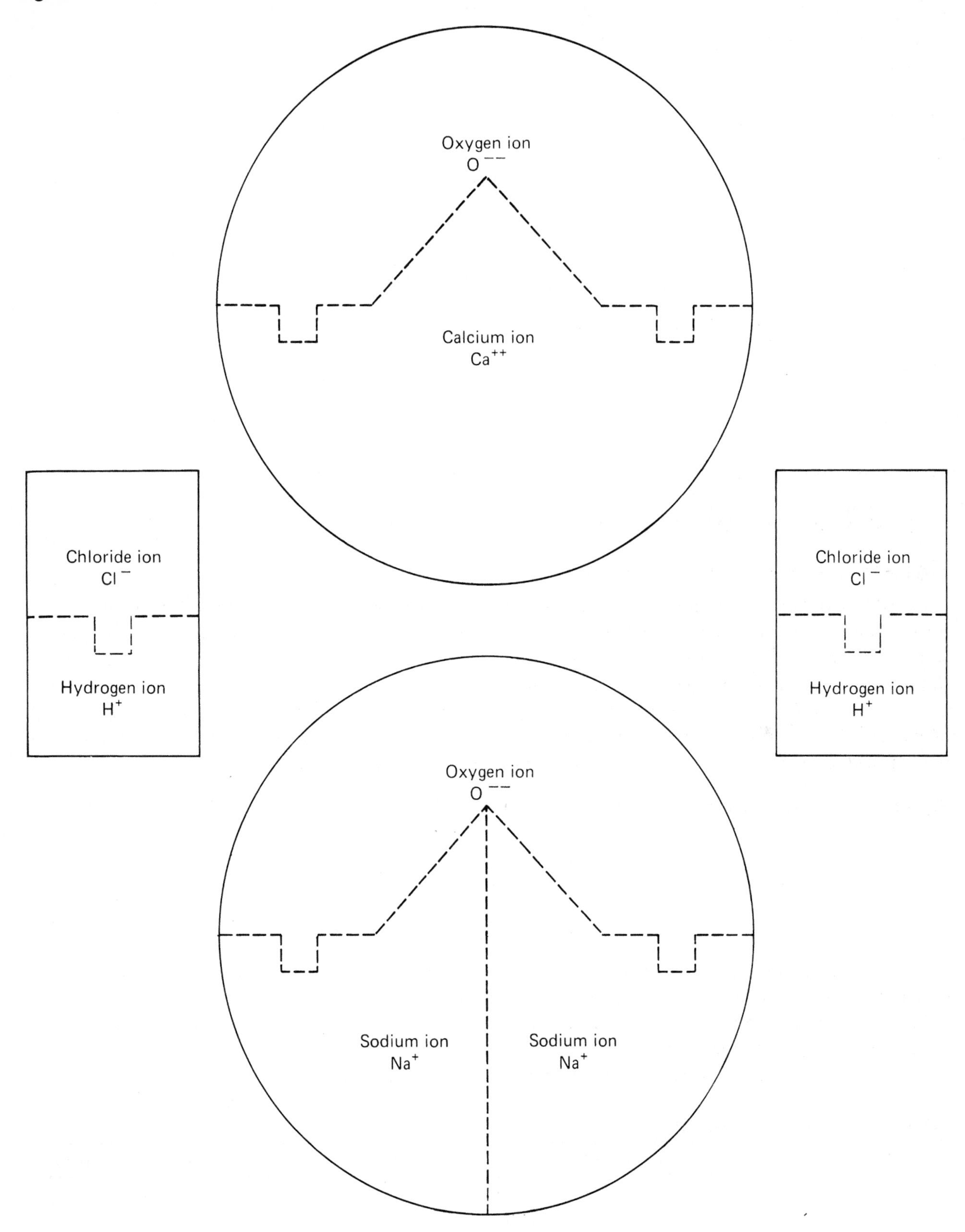

Directions: Use the information about the numbers of electrons atoms can share from the periodic table (table 2.1), and diagram the arrangement of the atoms for the molecules listed in the following space.

Simple Molecules

1. Methane (one carbon atom and four hydrogen atoms): CH_4

2. Ammonia (one nitrogen atom and three hydrogen atoms): NH_3

3. Water (one oxygen atom and two hydrogen atoms): H_2O

More Complex Molecules

Now let's look at some molecules that have a more complex structure. To diagram the arrangement of atoms in these molecules, use the following procedure.

a. Begin with the carbon atoms and bond them together in a chain or a ring.

b. Next, add the nitrogens if any are called for.

c. Then, add the oxygens, if any.

d. Finally, count the number of electrons in the whole molecule that are still available for bonding. (If this number is equal to the number of hydrogen atoms called for, simply add one hydrogen to each bondable point.) If there are too few hydrogens to complete available bonds, find two free bondable electrons on adjacent atoms and have them form a second bond between themselves (double bond). Now count the available bonding points—the number of hydrogens called for should equal the number of bondable electrons—and simply add the hydrogens where they can share electrons.

4. Ethane (two carbon atoms and six hydrogen atoms): C_2H_6

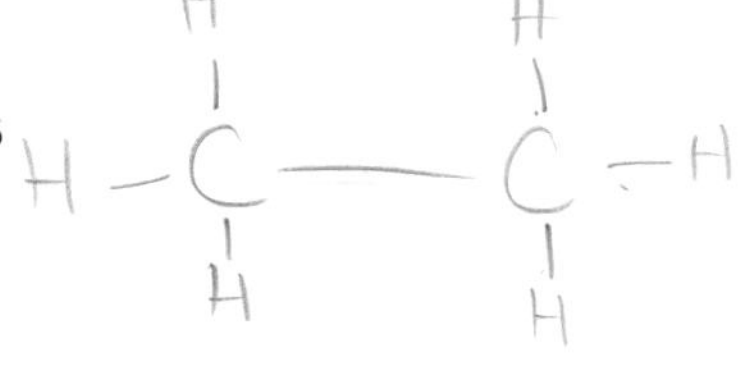

5. Ethyl alcohol: C_2H_5OH

6. Ethene (two carbon atoms and four hydrogen atoms): C_2H_4

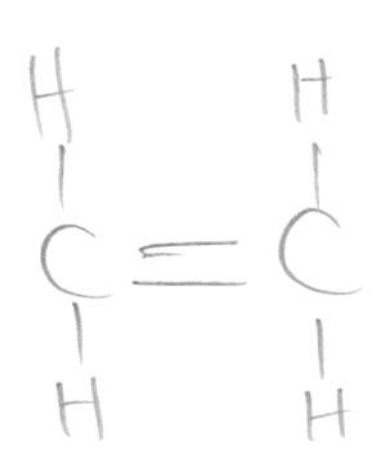

7. Aldehyde (one carbon atom, two hydrogen atoms, and one oxygen atom): CH_2O

8. Acetic acid: CH_3COOH

9. Amino acid: (Add COOH and NH_2 to the carbon skeleton. The "R" is used to refer to a number of different possible combinations of atoms. It could be an "H," a CH_3, or another combination of atoms.)

```
  H
  |
 -C-
  |
  R
```

10. Sugar: $C_6H_{12}O_6$

If molecular stick models are available, your instructor will help you demonstrate the three-dimensional arrangement of atoms in the organic compounds listed previously.

Preview: Part 2

Now that you understand something about atoms and chemical bonds, you will perform simple experiments to measure pH and observe chemical reactions.

Caution: This exercise requires the use of glassware and various chemicals; therefore, exercise care when moving around the work area and when handling the materials.

During this lab exercise, you will

1. determine and interpret the pH number of several solutions.
2. mix reactants and observe the results of chemical reactions.

Chemical Testing and Reactions

Acids, Bases, and pH

Acids are molecules that release hydrogen ions (H^+) when dissolved in water. A **hydrogen ion (H^+)** is a hydrogen atom that has lost its electron. **Bases** are substances that remove hydrogen ions from solution. A common base is the **hydroxide ion (OH^-).** It is frequently important to know if a solution is an acid or a base, and a number of different methods have been developed to test solutions for their acidity or alkalinity. All of these systems rely on a scale known as the **pH scale,** which is a measure of the number of hydrogen ions present in a solution. Pure water has both **hydrogen ions (H^+)** and hydroxide ions (OH^-) present in equal numbers, because water molecules dissociate ($HOH \rightarrow H^+ + OH^-$). Solutions with equal numbers of hydroxide and hydrogen ions are called neutral solutions. If there are more H^+ than OH^-, then the solution is an acid. If there are more OH^- than H^+, then it is a base (alkaline). The pH scale shown in figure 2.8 indicates the range of acidity and alkalinity that can exist. There are a couple of important things that you should know about the pH scale. First, it is a logarithmic scale, which means that the difference between any two numbers on the scale is a difference of 10 times. Second, it is an inverse scale, which means that, the smaller the pH number, the more hydrogen ions (H^+) are present. Therefore, a solution that has a pH of 5 is 100 times more acidic than a solution with a pH of 7.

Figure 2.8 pH scale.

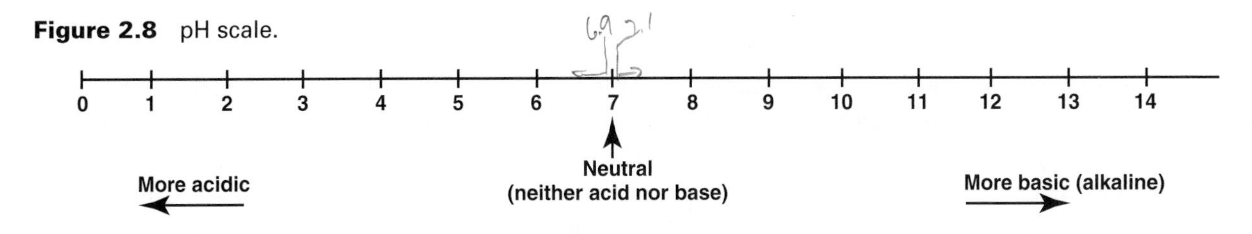

Refer to figure 2.8 and answer the following questions.

1. Which of the following pH values indicates the greater concentration of acid?
   pH 1 or pH 4 __________
   pH 5 or pH 3 __________
2. Which of the following pH values indicates the greater concentration of base?
   pH 8 or pH 10 __________
   pH 7 or pH 11 __________
3. What does a pH of 7 indicate? Neutral

Determining pH

Many kinds of materials change color as the pH of a solution changes. We can make use of this property to determine the pH of unknown solutions. You will find five unknown solutions labeled A through E on the worktable.

1. Using a pair of forceps, dip a piece of pHydrion paper into solution A. Compare the color of the pHydrion paper with the information on the paper dispenser chart and record the pH in table 2.2.
2. Place five drops of solution A in a test tube and add two drops of bromthymol blue. Record the color of the solution in table 2.2.

3. Place five drops of solution A in a clean test tube and add two drops of phenolphthalein. Record the color of the solution in table 2.2.
4. Repeat these three procedures with solutions B, C, D, and E, and complete table 2.2.

Table 2.2 Unknowns

Solution	pHydrion Paper	Color After Adding Bromthymol Blue	Color After Adding Phenolphthalein	pH Meter Readings
A				
B				
C				
D				
E				

Use the information from Table 2.2 to answer the following questions.

1. Bromthymol blue changes color when mixed with an acid. What color does it become? ____________________
2. What color is bromthymol blue when in a base? ____________________
3. What color would you expect bromthymol blue to be if it were in a neutral solution? ____________________
4. Phenolphthalein changes to ____________________ in the presence of ____________________.
5. What is the advantage of using pHydrion paper rather than bromthymol blue or phenolphthalein? ____________________

   ____
6. Which reagent (bromthymol blue or phenolphthalein) could be considered to be an acid indicator? ____________

Using a pH Meter

Measuring pH using the methods and materials already described has several drawbacks. Some of these drawbacks are (1) the solutions to be tested may be altered by the addition of indicator chemicals; (2) the accurate determination of a color change is difficult; and (3) a precise determination of pH is almost impossible. The use of an electronic pH metering machine is a better method of determining pH. The pH metering machine can be thought of as a battery that can develop different electrical potentials, depending on the hydrogen ion concentration of the solution in which it is placed. The solution serves as a source of the electrical potentials because of its ionic concentration. Even though these electrical potentials are relatively weak, this extremely sensitive machine is able to detect them and display them on a pH scale.

Your instructor will demonstrate the use of a pH meter.

1. Test the pH of solutions A through E using the pH meter and record your results in table 2.2. Compare the accuracy of the four methods used to determine pH.
2. Determine the pH of several other common solutions using the pH meter. These can include soft drinks, coffee, tap water, and detergents. Record your results in table 2.3.

Table 2.3 pH of Common Solutions

Common Solutions	pH Meter Readings

Caution: Read labels carefully and use caution when mixing all chemicals. Your instructor will provide instructions on the proper disposal of solutions.

Introduction

A chemical reaction occurs when atoms or molecules react with one another in such a way that chemical bonds are broken and new molecular combinations are made as new bonds are formed. In many cases, there is physical evidence that a reaction has taken place. This evidence might be a visible change, such as the production of a gas that bubbles off, a color change, the production of heat, or the development of an insoluble material that settles to the bottom of the container. As reactions occur between two chemical reactants, the chemical bonds generally rearrange to form a more stable, longer-lasting end product.

Chemical equations are shorthand statements used to represent chemical reactions. The **reactants,** which are changed, are shown on the left and the new substances formed **(products)** are on the right. The arrow indicates the direction of the chemical transformation. For example, hydrogen reacts with oxygen to form water:

$$2\ H_2 + O_2 \longrightarrow 2\ H_2O\ \text{(water)}$$

Reactants Products

Look carefully at the way the equation is written. Because hydrogen and oxygen exist as molecules consisting of two atoms, rather than as individual H and O atoms, we must write them as such in the equation.

Now, count the number of atoms on each side of the reaction arrow.

1. How many H atoms are on the left? ______________________
2. How many H atoms are on the right? ______________________
3. How many O atoms are on the left? ______________________
4. How many O atoms are on the right? ______________________

When the number and kinds of atoms on each side of the equation are the same, it is a **balanced equation.** All chemical equations must be balanced, because atoms are neither created nor lost during a reaction.

Examples

Consider the following reactions and equations:

1. $NaCl + AgNO_3 \longrightarrow NaNO_3 + AgCl$
   a. Mix a couple of drops of sodium chloride solution and silver nitrate solution in a test tube or watch glass.
   b. What physical changes do you observe as evidence that a chemical reaction has taken place?

   c. Complete the equation by writing out the names of the products.

   Sodium chloride + silver nitrate ⟶ ______________________ + ______________________

2. Mix a couple of drops of sodium iodide solution with a couple of drops of lead nitrate solution.
   a. What is the physical evidence that a reaction has taken place?

   b. Balance the equation by inserting the appropriate numbers in the spaces provided.

   $\_\_\_\_\ NaI + Pb(NO_3)_2 \longrightarrow PbI_2 + \_\_\_\_\ NaNO_3$

   ______________ + ______________ ⟶ ______________ + ______________

   c. Fill in the names of the reactants and the products below the chemical formulas.

3. Place a small amount of sodium bicarbonate in a test tube and add a couple of drops of hydrochloric acid.
   a. What physical evidence indicates that a reaction has taken place?

   b. Complete the equation by writing in the names and the chemical symbols for the end products that are not listed.

      $HCl + NaHCO_3 \longrightarrow CO_2 +$ ________________ + ________________

   c. Is the equation balanced? ________________________

2 Atoms and Molecules

Name ______________________________ Lab section ____________

Your instructor may collect these end-of-exercise questions. If so, please fill in your name and lab section.

End-of-Exercise Questions

Quiz

1. List several differences between ionic and covalent bonds.

   Ionic bonds steal electrons while covalent bonds share electrons. Ionic bonds change the charge of the atom and covalent bonds don't.

2. What information can be obtained from the periodic table of elements?

   The mass of the atom Its Atomic #
   Its Symbol
   The # of proton, neutrons, & electrons

3. What do you need to know to diagram an atom?

   How many electrons are available to bond.

4. What does a single straight line ($—CH_3$) extending from an atomic symbol represent?

   Represents an open bond an any atom can attach to it.

5. How does one know how many covalent bonds a particular atom will form?

   Depending on the unpaired atoms of its outer most shell.

6. Which groups of elements on the periodic table tend to form ions?

7. What is the difference between —OH and OH^-?

-OH is a covalent bond

OH^- is an Ionic bond

8. Define pH.

A system that scales a solutions acidity or basic

9. Explain how one might use bromthymol blue and phenolphthalein to test the pH of water in a swimming pool.

10. List three kinds of physical changes that indicate a chemical reaction has taken place.

Color Change
Bubble
Precipitate
Tempeture change

LABORATORY

3 Diffusion and Osmosis

+ Safety Box

- Laboratory thermometers do not need to be shaken down; they automatically register higher or lower temperatures. If you accidentally break a thermometer, dispose of the broken glass and mercury as indicated by your instructor.

Objectives

Be able to do the following:

1. Define these terms in writing.

    - diffusion
    - direction of net movement
    - osmosis
    - dynamic equilibrium
    - relative concentration
    - solvent
    - selectively permeable membrane
    - concentration gradient
    - solute
    - solution

2. Explain why a particular material diffuses in a particular direction.
3. Determine the net direction of diffusion.
4. Differentiate between diffusion and osmosis.
5. Describe the influence of temperature on the rate of osmosis.
6. Describe the influence that varying the concentration of solute and solvent has on the rate of osmosis.

Introduction

Although you may not know what diffusion is, you have experienced the process. Can you remember walking into the front door of your home and smelling a pleasant aroma coming from the kitchen? It was diffusion of molecules from the kitchen to the front door of the house that allowed you to detect the odors. **Diffusion** is the net movement of molecules from an area of greater concentration to an area of lesser concentration until the concentration everywhere is the same. The movement in one direction minus the movement in the opposite direction determines the **direction of net movement.** To better understand how diffusion works, let's consider some information about molecular activity.

The molecules in a gas, a liquid, or a solid are in constant motion because of their kinetic energy. Moving molecules are constantly colliding with each other. These collisions cause the molecules to move randomly. The higher the concentration of molecules in one region, the greater the number of collisions. Some molecules are propelled into the less concentrated area and others are propelled into the more concentrated area. Over time, however, there will be more collisions in the highly concentrated area, resulting in more molecules being propelled into the less concentrated area. Thus, the net movement of molecules is always from more tightly packed areas to less tightly packed areas.

Diffusion occurs when there is a difference in concentration from one region to another or from one side of a membrane to another (figure 3.1a). A difference in the concentration of molecules over a distance is called a **concentration gradient.** When the molecules become uniformly distributed, as in figure 3.1b, that have reached **dynamic equilibrium,** in which the number of molecules moving in one direction is balanced by the number moving in the opposite direction. It is *dynamic* because molecules continue to move, but, because motion is equal in all directions and there is no net change in concentration over time, *equilibrium* exists. The process of diffusion occurs in both living and nonliving systems. Biologically speaking, diffusion is responsible for the movement of a large number of substances, such as gases and small, uncharged molecules, into and out of living cells.

Figure 3.1 Molecular movement.

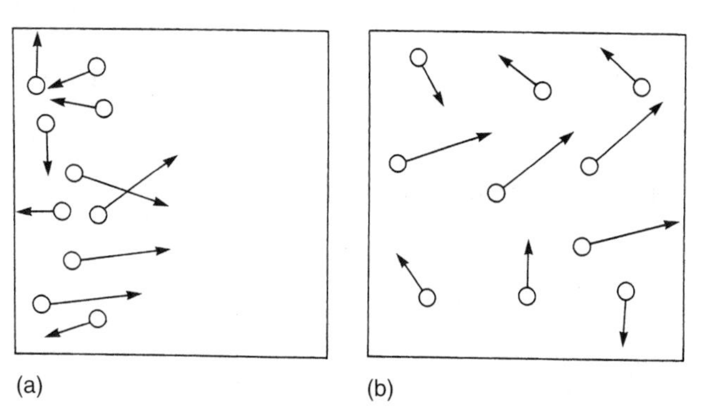

The direction of diffusion is always from where there are more molecules to where there are fewer. This is similar to the scattering of a crowd of people leaving a theater. Many of the individuals move from the theater to the outside, but some go back to retrieve their gloves or popcorn. The net movement, however, is the movement of the individuals leaving the theater minus the movement of those returning.

Imagine that your instructor opens a bottle of ammonia in a corner of the room. The bottle would have the highest concentration of ammonia molecules in the room; the individual ammonia molecules would move from this area of highest concentration to where they are less concentrated (figure 3.2).

Figure 3.2 Diffusion of ammonia.

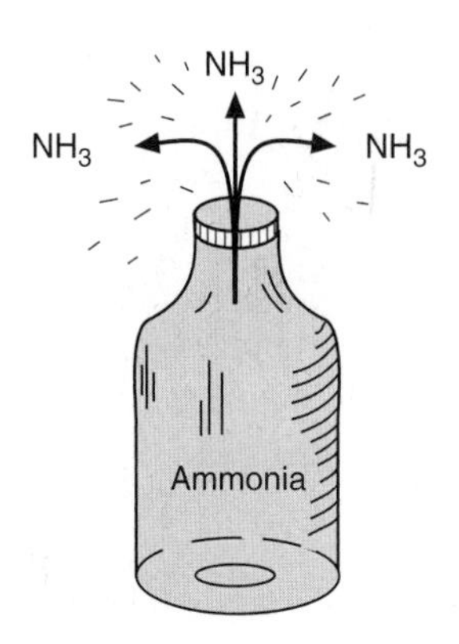

Although you could not actually see this happening, ammonia molecules would leave the bottle and move throughout the air in the room because of molecular movement. You could detect this by the odor of the ammonia. If you compared the relative number of ammonia molecules in the bottle to those dispersed in the room, you would be dealing with what is called relative concentration. **Relative concentration** compares the amount of a substance in two locations. Whenever there is a difference in concentrations of a substance, you can predict the direction that most of the molecules will move. You can predict that, when the bottle is first opened, ammonia molecules will move from the area of higher concentration (the bottle) to the region of lower concentration (the air in the room). Soon, however, the molecules of ammonia will mix with the air molecules in the room. Because the ammonia molecules are moving randomly, some of them will move from the air back into the bottle.

As long as there is a higher concentration of ammonia molecules in the bottle, more of them move out of the bottle than move in. One way of dealing with the direction of movement is to compare the number of molecules leaving the bottle with the number reentering the bottle. This is called the net amount of movement. The movement in one direction minus the movement in the opposite direction is the direction of net movement. If, for example, 100 molecules of ammonia leave the bottle and 10 reenter during that time, the net movement is 90 molecules leaving the bottle. Ultimately, the number of ammonia molecules moving out of the bottle will equal the number of ammonia molecules moving into it. When this point is reached, the ammonia molecules are said to have reached dynamic equilibrium.

When several kinds of molecules are present, consider only one case of diffusion at a time, even though several different types of molecules are moving. For example, consider the exchange of gases between the lungs and blood. In the lungs, a series of tubes transports gases. These tubes divide into smaller and smaller branches and eventually end at a series of small *alveolar sacs.* Adjacent to these sacs are a number of capillaries containing blood. By the process of diffusion, there is an exchange of oxygen and carbon dioxide between the alveolar sacs and the blood in the capillaries (figure 3.3).

Figure 3.3 Diffusion of oxygen and carbon dioxide.

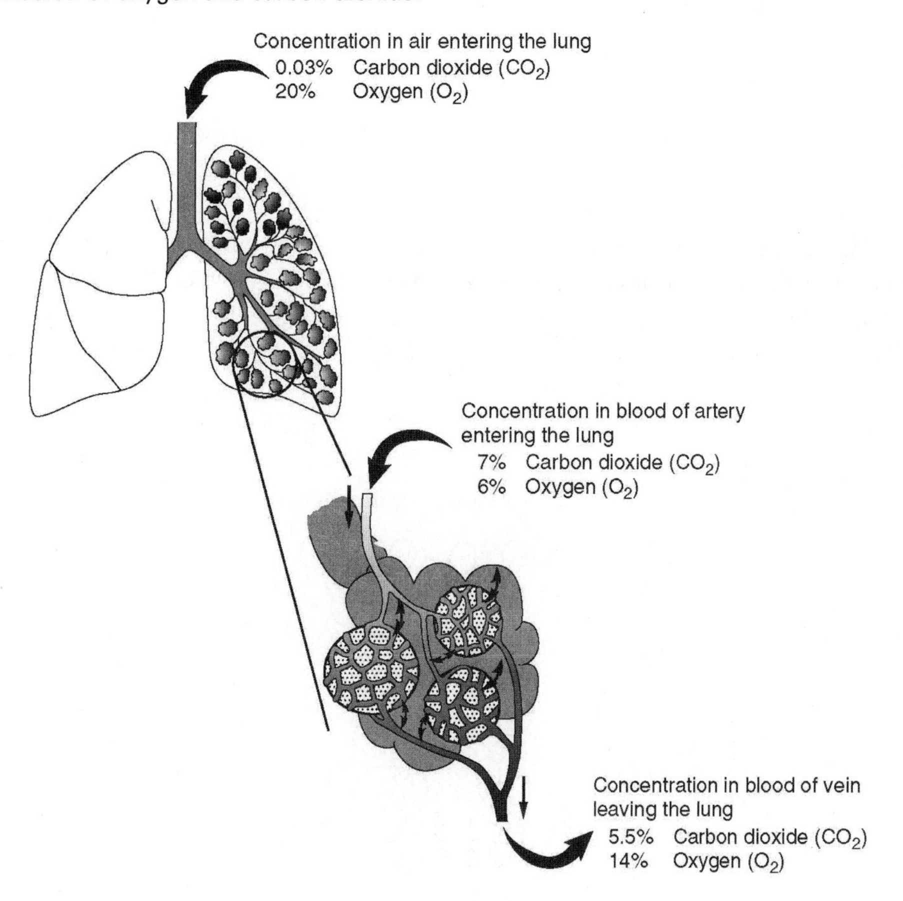

1. Is the direction of net movement of carbon dioxide molecules in figure 3.3 from the blood to the lungs or from the lungs to the blood? Explain your answer.

2. Is the direction of net movement of oxygen molecules in figure 3.3 from the blood to the lungs or from the lungs to the blood? Explain your answer.

Another example of diffusion is sugar *dissolving* in water. When sugar molecules and water molecules mix, a *solution* is created. A **solution** is any mixture where two or more different types of molecules are evenly dispersed throughout the system.

3. Draw an arrow on figure 3.4 to show the net direction of sugar movement.

Figure 3.4 Sugar diffusion.

Figure 3.5 shows a selectively permeable membrane. A **selectively permeable membrane** is a thin sheet of material that selectively allows certain molecules to cross but prevents others from crossing. *The membrane in this figure is permeable only to water molecules.* Water molecules may freely diffuse across the membrane, but other types of molecules cannot. The diffusion of water across a selectively permeable membrane is called **osmosis.** On each side of the selectively permeable membrane in figure 3.5 is a chloride solution. A solution is characterized by the dissolved substance called the **solute.** Chloride in this example is the solute. The substance in which the solute is dissolved is called the **solvent.** In figure 3.5 and in biological systems, water is the solvent.

Figure 3.5 Diffusion of water.

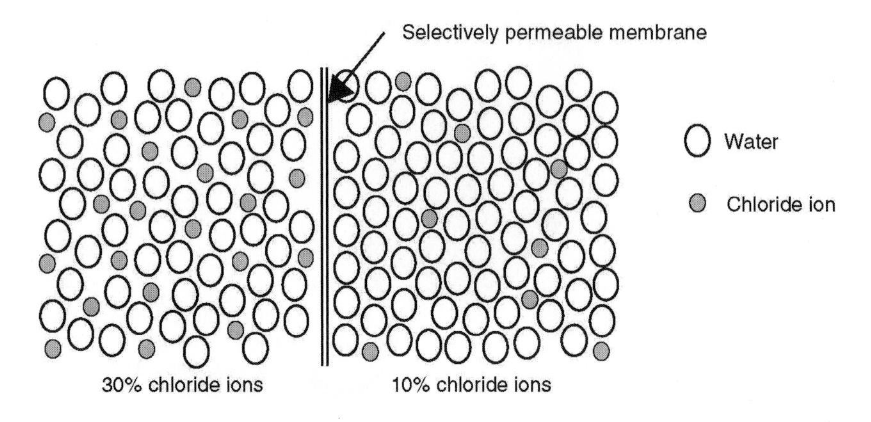

4. What is the percentage of solute in the left side of the container (figure 3.5)? ______________________
5. What is the percentage of solvent in the left side of the container? ______________________
6. What is the percentage of solute in the right side of the container? ______________________
7. What is the percentage of solvent in the right side of the container? ______________________
8. Where is the water in higher concentration—the left or right side? ______________________
9. Draw an arrow to indicate the net direction of movement of the water molecules.

In each of the previous examples, the net movement was a result of diffusion of molecules from a place of higher concentration to a place of lower concentration. The rate at which diffusion occurs is related to the amount of energy the molecules have and the degree of difference between the areas of high and low concentration. Adding energy doesn't change relative concentrations, nor does it influence the direction of diffusion. It merely affects the rate at which diffusion occurs. Molecules with greater kinetic energy move faster, causing diffusion to happen more quickly.

Preview

The kinetic molecular theory states that all substances are made up of molecules that occupy space and are constantly in motion. This exercise helps you examine some phenomena related to this motion of molecules.
During this lab exercise, you will

1. set up a demonstration of osmosis under a variety of temperature conditions and determine how temperature influences the rate of osmosis.
2. set up a demonstration of osmosis using a variety of concentration gradients and determine how concentration differences influence the rate of osmosis.
3. graph the results of the osmosis demonstrations.

Procedure

Osmosis: Effect of Temperature

Construct a hypothesis about the relationship between temperature and the rate of osmosis and write it in the following space.

Test your hypothesis by doing the following experiment.
Working in groups, prepare three sacs to demonstrate osmosis.

1. Obtain three pieces of dialysis tubing (sausage casing) and soak them in tap water for about 1 minute.
2. Form each of the pieces of tubing into a tubular bag. Shake off the excess water, twist and fold over one end of the dialysis tubing, and securely tie it with a piece of string.
3. Fill each tubular bag about half full with clear Karo syrup. Leave room for a small pocket of air. Twist the open end of each bag, fold the end over, and tie it. After rinsing each bag, pat it dry and cut off any excess string.
4. Use a balance to obtain an initial weight for each bag. Record your data in table 3.1.
5. a. Place one bag in a beaker of water at room temperature (approximately 20°C).
   b. Place the second bag in a waterbath heated to 40°C.
   c. Place the third bag in a beaker of ice water (approximately 0°C) (figure 3.6). Record the exact temperature of the water in each beaker. Make certain that each bag is completely covered with water.
6. After 5 minutes, remove each bag and gently squeeze to assess any changes in firmness. Observe the size, shape, and firmness of each bag. Gently pat each bag dry and weigh. Record your results in table 3.1.
7. Return each bag to its appropriate container of water for another 5-minute interval. Repeat your observations and measurements every 5 minutes. (Measurements should be taken at 0, 5, 10, 15, and 20 minutes.)
8. To more easily visualize the effect of temperature on osmosis, use the data collected in table 3.1 to construct a graph of the effect of temperature on the rate of osmosis. If you do not have three different colored pencils, use a solid line for the hot water, a dashed line for room-temperature water, and a dotted line for the ice water.

Figure 3.6 Osmosis: effect of temperature setup.

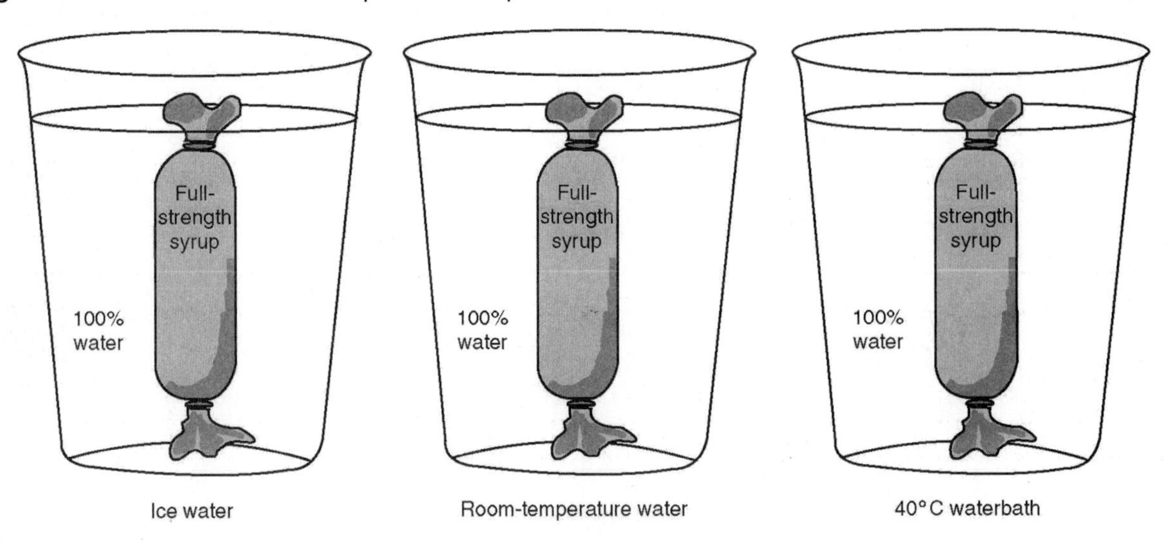

Table 3.1 Effect of Temperature on the Rate of Osmosis

Temperature	Initial Weight	5 Minutes	10 Minutes	15 Minutes	20 Minutes
10:25 Warm water	24.3	25.8	26.6	27.9	28.9
10:18 Tap water	22.4	22.7	22.8	23.2	23.5
10:30 Ice water	26.1	26.7	27.2	27.7	28.1

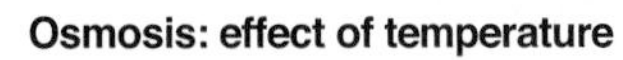

Osmosis: effect of temperature

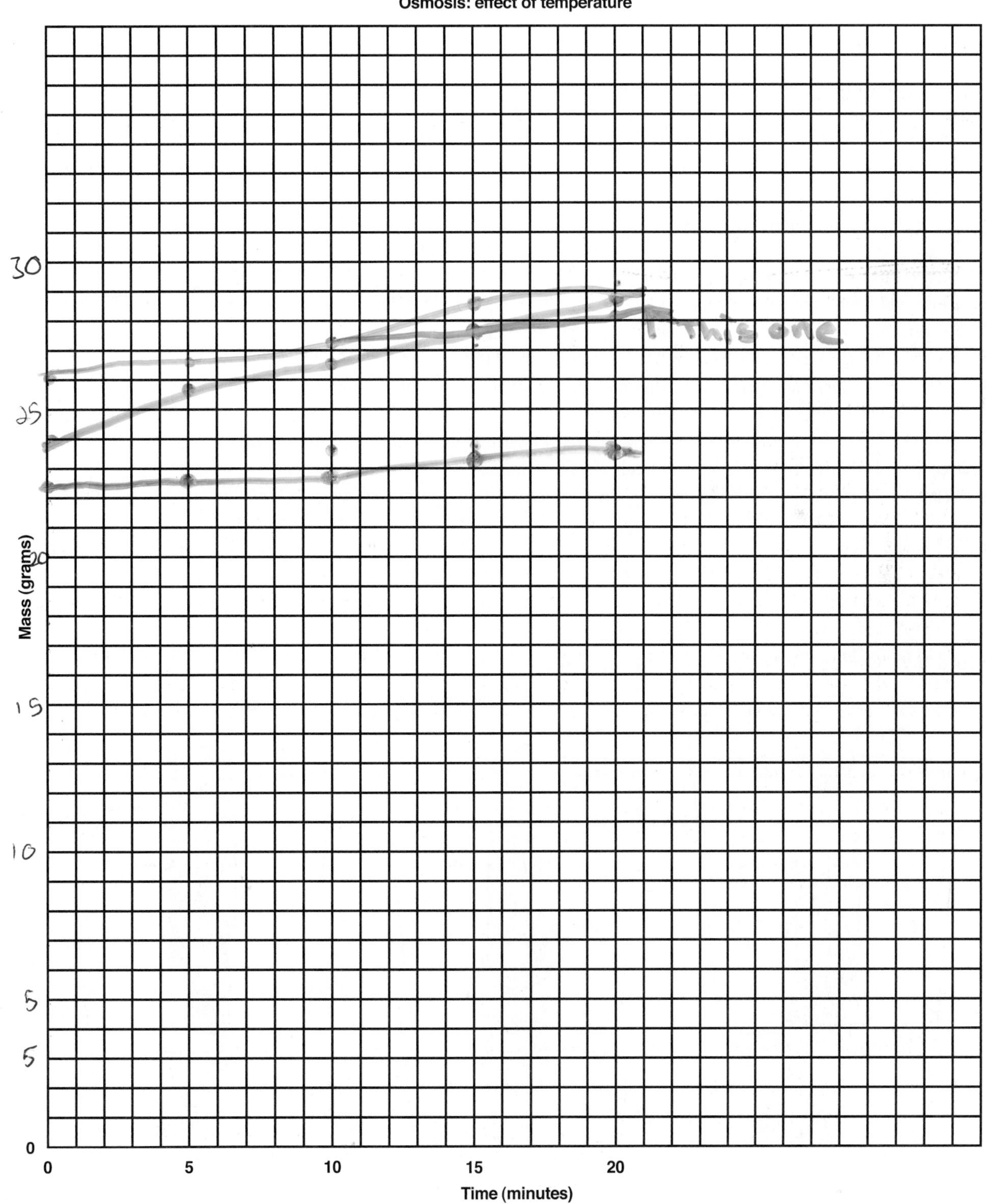

9. Publish your data by drawing your graph on a blackboard, an overhead projector, or similar equipment.
10. Compare and discuss your results with others in the class.
11. What are your conclusions? Did you support or disprove your hypothesis? Explain.

Osmosis: Effect of Concentration

Construct a hypothesis about the relationship between the concentration of a syrup solution and the rate of osmosis and write it in the following space.

Test your hypothesis by doing the following experiment.

1. Repeat the construction of the selectively permeable bags (see instructions under "Osmosis: Effect of Temperature"). Prepare three bags but fill each with a different concentration of clear Karo syrup according to the following specifications (figure 3.7):
   Bag 1: one part syrup to three parts water (2.5-mL syrup; 7.5-mL water)
   Bag 2: one part syrup to one part water (5-mL syrup; 5-mL water)
   Bag 3: full-strength syrup (10-mL syrup)
2. Rinse and gently pat the bags dry.
3. Record the initial weight of each bag.
4. Place each bag in a container of room-temperature tap water.
5. Check their weight and firmness at 5-minute intervals for a period of 20 minutes.
6. Record all data in table 3.2, and graph your results.
7. Publish your data by drawing your graph on a blackboard, an overhead projector, or similar equipment.
8. Compare and discuss your results with others in the class.
9. What are your conclusions? Did you support or disprove your hypothesis? Explain.

Figure 3.7 Osmosis: effect of concentration setup.

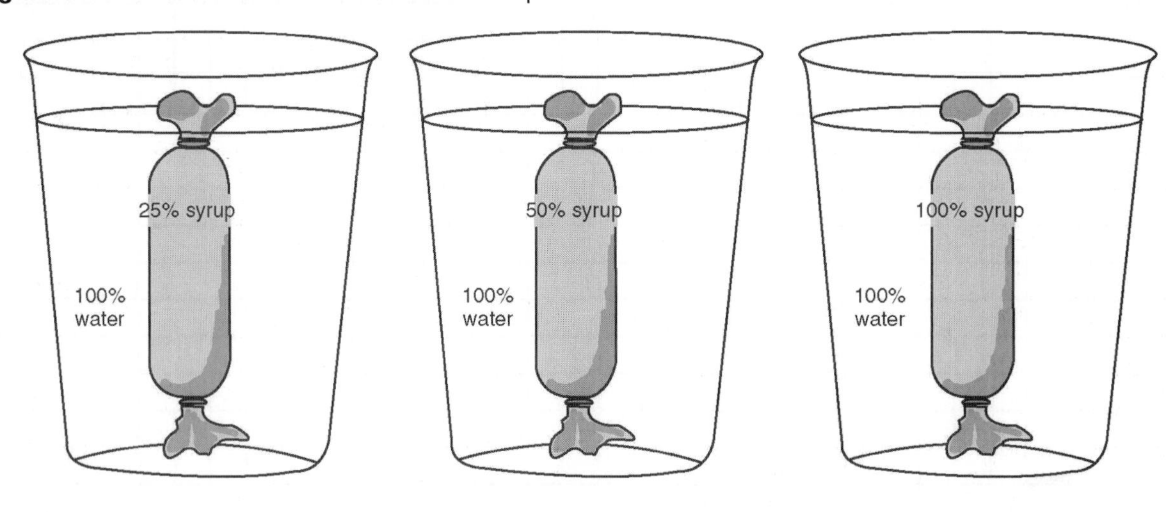

Table 3.2 Effect of Concentration on the Rate of Osmosis

Concentration	Initial Weight	5 Minutes	10 Minutes	15 Minutes	20 Minutes
25%	16.9	17.4	17.9	18.4	18.7
50%	21.6	22.	22.3	22.7	23.7
100%					

10:51

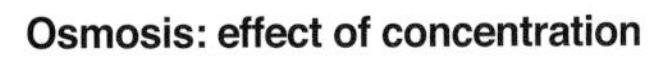

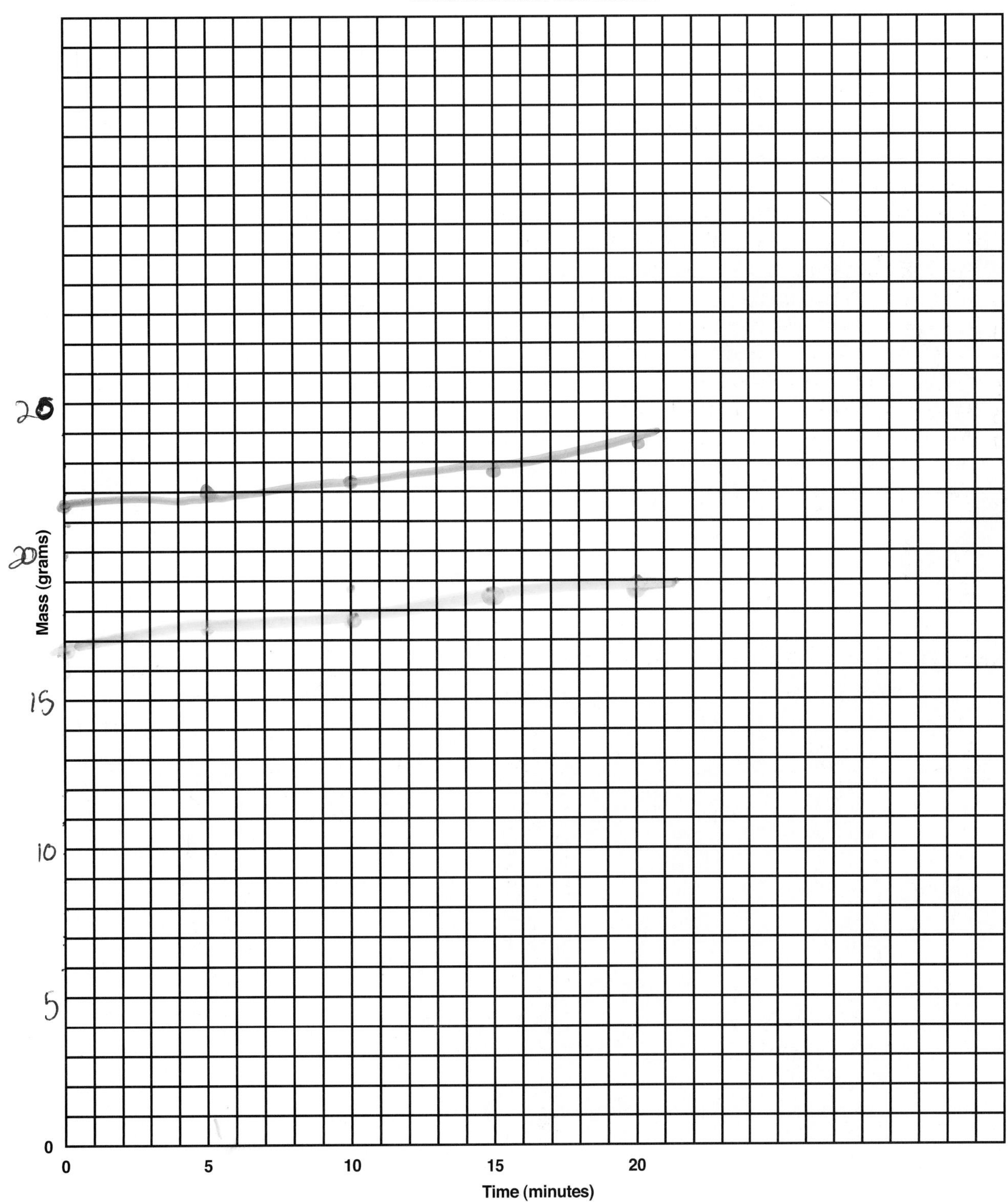

3 Diffusion and Osmosis

Name ______________________________ Lab section ____________

Your instructor may collect these end-of-exercise questions. If so, please fill in your name and lab section.

End-of-Exercise Questions

1. In a perfectly tied and unbroken bag, should we see evidence of sugar molecules passing through the "membrane"? Qualify your answer in terms of how selective permeability operates.

   NO.
   The sugar molecules are too big to go through the membrane

2. From your graph of the influence of temperature on the rate of osmosis, what can you conclude about the effects of temperature on the rate of osmosis? Hint: How are temperature and molecular motion related?

   Increase temp you increase the rate of osmosis

3. Was dynamic equilibrium reached in any of the syrup demonstrations? Explain your answer.

   The syrup can't pass through the bag

4. How do differences in concentration affect the rate of osmosis? Why is there a difference?

   Increase conc. you increase rate of osmosis

5. Why does a good cook wait to put the salad dressing on a salad until just before serving? Answer by explaining what happens to the cells in a lettuce leaf when the dressing is added.

6. Human cells contain 0.9% solutes (dissolved materials). Therefore, there is 99.1% water in these cells. The Pacific Ocean contains 3.56% salt. Although this seems like a silly question, how much (%) water is in the ocean? You are cast adrift on this ocean. What would happen to your cells if you were to drink the salt water? .

7. Your younger brother just put your favorite saltwater fish into his freshwater aquarium. Predict what will happen to the fish, its cells, and your younger brother.

LABORATORY

Structure of Some Organic Molecules 4

+ Safety Box

- No unusual hazards are associated with this laboratory experience. Please follow standard laboratory safety procedures.

Objectives

Be able to do the following:

1. Define these terms in writing.

   functional group
   organic chemistry
   carbon skeleton
   hydrolysis
   covalent bond
   hydrogen bond
   dehydration synthesis
   peptide bond

2. Recognize the following functional groups by structural formula, chemical formula, or name:

   carboxylic acid group
   methyl group
   phosphate group
   ketone group
   alcohol group
   amino group
   aldehyde group

3. Recognize a carbon skeleton and identify by name the functional groups attached to it.
4. Identify the following organic molecules by their structural formulas and/or their chemical formulas:

   simple sugar (monosaccharide)
   lipid
   hydrocarbon
   phospholipid
   fatty acid
   amino acid
   unsaturated fatty acid
   purine
   complex carbohydrate
   fat
   steroid
   glycerol
   protein
   saturated fatty acid
   pyrimidine
   nucleotide

Introduction

Organic molecules are interesting because they come in many shapes and sizes. The one consistent characteristic of organic molecules is that they contain carbon atoms; virtually all organic compounds also include hydrogen atoms and one or more **functional groups** that give the molecules specific chemical properties.

Most organic molecules are large and consist of a backbone of carbon atoms arranged in chains or rings. It was originally thought that organic molecules could be constructed only by living cells. In 1923, however, a young medical student named Friedrich Wohler was the first to synthesize an organic compound, urea (NH_2—CO—NH_2), in the laboratory. The discovery that organic molecules could be synthesized without the aid of living cells required that an alternative definition of organic chemistry be developed.

The next distinction proposed was that any molecule that contained carbon could be considered organic. This distinction proved to be better but had the drawback that some very simple molecules, such as carbon monoxide (CO) and carbon dioxide (CO_2), were simple in structure and similar in behavior to inorganic molecules yet contained a carbon atom. Today, we define **organic chemistry** as the study of carbon compounds. This allows for flexibility when considering small, simple compounds, such as CO and CO_2.

The purpose of this exercise is to allow you to become familiar with the way in which formulas for organic molecules are written, how their atoms are arranged, and how they are logically grouped by their structural similarities.

Recall that carbon is an atom that is not likely to form ionic bonds. It is much more likely to share its electrons with other atoms as their orbitals overlap. This type of bonding is called **covalent**. You may want to review the section on the formation of these bonds in your textbook before beginning this activity. Carbon has 4 electrons in its outermost energy level. There are 2 electrons in an inner orbital, but they are unavailable to bond, so we generally dismiss them. The 4 electrons in the second energy level are thought to be as far apart from each other as possible.

1. To visualize this arrangement of electrons, use a spherical piece of clay and four toothpicks to show where the bondable electrons of carbon are located. Stick three toothpicks into the clay sphere so they form a tripod. Now, insert the fourth toothpick vertically on the top of the sphere. These four toothpicks are as far apart from each other as they can be and are similar in arrangement to the electrons in the second energy level of a carbon atom.

   It is very difficult to constantly diagram the atoms as three-dimensional models, so we usually draw the carbon atom with the bondable electrons at the four major compass points: N, S, E, and W. Most of the time, picturing the atoms as flat structures is sufficient to tell you how the bonds are formed.
2. Nitrogen has 5 electrons in its second energy level, 2 in the *2s* orbital, and 3 in the *p* orbitals: 1 each in the *2px, 2py,* and *2pz* orbitals. Therefore, nitrogen has 3 electrons available to form covalent bonds. Place three toothpicks in a clay sphere to help you visualize this arrangement. Ask your instructor for help if you are unsure of the electron placement.
3. Oxygen has 2 electrons available for covalent bonds, and hydrogen has only 1 electron to share. Most of the organic molecules that we will work with have various combinations of carbons, nitrogens, oxygens, and hydrogens. Complete table 4.1 to remind yourself of the bonding behavior of these four atoms.

Table 4.1 Four Most Common Atoms in Organic Compounds

Name	Symbol	Number of Outer Electrons	Bonding Capacity
Carbon	C	4	4
Hydrogen	H	1	1
Nitrogen	N	5	3
Oxygen	O	2	2

As you do this lab, use a straight line between two atoms to represent a pair of electrons that are shared. Note that there are 2 electrons, 1 from each atom, represented by this line. You might think of it as a handshake between two individuals. The handshake requires two hands, one from each of the individuals. Sometimes, we want to indicate that an atom has an electron that it is willing to share, but no other atom is sharing it. This is like a person with a hand outstretched, waiting to shake hands if anyone else wants to be greeted. A straight line from an atom with nothing on the other end of it is 1 electron available to be shared.

The sketches of the following incomplete molecules have several straight lines. Some represent 2 electrons, some 1. Place a "2" or a "1" on each line to indicate if the line represents 2 shared electrons or an atom willing to share an electron.

```
    H
    |                    |   |
  — C — H            H — C — C —
    |                    |   |
                         H   H
```

The next two sketches contain a double bond. If a single straight line between two atoms represents 2 shared electrons, two such lines between two atoms represent 4 shared electrons. Place a "4" on the double bonds, a "2" on each line that represents a single bond between two atoms, or a "1" on lines that represent unpaired electrons in the sketches.

```
    H     O                O
    |    //               //
H — C — C               — C
    |    \                 \
    H     H                 H
```

The unique characteristic of organic molecules is the chain, or ring, of carbon atoms. The molecule that follows has a 3-carbon chain in the center. It is commonly called propane.

```
    H   H   H
    |   |   |
H — C — C — C — H
    |   |   |
    H   H   H
```

Propane

This chain of carbon atoms has only hydrogen atoms bonded to it. Such molecules that are composed of only carbons and hydrogens are called **hydrocarbons.** This arrangement of carbon and hydrogen is the basic structural component of many organic molecules. We call this carbon chain the **carbon skeleton,** or backbone, of the molecule. Suppose we take 1 of the carbons (and its hydrogens) off the chain. The carbon with 3 hydrogens and 1 electron available for bonding is indicated in two ways:

```
                    H
                    |
— CH3      or     — C — H
                    |
                    H
```

Methyl group

This arrangement of atoms is termed a **methyl group.** Another way to look at the 3-carbon propane molecule is to think of it as having a central carbon atom with a methyl group on each side. Now, let's replace the methyl group with an amino group. The **amino group** is composed of 1 nitrogen with 2 hydrogens attached and 1 electron available for bonding. An amino group can be indicated as follows:

```
                      H
                     /
— NH2      or     — N
                     \
                      H
```

Amino group

Sketch a 2-carbon skeleton. It should be composed of 2 carbon atoms, 5 hydrogens, and 1 bondable electron.

Now, attach an amino group to the bondable electron of the carbon skeleton you just drew. Label your sketch with the terms *carbon skeleton* and *amino group.*

A carbon bonded to an OH group and double-bonded to an oxygen (O) forms a **carboxylic acid group.** Following is a 2-carbon skeleton with a carboxylic acid group on it. Draw a square around the carbon skeleton and label it, and then draw a circle around the acid group and label it.

Carboxylic acid

A carbon skeleton that has a carbon with 2 hydrogens and 1 OH group attached to it is termed an **alcohol.** The OH group can be attached anywhere in the carbon chain. Draw a 3-carbon skeleton below, with hydrogens at every available bonding point. The formula should be C_3H_8.

Now, draw the same 3-carbon skeleton and replace 1 of the hydrogens on the middle carbon with an OH group (an oxygen bonded to a hydrogen).

3-carbon skeleton full of hydrogen

3-carbon skeleton with an alcohol functional group

Highlight the alcohol group in your last sketch by drawing a circle around it and labeling it *alcohol group.* Look back at the carboxylic acid group on the carbon skeleton. Note that it has an OH group as a part of the acid group. However, it doesn't act like an alcohol because the double-bonded oxygen together with the OH group causes the OH to have different properties. It is important that you recognize the functional groups, because their presence is responsible for the behavior of the organic molecules.

So far, we have identified the following functional groups; a *methyl group,* an *amino group,* a *carboxylic acid group,* and an *alcohol group.* In addition to these four, you need to recognize a *phosphate group,* an *aldehyde,* and a *ketone group.*

A **phosphate group** is easily recognized because it has phosphorus as a central part of the group. The formula for the phosphate is $H_2PO_4^-$:

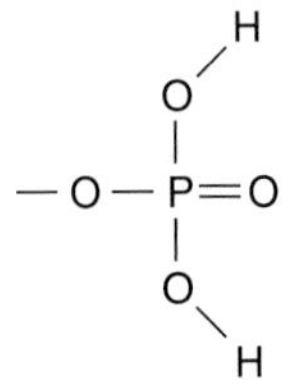

Phosphate group

The oxygen at the left side of the molecule has a bondable electron available to share with a carbon of a carbon skeleton. Therefore, the phosphate group can be attached to any point on the carbon skeleton where there is an electron available to form a covalent bond. (You may remember from a previous discussion of phosphate that it frequently forms phosphoric acid by bonding to a hydrogen at the bondable point. The formula for phosphoric acid is H_3PO_4.)

Sketch a 2-carbon skeleton and add a phosphate group to the carbon on the right end of the molecule. Draw a square around the carbon skeleton and label it; then draw a circle around the acid group and label it.

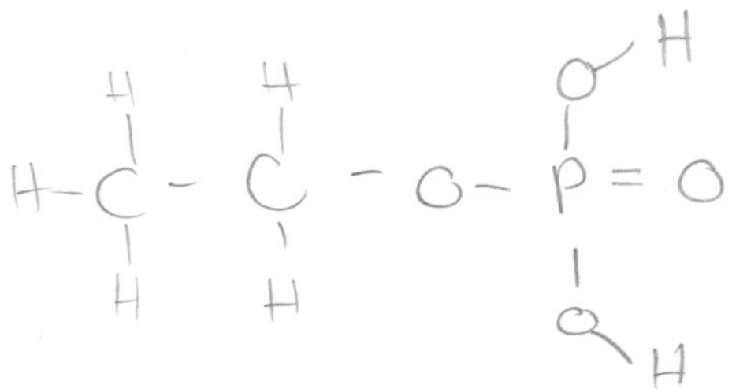

2-carbon skeleton with a phosphate group

The two final groups that you need to recognize are aldehydes and ketones. These both have double-bonded oxygens. The difference between them is the location of the double-bonded oxygen within the carbon chain. If the oxygen is on the end of a carbon skeleton, it is an **aldehyde;** if the oxygen is on a carbon in the middle of a carbon skeleton, then the group is a **ketone group.** Note the following two molecules. Circle the two double-bonded oxygens. Label the one on the left *aldehyde* and the one on the right *ketone.*

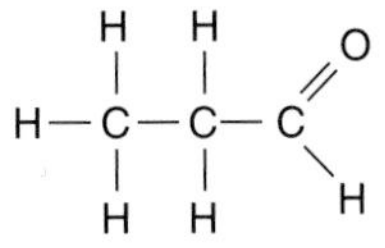

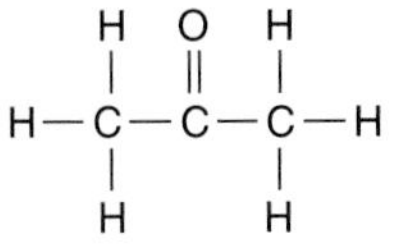

Following are several carbon skeletons with functional groups on them. Identify as many functional groups as you can. There may be more than one functional group per carbon skeleton.

build These
B 1-6 pg 21

$H_3C - CH_2 - CH_3$

$H_2N - CH_2 - CH_3$

$H_3C - C(=O) - CH_3$

$H_3C - CH_2 - C(=O) - O - H$

$H_3C - CH_2 - C(=O) - H$

$H_3C - CH_2 - CH_2 - O - H$

$HO - CH_2 - CH(OH) - CH_2 - OH$

$H_3C - CH_2 - COOH$

$H_3C - CH_2 - C(=O) - H$

Draw a carbon skeleton composed of 6 carbons and 14 hydrogens.

6-carbon skeleton full of hydrogen

Now, draw this carbon skeleton composed of 6 carbons, but this time have the skeleton form a ring with the sixth carbon attaching to the second carbon. How many hydrogens will share all the available bondable electrons in this molecule? This is easy if you first bond the carbons together and draw lines from each carbon to show the remaining bondable electrons. Put the 12 hydrogens on the molecule now. There is a methyl group on this molecule. Draw a circle around it and label it *methyl group.*

6-carbon skeleton *ring* full of hydrogen

Draw the same molecule three more times, and on the first one replace the methyl group with a methanol (a carbon with one alcohol group and 2 hydrogens). On the second, replace the methyl group with an aldehyde, and on the third, replace one of the hydrogens on the methyl group with an amino group and another of the hydrogens from the methyl group with a carbon attached to a carboxylic acid group. If you have difficulty with these, ask your instructor for help.

6-carbon skeleton with a methanol

6-carbon skeleton with an aldehyde

6-carbon skeleton plus an amino group plus a carboxylic acid group

Four Types of Organic Molecules Common in Living Things

Organic molecules are frequently categorized into classes based on their structures. The four classes of organic molecules we will examine in this exercise are the carbohydrates (simple and complex); the lipids, which can be subdivided into fat, phospholipids, or steroids; the proteins, which are composed of amino acids; and the nucleic acids, which are constructed from RNA and DNA nucleotides.

Carbohydrates

Carbohydrates have the general formula $C_xH_{2x}O_x$. Because it appeared to early scientists that $C(H_2O)$ described the molecule, they thought of these molecules as being composed of carbon to which water had been added. Hence, the name *carbohydrate* (watered carbon) was appropriate.

$C_6H_{12}O_6$ is a carbohydrate, and so is $C_3H_6O_3$. These molecules are sketched below, with carbon skeletons in a ring. By convention, organic chemists commonly abbreviate —O—H to —OH.

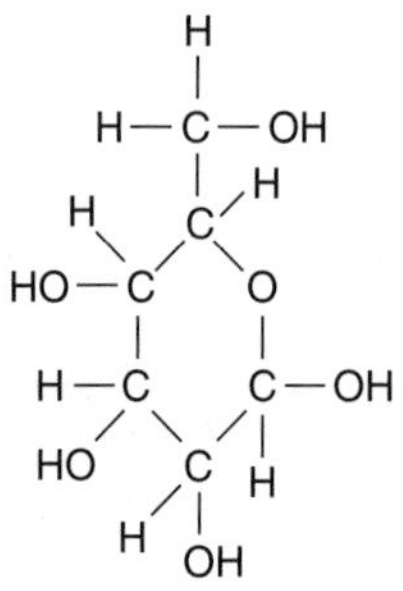

$C_6 H_{12} O_6$

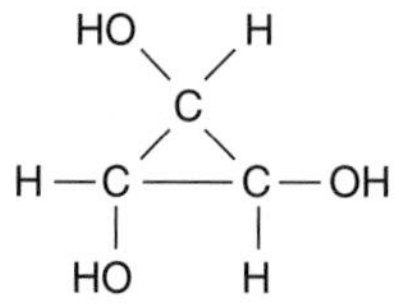

$C_3 H_6 O_3$

Complex carbohydrates, such as starch and cellulose, can be broken down into simple sugars known as **monosaccharides.** Simple sugars can be joined together to form disaccharides and polysaccharides. Note the highlighted atoms on the following molecules. The joining of two simple sugars involves interactions between OH groups on the molecules. The bond between the oxygen and carbon of one is broken, and the bond between the hydrogen and oxygen of the second OH group is broken. These fragments, OH from one sugar and H from the other, form a molecule of water. This causes a bondable electron to be available on each of the sugars: the one on the oxygen, which lost its hydrogen, and the one on the carbon, which lost its OH group. These two bondable electrons form a covalent bond and hold the two remaining parts of the sugars together. Look closely at the following sketch. Note the portions labeled sugar #1, sugar #2, attachment bond, and water. The arrow in the sketch represents a reaction: The two molecules on the left, sugars #1 and #2, have become the **disaccharide** and water on the right.

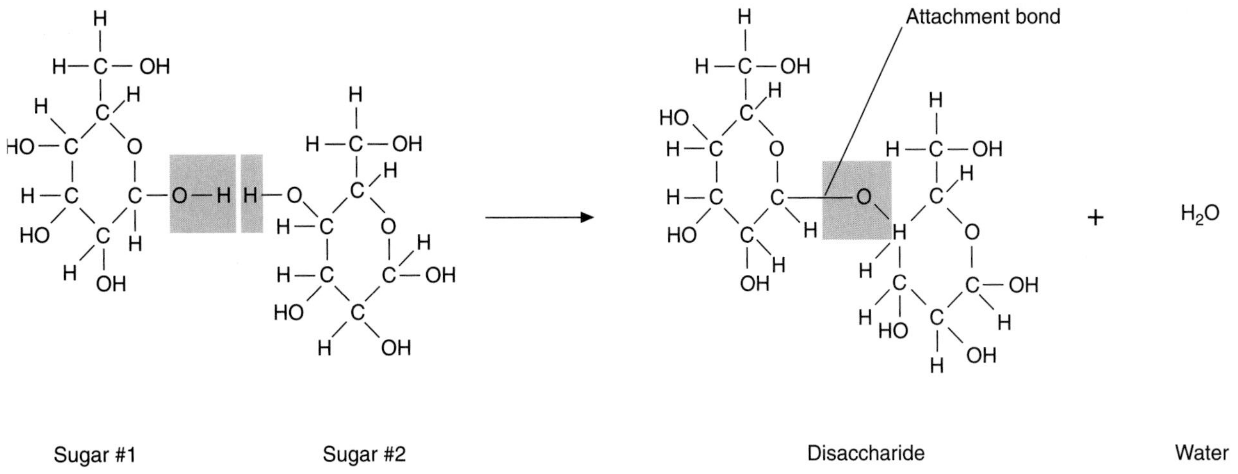

Attachment bond

$+$ H_2O

Sugar #1 Sugar #2 Disaccharide Water

A sugar has the formula $C_xH_{2x}O_x$. Does the sugar on the right have this type of formula? Count the carbons, hydrogens, and oxygens in the whole molecule and write them in the following manner:

C H O

The new disaccharide is 1 oxygen and 2 hydrogens short of a carbon-hydrogen-oxygen ratio of 1:2:1. This is typical of a complex carbohydrate. When two simple sugars (which have the expected ratio of carbon, hydrogen, and oxygen) react with each other and form a larger sugar molecule, they lose a water molecule from between them. Therefore, the complex sugar molecule is missing 2 hydrogens and 1 oxygen. This reaction is called a **dehydration synthesis reaction** because a larger molecule is being synthesized, and a water is being removed. Sometimes, it is termed a **condensation reaction** because two small molecules have been consolidated into one larger molecule.

You will be able to classify any carbohydrate now as one that is simple or complex by determining if it fits the pattern of equal numbers of carbons and waters, or if it is missing a water. What if you attach three simple sugars together? The resulting molecule is missing two waters, one between sugars #1 and #2 and one between sugars #2 and #3.

Following is a list of chemical formulas of carbohydrates. Beside each, put an *S* if it is a simple carbohydrate (simple sugar) or a *C* if it is a complex carbohydrate.

$C_3H_6O_3$ $C_6H_{12}O_6$ $C_4H_8O_4$ $C_{18}H_{30}O_{15}$
$C_5H_{10}O_5$ $C_9H_{14}O_7$ $C_7H_{14}C_7$ $C_{15}H_{26}O_{13}$

A simple sugar is one *saccharide* unit by itself, or ***mono***saccharide. Two sugars bonded together form a ***di***saccharide, and three sugars joined form a ***tri***saccharide. In the preceding list, behind each of the carbohydrates, put a *Mono, Di,* or *Tri* to indicate how many sugars are attached. (Remember that you can easily tell this by determining how many waters are missing.)

On page 50 are some structural formulas of carbohydrates. Label each as an *S* (simple carbohydrate) or *C* (complex carbohydrate), and then label the *monosaccharides, disaccharides,* and *trisaccharides.* Note that each of the sugars, when attached to another sugar, is bonded through an oxygen linkage. This is because the water that formed did so by removing an H and OH. Soon you will be looking at a structural formula and keying right in on the oxygen that links these two parts together.

Lipids

Lipids are organic molecules characterized by the fact that they are insoluble in water. Although they are composed of carbon, hydrogen, and oxygen, the amount of oxygen is much less than the amount of carbon. You will not confuse these two formulas if you remember that the name *carbohydrate* tells you it is "watered carbon" and implies nearly equal amounts of carbon and oxygen, whereas the lipid has much less oxygen.

Lipids are subdivided into three groups: *fats, phospholipids,* and *steroids.* **Fats** are lipid molecules composed of two kinds of building blocks: a *glycerol* and *three fatty acids.* The **glycerol** molecule is a 3-carbon skeleton with three alcohol groups, one on each carbon. Its chemical formula is $C_3H_5(OH)_3$. Sketch this molecule:

1. ______ ______

2. ______ ______

3. ______ ______

4. ______ ______

5. ______ ______

6. ______ ______

7. ______ ______

8. ______ ______

A **fatty acid** is a long carbon skeleton, usually 10 or more carbons with a *carboxylic acid group* at one end. Draw a circle around the acid group in this *stearic acid* molecule:

```
O        H   H   H   H   H   H   H   H   H   H   H   H   H   H   H   H   H
 \\      |   |   |   |   |   |   |   |   |   |   |   |   |   |   |   |   |
   C  —  C — C — C — C — C — C — C — C — C — C — C — C — C — C — C — C — C — H
 /       |   |   |   |   |   |   |   |   |   |   |   |   |   |   |   |   |
HO       H   H   H   H   H   H   H   H   H   H   H   H   H   H   H   H   H
```

Stearic acid

Notice in the stearic acid that the carbon skeleton does not have any double bonds. It has as many hydrogens as it can possibly hold. We say that this particular fatty acid is **saturated** with hydrogens. In other words, it is completely *hydrogenated.*

Following is the structural formula of an **unsaturated fatty acid**—*oleic acid.* Draw a circle around the *carboxylic acid group* and then put an arrow near the points on the carbon skeleton where it is unsaturated (where double bonds form between the carbons).

```
O        H   H   H   H   H   H   H   H   H   H   H   H   H   H   H   H   H
 \\      |   |   |   |   |   |   |   |   |   |   |   |   |   |   |   |   |
   C  —  C — C — C — C — C — C — C — C = C — C — C — C — C — C — C — C — C — H
 /       |   |   |   |   |   |   |           |   |   |   |   |   |   |   |
HO       H   H   H   H   H   H   H           H   H   H   H   H   H   H   H
```

Oleic acid

A glycerol can react with three fatty acids to form a fat. The reactions are dehydration synthesis reactions, and they occur between the OH group of the glycerol and the OH portion of the acid group on the fatty acid. Following is a glycerol with one dehydration synthesis reaction completed, a second fatty acid in position to react, and a space for a third fatty acid. Draw a rectangle around the glycerol and an oval around the fatty acid ready to attach. Add a third fatty acid and show it attached to the glycerol at the appropriate point.

```
    H           O   H   H   H   H   H   H   H   H   H   H   H   H   H   H   H   H   H
    |           ||  |   |   |   |   |   |   |   |   |   |   |   |   |   |   |   |   |
H — C — O — C — C — C — C — C — C — C — C — C — C — C — C — C — C — C — C — C — H    +    H — O
    |               |   |   |   |   |   |   |   |   |   |   |   |   |   |   |   |   |              \
    |               H   H   H   H   H   H   H   H   H   H   H   H   H   H   H   H   H               H
    |
    |
    |                       O   H   H   H   H   H   H   H   H   H   H   H   H   H   H   H   H   H
    |                       ||  |   |   |   |   |   |   |   |   |   |   |   |   |   |   |   |   |
H — C — O — H  H — O — C — C — C — C — C — C — C — C — C — C — C — C — C — C — C — C — C — H    +    ______
    |                           |   |   |   |   |   |   |   |   |   |   |   |   |   |   |   |   |
    |                           H   H   H   H   H   H   H   H   H   H   H   H   H   H   H   H   H
    |
    |
H — C — OH
    |
    H
```

Did you add a fatty acid that was saturated or unsaturated? ______________________________

A **phospholipid** contains glycerol and fatty acids, as fat molecules do, but differs from fats in that one fatty acid chain is replaced by a complex group of atoms that contains a phosphate group. Next is a sketch of a phospholipid. Compare it with a fat. Draw a circle around the portion of the structural formula that makes it a phospholipid rather than a fat.

```
    H         O   H   H   H   H   H   H   H   H   H   H   H   H   H   H   H   H   H
    |         ||  |   |   |   |   |   |   |   |   |   |   |   |   |   |   |   |   |
H — C — O — C — C — C — C — C — C — C — C — C — C — C — C — C — C — C — C — C — C — H   +  H — O
    |             |   |   |   |   |   |   |   |   |   |   |   |   |   |   |   |   |              \
    |             H   H   H   H   H   H   H   H   H   H   H   H   H   H   H   H   H               H
    |
    |         O   H   H   H   H   H   H   H   H   H   H   H   H   H   H   H   H   H
    |         ||  |   |   |   |   |   |   |   |   |   |   |   |   |   |   |   |   |
H — C — O — C — C — C — C — C — C — C — C — C — C — C — C — C — C — C — C — C — C — H   +  H — O
    |             |   |   |   |   |   |   |   |   |   |   |   |   |   |   |   |   |              \
    |             H   H   H   H   H   H   H   H   H   H   H   H   H   H   H   H   H               H
    |
    |         O       H   H      CH3
    |         ||      |   |     /
H — C — O — P — O — C — C — N                        +                              H — O
    |         |       |   |     \                                                          \
    H         O       H   H      CH3                                                        H
              |
              H
```

When fats or phospholipids are synthesized, they react with their component parts, and water is removed from between them. When they are digested, or broken down, the reverse occurs. When water is added to a large molecule to break it into smaller component parts, the reaction is called **hydrolysis.**

Steroids are complex lipids composed of four interlocking carbon rings. Following are three steroids. Write the chemical formula for each below its structural formula.

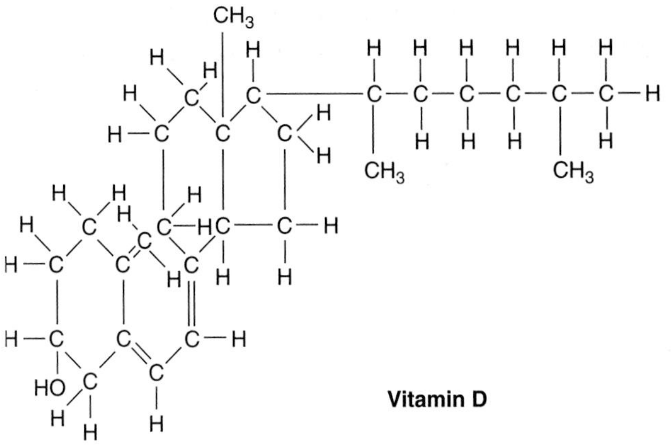

Vitamin D

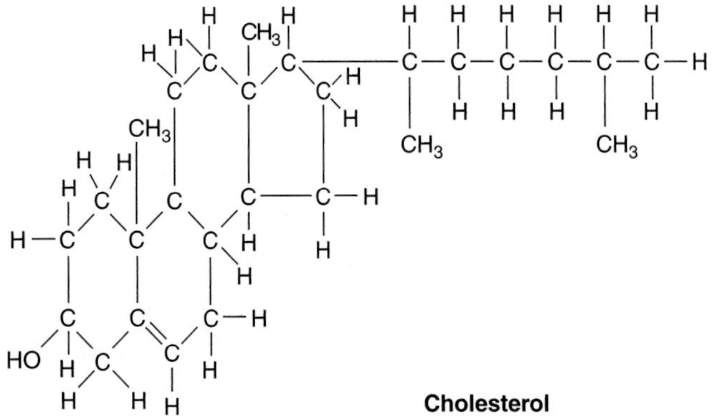

Cholesterol

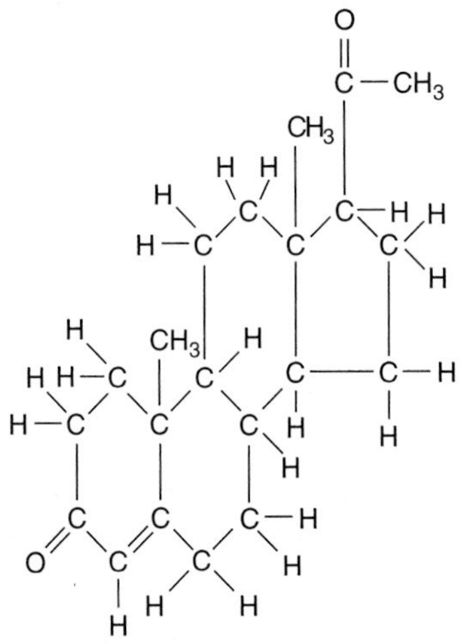

Progesterone

Proteins

Proteins are composed of building blocks called **amino acids.** Each amino acid is composed of a carbon skeleton with an *amino group* and a *carboxylic acid group* attached. The remainder of the carbon skeleton varies from one amino acid to the next. In each of the following two sketches, draw a circle around the amino group, a triangle around the carboxylic acid group, and a square around the carbon skeleton:

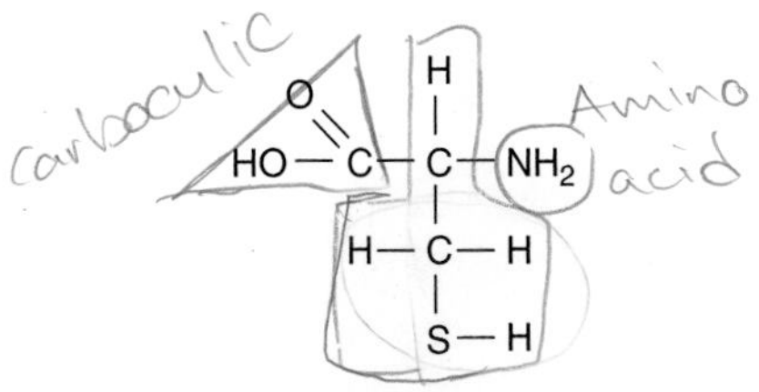

Cysteine

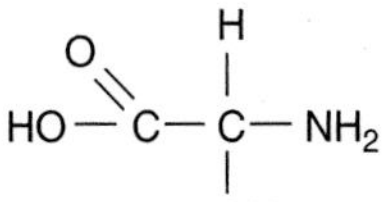

Alanine

When two amino acids react with each other, water is removed between them. They react by removing a hydrogen from the amino group and an OH from the carboxylic acid. Notice that, when this happens, there is no oxygen left behind to hold the two molecules together.

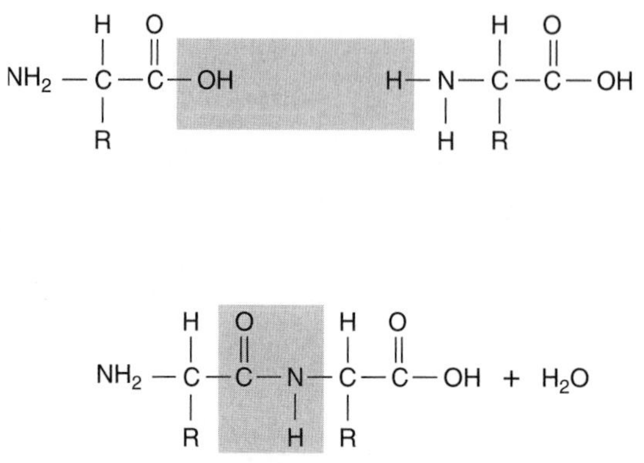

```
        H  O                      H  O
        |  ||                     |  ||
NH2 — C — C — OH          H — N — C — C — OH
        |                     |   |
        R                     H   R
```

```
          H  O     H  O
          |  ||    |  ||
NH2 — C — C — N — C — C — OH  +  H2O
          |       |   |
          R       H   R
```

They bond directly from the nitrogen of the amino group to the carbon of the carboxylic acid group. This bond, which occurs only between amino acids, is called a **peptide bond.** A molecule that consists of several amino acids joined by peptide bonds is called a *polypeptide.* Draw ovals around the peptide bonds in the following sketch:

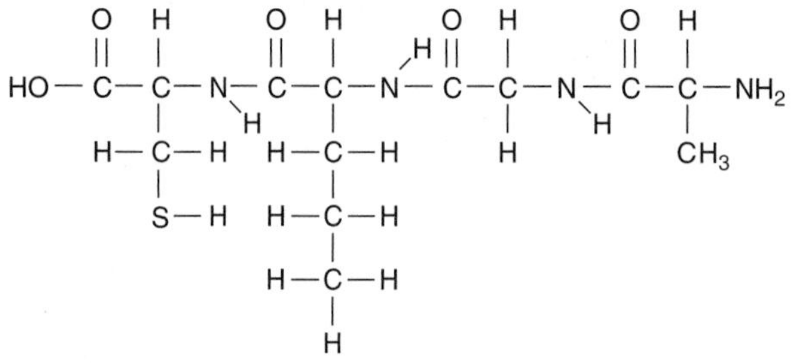

```
      O  H        O  H          O  H        O  H
      || |        || |        H ||  |        || |
HO — C — C — N — C — C — N — C — C — N — C — C — NH2
         |    H      |              |    H      |
     H — C — H   H — C — H          H          CH3
         |           |
     S — H       H — C — H
                     |
                 H — C — H
                     |
                     H
```

Polypeptide

Draw a dipeptide composed of cysteine and alanine in the following space.

Nucleic Acids

The last group of complex organic molecules we will study are the **nucleic acids.** Nucleic acids are composed of subunits called **nucleotides.** There are eight different nucleotides. Each nucleotide is constructed of a *phosphate group,* a *sugar,* and a *nitrogenous base.*

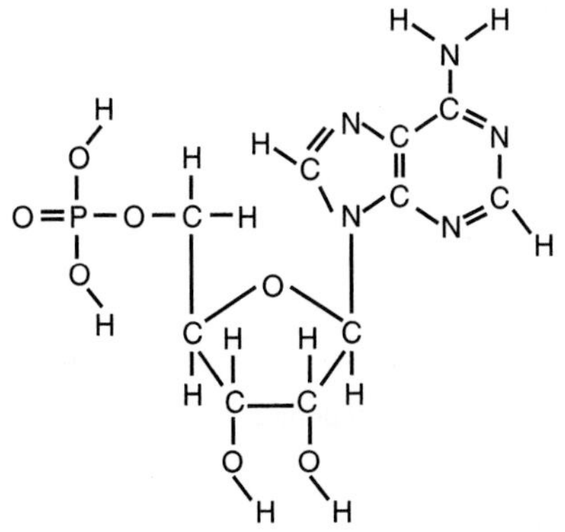

Nucleotide

Two kinds of 5-carbon sugars can be part of a nucleotide: *ribose* and *deoxyribose.* Nucleotides containing ribose are RNA nucleotides, and nucleotides containing deoxyribose (lacking one oxygen atom) are DNA nucleotides. Look at the carbon atom in the lower right-hand portion of the ribose and deoxyribose to see how they differ. Label the nucleotide above as either a ribose or deoxyribose nucleotide.

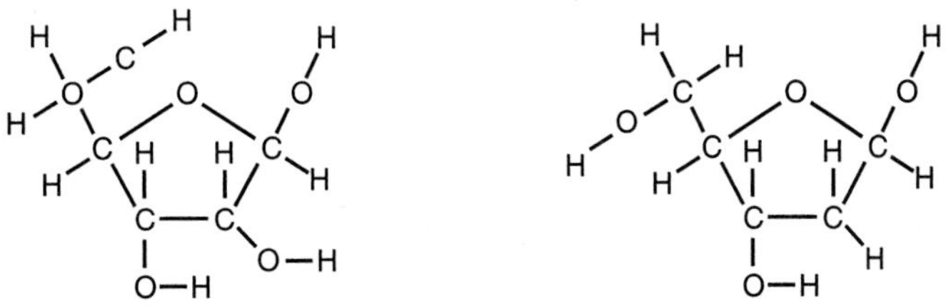

Ribose

Deoxyribose

There are five different nitrogenous bases that can be part of the nucleotide structure: *adenine, guanine, cytosine, thymine,* and *uracil.* Cytosine, thymine, and uracil are single-ring molecules called **pyrimidines.** Adenine and guanine are larger double-ring molecules called **purines.** DNA nucleotides can contain adenine, guanine, cytosine, or thymine. RNA nucleotides contain adenine, guanine, cytosine, or uracil.

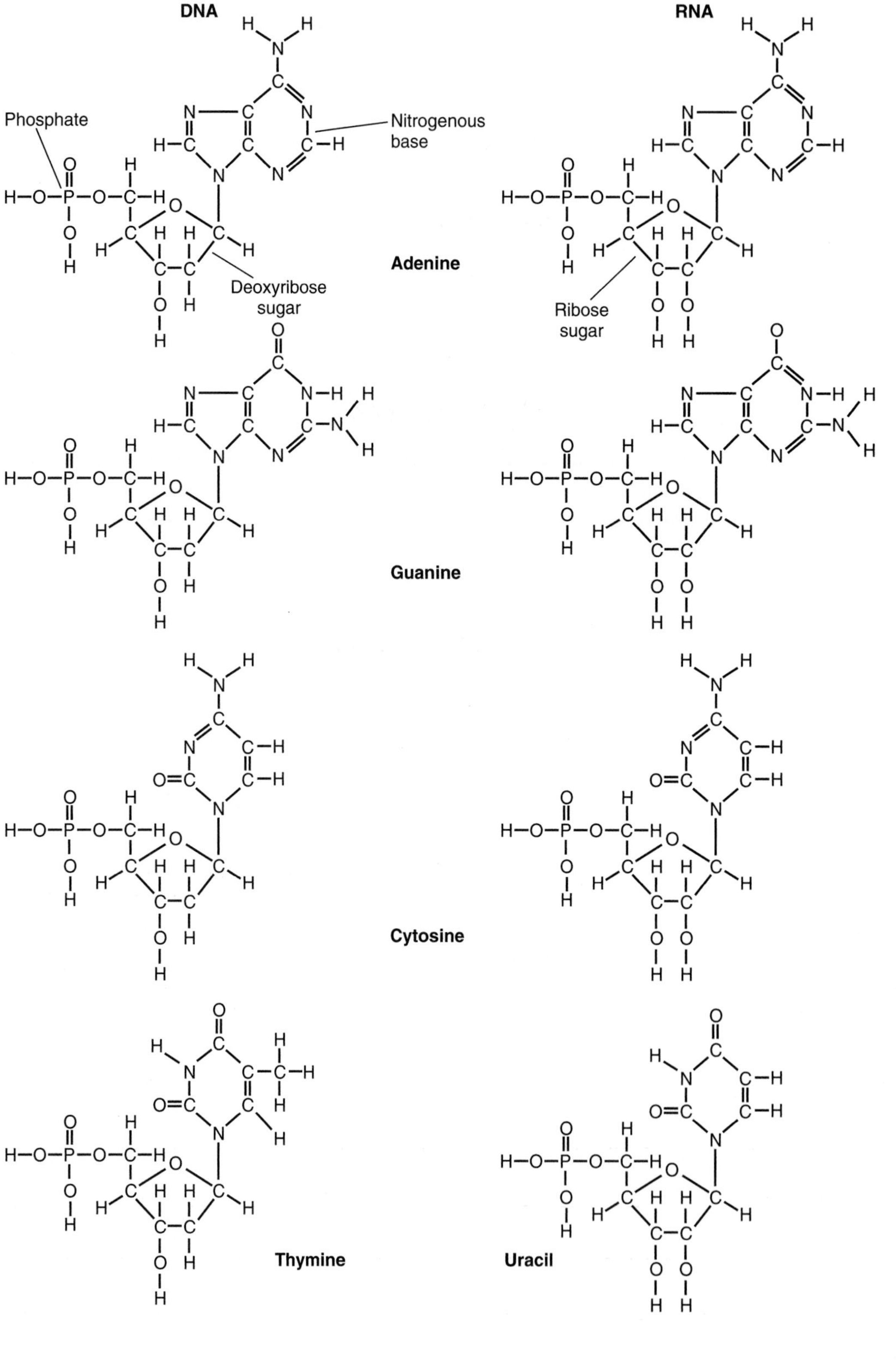

Each nitrogenous base fits with specific nitrogenous bases of other nucleotides. Guanine fits with cytosine, and adenine fits with either thymine or uracil. Individual nucleotides combine by reactions that occur between the sugar subunit of one nucleotide and the phosphate group of a second nucleotide. Several such reactions result in long nucleotide chains. The sugars and phosphates form the backbone of the nucleotide chain. With arrows, indicate the locations on the nucleotide chain above where dehydration synthesis reactions have occurred between nucleotides.

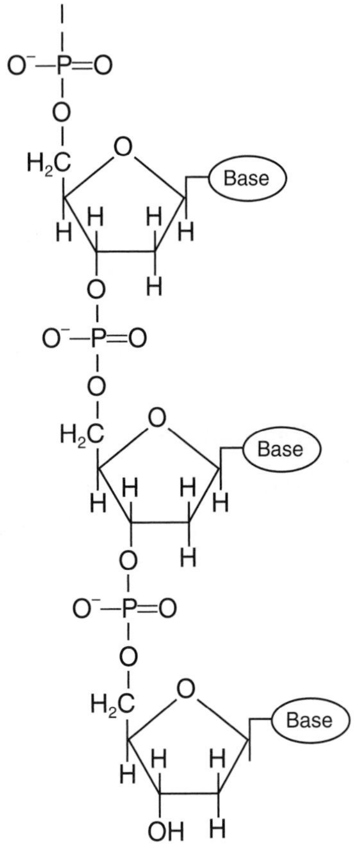

Review Problems

Review the terms *carbon skeleton, functional group, dehydration synthesis,* and *hydrolysis.* Make sure that you can use them correctly to describe what is being sketched in a structural formula.

Following is a list of chemical formulas. They are of simple carbohydrates, complex carbohydrates, alcohols, aldehydes, ketones, carboxylic acids, fatty acids, and amino acids. Identify each as completely as possible.

1. $C_6H_{10}O_5$ ______________________

2. $C_3H_5(OH)_3$ ______________________

3. $C_3H_7(OH)$ ______________________

4. $CH_3—CO—CH_3$ ______________________

5. $CH_3—CH_2—CHO$ ______________________

6. $CH_3—CH_2—COOH$ ______________________

7. $CH_3—(CH_2)_{17}—COOH$ ______________________________

8. $COOH—CH_2—CH_2—NH_2$ ______________________________

9. $CH_3—CH—NH_2$ with COOH attached to the CH (drawn below it) ______________________________

10. $COOH—CH_2—NH_2$ ______________________________

11. $C_{15}H_{22}O_{11}$ ______________________________

12. $C_6H_{12}O_4$ ______________________________

13. $C_9H_{14}O_7$ ______________________________

Summary of Hydrolysis and Dehydration Synthesis Reactions

Macromolecules, large organic molecules, are formed by the removal of water from two adjacent functional groups. This process is called dehydration synthesis. A macromolecule can in turn be broken down by the addition of water between two adjacent subunits. This process is called hydrolysis.

Dehydration Synthesis:

H—*subunit*—OH + H—*subunit*—OH ⟶ H—*subunit*—*subunit*—OH + HOH

Hydrolysis:

H—*subunit*—*subunit*—OH + H_2O ⟶ H—*subunit*—OH + H—*subunit*—OH

What can you say about the equations for these two reactions that will help you remember which is which?

4 Structure of Some Organic Molecules

Name ______________________________ Lab section ____________

Your instructor may collect these end-of-exercise questions. If so, please fill in your name and lab section.

End-of-Exercise Questions

1. What does a straight line extending from an atom (—H) represent?

   an open single covalent bond

Quiz

2. What does a straight line connecting two atoms (H—O) represent?

   Covalent bond (single)

3. Identify the following functional groups:

   $—CH_3$ ______________________ $—NH_2$ ______________________

   —COOH ______________________ —OH ______________________

   —COH ______________________ —C(=O)— ______________________

   $—PO_4^{-2}$ ______________________

4. List by symbol the types of atoms that can be found in

   carbohydrates ______________________.

   lipids ______________________.

   proteins ______________________.

   nucleic acids ______________________.

5. Even though carbohydrates and lipids are composed of the same three types of atoms, list at least two ways in which they differ. In other words, how do you distinguish a carbohydrate from a lipid?

   Lipids have less oxygen than carbohydrates

   Carbohydrate are soluable in water but lipids are not

6. Label the following reactions as either dehydration synthesis or hydrolysis:

   a. Monosaccharide + monosaccharide ⟶ disaccharide + water

   dehydration synthesis

   b. $C_{12}H_{22}O_{11} + H_2O \longrightarrow C_6H_{12}O_6 + C_6H_{12}O_6$

   hydrolosis

7. *Catabolic* reactions, such as digestion, result in the breakdown of larger organic compounds into smaller subunits. *Anabolic* reactions involve the buildup of large macromolecules from smaller building blocks. Dehydration synthesis must be associated with ______________________,

   and hydrolysis is associated with ______________________.

8. What is the difference between saturated fat and unsaturated fat?

Unsaturated double bonds less than the number H
Saturated no doube bonds & maximum # of H

9. How do phospholipids differ from fats?

They contain a phosphate group.
Phospholipids are watersoluble

10. Name a distinguishing feature of steroids.

Composed of four interlocking rings

11. Explain why the name *amino acid* is an appropriate one for the building blocks of protein molecules.

Come from an amino group & a carbocylic acid.

12. What are the components of a nucleotide?

Phosphate group
sugar nitrogenous base

LABORATORY

5 The Microscope

Safety Box

- A microscope is very expensive to replace; therefore, be particularly careful when handling one. Use both hands to carry it around the room. Hold it upright, so that the ocular lens does not slip out. Use only clean, dry lens paper to clean dust off the glass lenses. Do not use wet paper, paper towels, or other materials that may scratch these lenses.
- Slides and coverslips are glass. Be careful not to cut yourself when using them. Carefully clean slides before you use them, because dust and fingerprints interfere with your ability to see a specimen.
- Dispose of broken glass or organic materials as indicated by your instructor.

Objectives

Be able to do the following:

1. Define these terms in writing.

   - coarse adjustment knob
   - fine adjustment knob
   - stage
   - iris diaphragm
   - ocular lens
   - objective lens
   - parfocal capability
   - magnifying power
   - field of view
   - depth of field
   - resolving power
   - slide
   - coverslip
   - wet mount slide
   - scanning lens
   - compound microscope
   - plane of focus
   - oil-immersion lens
   - low-power lens
   - high-power lens
   - condenser

2. When given a slide, coverslip, and specimen, construct a temporary wet mount and focus on the specimen in the center of the field of view, using low or high power.
3. Describe in writing or draw what is viewed through the microscope.
4. Determine the total magnification of a set of lenses, given the magnification of each lens.
5. Describe the function of the various parts of a microscope.

Introduction

Because biological study includes the microscopic examination of one-celled organisms and the cells and tissues of multicellular organisms, it is important to learn the correct procedures for efficient operation of a light microscope. In addition to giving you an opportunity to learn proper microscope technique, this exercise also gives you a chance to practice using a microscope.

Preview

During this lab exercise, you will

1. identify and name the parts of a light microscope and describe the functions of the various parts.
2. determine the total magnification of a set of lenses in a compound microscope.
3. focus on a practice slide.
4. focus on crossed hairs in a temporary wet mount to determine the depth of field of a set of lenses.

Procedure

Carefully carry your assigned microscope to your work space using both hands. One hand should hold the microscope by the arm and the other hand should support the microscope base.

Refer to figure 5.1 and table 5.1 to familiarize yourself with the operation and function of each microscope part.

Magnification

The **compound microscope** is a device that uses two sets of lenses to increase the apparent size of objects. The two lenses are known as the **ocular lens** (the lens you look through) and the **objective lens,** which is near the stage. The **magnifying power** (how much it magnifies) of each lens is marked on its tubular housing. Simply multiply the magnifying power marked on the ocular lens housing times the value marked on the objective lens housing to determine how many times your specimen is enlarged. Notice that your ocular lens magnification is 10. If the **low-power lens** is also marked 10, the total low-power magnification is $10 \times 10 = 100$.

Microscopes often have additional objective lenses—namely, a **scanning lens,** which typically has a magnifying power of 4 and is used for initial viewing of the specimen; a **high-power lens,** which typically has a magnifying power of 40 or 45; and an **oil-immersion lens,** which typically has a magnifying power of 100. As the magnifying power increases, the lenses get longer. *Use of the oil-immersion lens requires special training, so do not use it unless instructed to do so by your instructor.* Improper use could cause severe and costly damage to the oil-immersion lens.

Calculate the magnification of your microscope when the high-power objective lens is used. ______________

Resolving Power

Resolving power is a measure of lens quality. Quality lenses have a high **resolving power,** which is the capacity to deliver a clear image in fine detail. If a lens has a high magnifying power but a low resolving power, it is of little value. Although the image may be large, it is not clear enough to show fine detail.

Another factor that influences resolving power is the cleanliness of the lenses. Dirt, water, or oil on the lens may scatter light and reduce the effective resolving power of the microscope. Therefore, lenses should always be kept clean. *Use only lens paper to clean the lenses.*

Figure 5.1 Light microscope components and their functions. (Courtesy of Leica, Inc.)

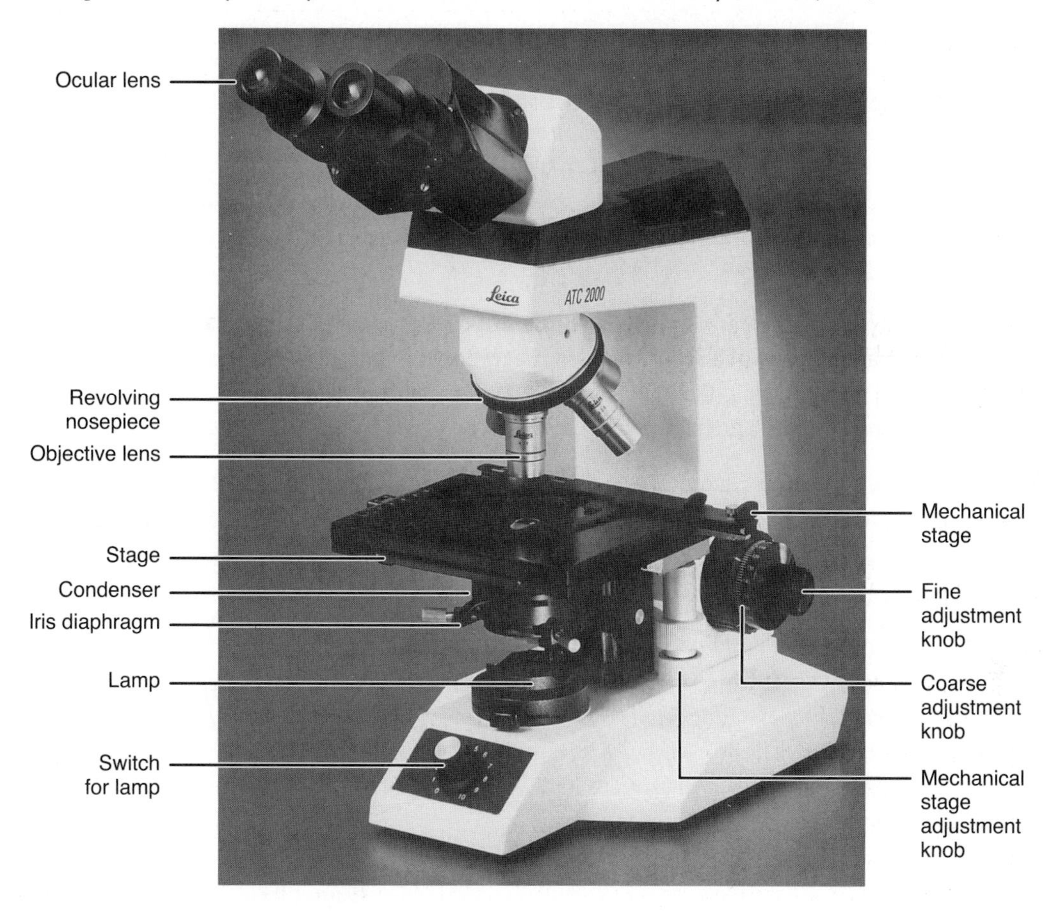

Table 5.1 Parts and Function of a Light Microscope

Part	Function
Ocular lens (eyepiece with pointer)	Lens through which you view magnified specimen; pointer may appear as a needle or as a curved line
Revolving nosepiece	Movable mount for selecting the objective lens that provides the magnification desired
Objective lens	Lens on revolving nosepiece, which accomplishes the initial magnification of the specimen
Stage	Flat work surface upon which the slide is placed
Iris diaphragm	Regulates the amount of light passing through the stage aperture and specimen
Lamp	Constant light source beneath the iris diaphragm
Mechanical stage	A microscope with a mechanical stage has a lever that is opened laterally *(never lifted)* to accept and secure the slides to the stage; control knobs are used for precise movement of a slide on the stage
Coarse adjustment knob	Gives initial focus on low power
Fine adjustment knob	Gives refined focus on low power, high power, and oil immersion
Condenser	Focuses light from the light source on the specimen on the slide

Field of View

You have already learned that lenses can have different magnifying powers, but it is also important to understand that each lens has a particular *field of view*. The **field of view** is the size of the area that the lens views. *The larger the magnifying power of an objective lens, the smaller the area viewed.* This is sometimes hard to appreciate, because to you—the observer—the size of the circle of light you see through the ocular lens appears the same for all powers of the objective lenses. When you switch from low power to high power, you are actually looking at the central portion of what was visible under low power. Therefore, it is important to *center the specimen on low power before making the switch to high power.* Figure 5.2 shows how the field of view differs when using low and high power.

Parfocal Capability

A feature of a good-quality microscope is its **parfocal capability.** This means that, when a specimen is in focus under low-power magnification, you can switch to high-power magnification and have the specimen remain in reasonably good focus. Usually, just a slight touch to the fine adjustment knob is all that is needed to sharpen the focus. Of course, it is imperative that your specimen be accurately centered just before you switch over to high power.

Figure 5.2 On high power an object is more highly magnified than on low power, but the field of view is smaller.

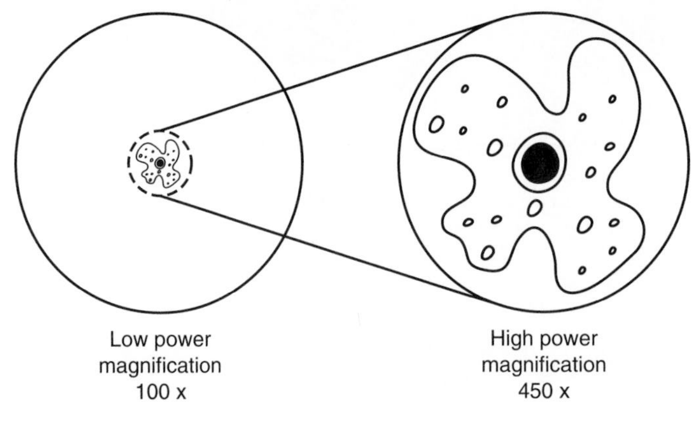

Viewing and Focusing

Before you attempt to view any specimen through the microscope, you must learn the correct use of its parts. Always use the following procedures when viewing objects through the microscope.

1. Carefully carry your assigned microscope to your work space using both hands. One hand should hold the microscope by the arm and the other hand should support the microscope base.
2. Make sure the microscope is plugged in. Turn on the light.
3. Rotate the low-power objective lens or scanning lens into position directly over the round opening in the flat stage of the microscope. Move it until it clicks into place.
4. Rotate the coarse adjustment knob so that the distance between the objective lens and the stage is at its maximum.
5. Center your specimen over the opening in the stage. Make sure that the slide is securely held in by the clips or the fingers of the mechanical stage.
6. Watch the stage and objective lens from alongside (not through) the microscope. Make the distance between the specimen and the low-power lens as small as possible.
7. While looking through the ocular lens, turn the coarse adjustment knob to move the objective lens away from the specimen until a part of the specimen comes into focus.
8. While looking through the ocular lens, center the specimen in the field of view.
9. The iris diaphragm is located below the stage; it regulates the amount of light passing through the specimen. Locate the lever that adjusts the iris diaphragm and move it so that you get a good image.
10. If you have an adjustable condenser below the stage, use the condenser knob to move the condenser up and down. Adjust the position of the condenser to make the image clearer. On low power, it is generally best to have the condenser positioned near the stage.
11. Switch to high power and sharpen the focus *with the fine adjustment knob only.*
12. If you are unable to find the specimen, switch back to low power and repeat steps 7, 8, and 9.
13. Keep both eyes open, even though only one is used in the monocular, compound microscope. After a short while, you can get accustomed to ignoring impressions coming from your free eye. If you have trouble at first, simply cover your free eye with your hand. Squinting leads to muscle fatigue and headaches.

Making a Wet Mount

A **wet mount slide** is made by placing the object in a drop of water on the slide and covering it with a thin glass coverslip. A coverslip must always be used. The use of a coverslip gives a flat surface to look through. If you don't use a coverslip, the water drop will form a curved surface and make viewing difficult. In addition, on high power, the heat from the lamp will cause water to evaporate and condense on the lens. A fogged lens is difficult to see through.

1. Make a **wet mount slide** by cutting out one of the words in figure 5.3 on page 65; your instructor may indicate a particular word.
2. Place the word on the slide, put one or two drops of water on the paper, and place a coverslip over the paper. If you place one edge of the coverslip against the glass slide and gently lower it into position, as shown in figure 5.4, you will not trap air bubbles, which interfere with your ability to see the object.

Figure 5.3 Making a wet mount using a word.

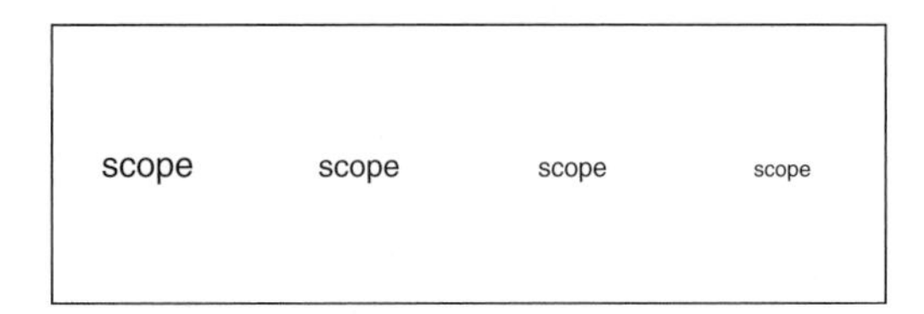

Figure 5.4 How to make a wet mount slide.

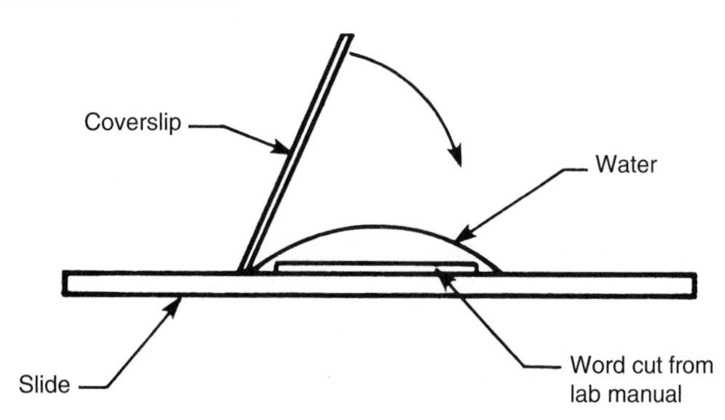

3. Place this slide on the microscope and examine it under low power. You should find that something comes into view with about a quarter of a turn of the coarse adjustment knob. Move the knob smoothly and slowly. If you cannot see anything, start over by returning the low-power lens to the position closest to the slide and trying again. If you still have trouble, ask your instructor for assistance. The first things you see are the fibers in the paper. You may also see the ink that forms the letters on the paper. If you do not see any letters, move your slide around until you do.
4. Adjust the amount of light by moving the lever connected to the iris diaphragm. Notice that there is an optimal position for this lever that allows you to see the letters on the paper clearly. If you have any problems locating something to look at, call your instructor.
5. Center a letter or portion of a letter in the center of the low-power field of view and switch to high power.
6. Focus using the fine adjustment knob.

7. Use a pencil to sketch the word as seen with (1) the unaided eye, (2) the low-power objective and, (3) the high-power objective of the microscope. The letters in your drawings should be oriented as they actually appear when viewed through the microscope.

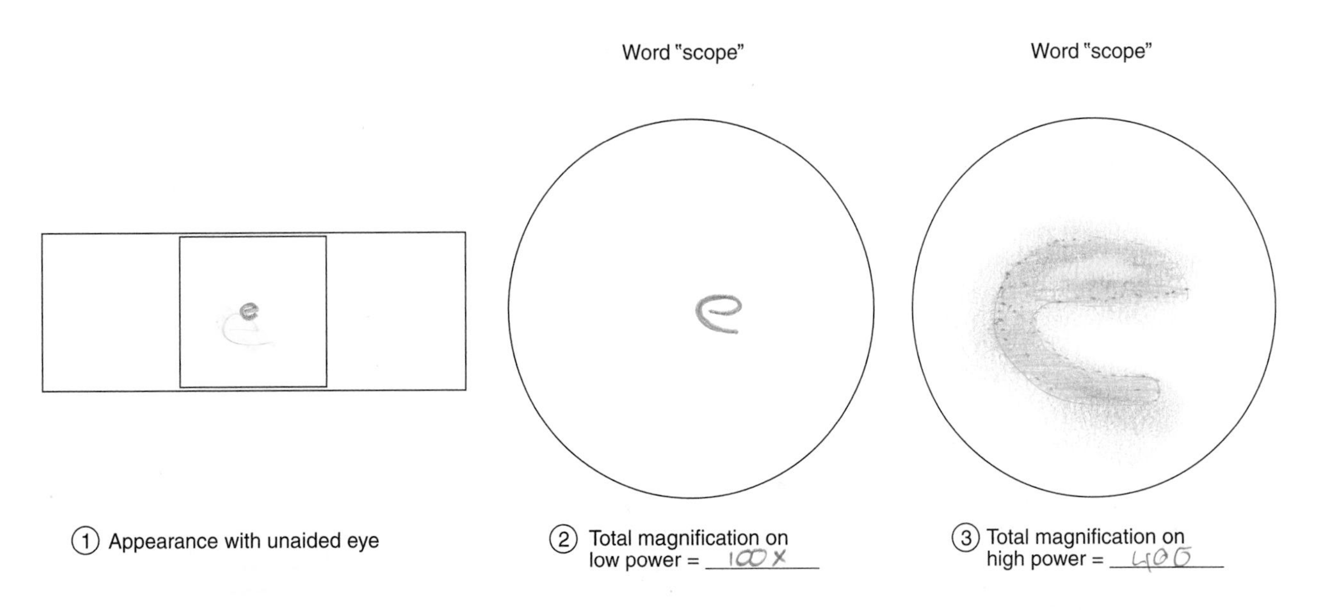

a. Compare your three drawings.
How do they differ in the orientation of the letters?

How are they different in regard to the number of letters visible?

b. Use the low-power lens and focus on a letter. While looking through the microscope, move the slide slightly to the right. In which direction does the image of the letter appear to move?

c. In what direction would you move the slide if a swimming specimen were leaving the field of view at the top of the slide and you wanted to continue looking at it?

Planes of Focus and Depth of Field

Lenses have a **plane of focus,** which is a position a specific distance from the lens where an object (or a portion of it) appears in sharp focus. You can actually move up and down through some specimens and focus on different levels of the specimen. The portion that is in focus at any given level is the portion in the plane of focus. This plane of focus has some depth to it. Therefore, it is a layer in space, a specific distance from the lens where objects are in sharp focus. The thickness of the layer that is in sharp focus is known as the **depth of field.** *High-power lenses have a much shallower depth of field than do low-power lenses.*

To experience microscopic depth of field, prepare a wet mount of two human hairs. If possible, use two different types of hair—a light hair and a dark hair, or a coarse hair and a fine hair.

1. Place a drop of water on the center of a clean slide.
2. Take a piece of hair (about 2 cm long) and place it lengthwise in the drop of water. Then, take a second piece of hair (again, about 2 cm long) and lay it over the first, so that they are crossed at right angles.
3. Focus on the crossed hairs under low power. You will probably see both hairs clearly at the same time (figure 5.5). They are both within the depth of field.
4. Move the slide so that the point where the hairs cross is in the center of the field of view and switch to high power. Use the fine adjustment knob to bring the bottom hair into focus. Note that only one hair is in sharp focus. It is within the depth of field. The other is blurry. It is still distinguishable as a hair but is not clearly focused. Focus upward with the fine adjustment knob to see that the bottom hair goes out of focus as the top hair comes into focus. Is the high-power depth of field greater or less than the low-power depth of field? ______________________

Figure 5.5 Plane of focus.

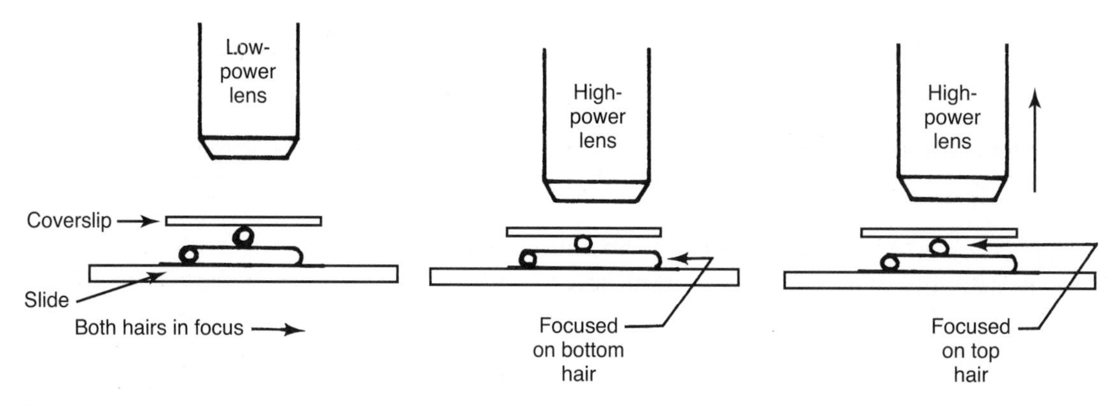

Further Practice

1. Obtain a prepared slide. These slides have been stained with particular stains so that specific structures are highlighted. If the slide is of a large object, it has been sliced into thin sheets, which were then mounted on the slide and a coverslip was permanently affixed to the top surface. Remember, your purpose is to practice looking at microscopic structures. Use the iris diaphragm, low power, and high power and move the slide around. Make drawings of your observations in the following space. Be sure to label all drawings with the name of the organism and the total magnification, as demonstrated in the following example. Make your drawings large, so that you can show details of the structure.

Title

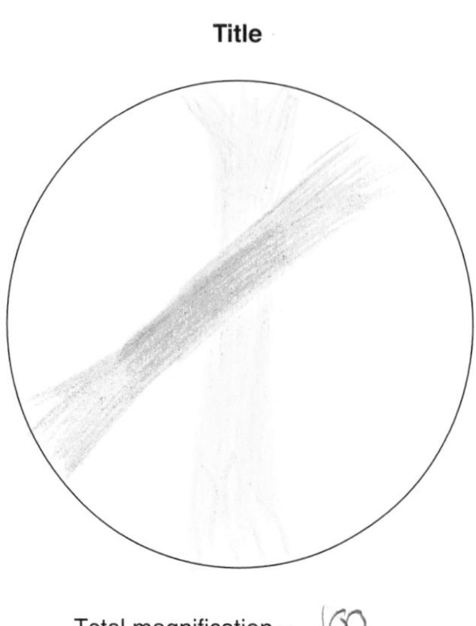

Total magnification = ___160___

2. Prepare wet mounts of the fresh specimens available. This variety of material could include such things as protozoa, cork, potato, algae, or microscopic animal life. Draw your observations here.

Title

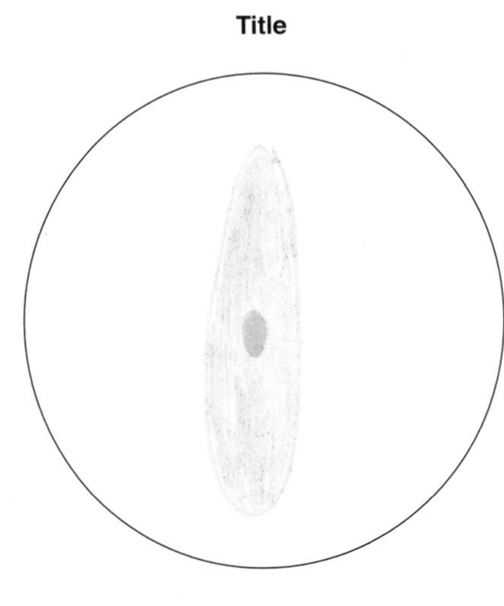

Total magnification = 400

Title

Total magnification = 100

3. When finished, clean and dry your slides and return them to the slide box. Because coverslips are fragile, your instructor will inform you how to clean and return them. If you break any slides or coverslips, DO NOT PUT THEM IN THE WASTEPAPER BASKET. Place broken glass in the container designated for the purpose.
4. When you are finished using your microscope, make sure you have removed any slides, cleaned any moisture from the stage or lenses, and positioned the nosepiece on low power before returning the instrument to its correct storage place.

5 The Microscope

Name ______________________________ Lab section ____________

Your instructor may collect these end-of-exercise questions. If so, please fill in your name and lab section.

End-of-Exercise Questions

1. Part of learning how to use the microscope is learning how to troubleshoot. When you have difficulty, what should you do? In each of the following situations, list possible causes of the problem and indicate what you should do to resolve it.

   a. You have just set up your microscope for use and turned it on. However, when you look through the ocular lens, it is dark—there is no light coming through the lens.
   Possible problem: lenses, iris diaphragm, lamp is off, not plugged in, bulb.

   Solution:

   b. Your microscope is focused on an object on the slide but not the object you are to observe.
   Possible problem: Not in field of view

   Solution: start over, and center it.

   c. The specimen you are examining is very thin and transparent. What adjustments could you make to the microscope to make it easier to see?

   Turn the light down

   d. You are looking at a specimen on low power and you cannot see the details that your instructor asked you to look at. What should you do? How is this done?

   center it turn it to high power

2. Complete the table, indicating the magnifying power of your microscope.

	Ocular Lens	Objective Lens	Total Magnification
Scan	10	4	40
Low	10	10	100
High	10	40	400
Oil immersion			

3. The following diagram illustrates the field of view as if you were using low-power magnification. Circle the part of the slide (numbers) you would see if you switched from low- to high-power magnification.

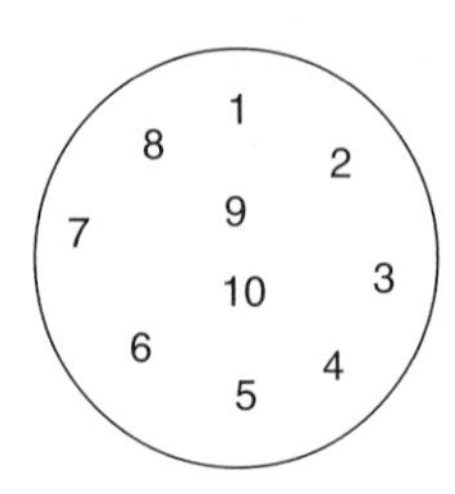

4. If you were using low power and wanted to look at the number 3 on high power, what should you do before you switch to high power? center it.

5. Label the following structures on the microscope drawing: lamp, fine adjustment knob, coarse adjustment knob, mechanical stage knob, condenser, revolving nosepiece, stage, ocular lens, objective lens, iris diaphragm.

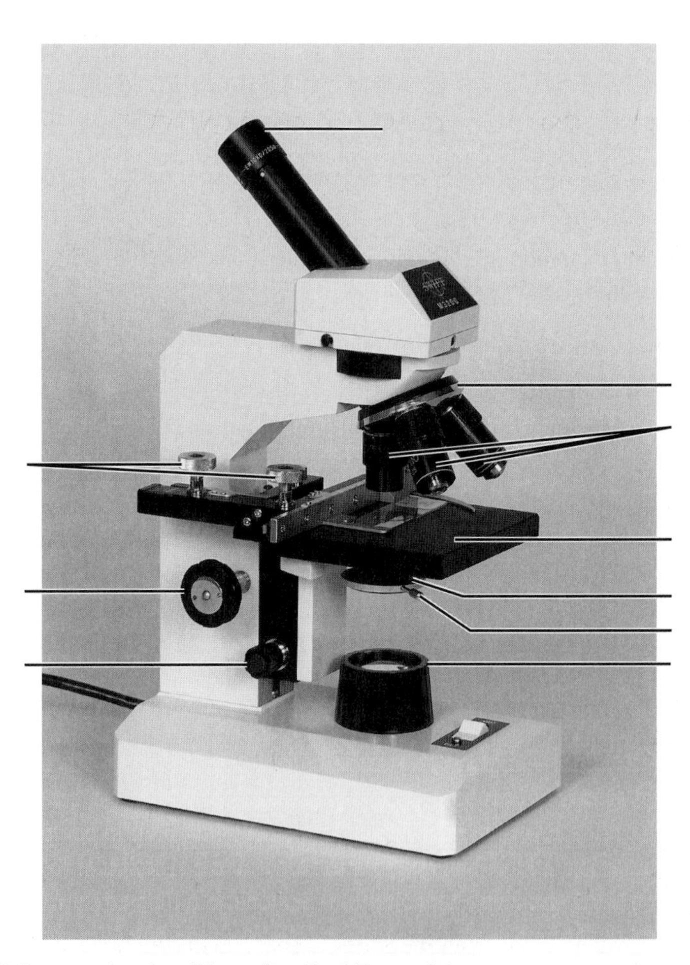

(© 1996 The McGraw-Hill Companies, Inc./Photo by Jim Womack.)

LABORATORY

Survey of Cell Types: Structure and Function

6

+ Safety Box

- A microscope is very expensive to replace; therefore, be particularly careful when handling one. Use both hands to carry it around the room. Hold it upright, so that the ocular lens does not slip out. Use only clean, dry lens paper to remove dust from the glass lenses. Do not use wet paper, paper towels, or other materials that may scratch these lenses.
- Slides and coverslips are glass. Be careful not to cut yourself when using them. Carefully clean slides before you use them, because dust and fingerprints interfere with your ability to see a specimen.
- Dispose of broken glass or organic materials as indicated by your instructor.
- Reagents, such as Lugol's solution or methylene blue, may permanently stain clothing.
- Materials containing human tissue must be treated with special care. Dispose of cheek epithelial cells in the special manner indicated by your instructor.

Objectives

Be able to do the following:

1. Define these terms in writing.

cell	stain	algae
cytoplasmic streaming	nucleus	protozoa
organelle	prokaryotic cell	pyrenoids
eukaryotic cell	cell wall	eyespot
cell membrane	nucleolus	food vacuoles
cytoplasm	vacuole	macronucleus
chloroplast	Protista	micronucleus
flagellum	cilia	bacteria
hyphae	contractile vacuole	spores
Fungi	Eubacteria	heterocysts
Plantae	Animalia	crosswall

2. Make a wet mount of any cellular material provided and focus on a cell.
3. Stain cells with either Lugol's solution or methylene blue.
4. Describe the three-dimensional structure of onion cells, cheek epithelial cells, *Elodea, Paramecium, Euglena,* and *Spirogyra.*

5. Diagram any cell observed through the microscope and indicate the three-dimensional structure of the cell.
6. Locate the following structures in an onion cell:

   a. cell wall b. cytoplasm c. nucleus d. cell membrane e. vacuole f. nucleoli

7. Locate the following structures in *Paramecium:*

   a. cell membrane b. food vacuoles c. contractile vacuole d. cilia

8. Locate the following structures in an *Elodea* cell:

   a. cell wall b. vacuole c. chloroplast

9. Locate the following structures in a cheek epithelial cell:

   a. cell membrane b. nucleus

10. Locate the following structures in a *Spirogyra* cell:

    a. cell wall b. chloroplast

11. Describe the FUNCTION of each of the following:

    a. nucleus b. cell wall c. food vacuole in protozoa d. cell membrane e. vacuole in plant
    f. chloroplast g. flagellum h. cilia i. eyespot j. contractile vacuole in protozoa
    k. pyrenoid l. heterocyst

12. List ways in which the cells of each of the following classifications of organisms differ from one another. List ways in which the cells of each of the following classifications are similar to one another.

    Animalia, Plantae, Protista, Fungi, Archaea, and Eubacteria

Introduction

The cell concept is basic to understanding the activities and characteristics of organisms. **Cells** are the smallest units of living things and are the units of structure and function of an organism. As functional units, they reflect the abilities of the organism as a whole. Some simple kinds of organisms consist of individual cells, but many of the organisms with which we are most familiar are multicellular. Multicellular organisms usually are composed of several different kinds of cells, each having specific characteristics that relate to its function. The various kinds of living things have been subdivided into three domains: Eubacteria, Archaea, and Eucarya. The cells of the Eubacteria and Archaea are small and simple, and they lack a nucleus. These cells are called **prokaryotic cells.** The cells of the Eucarya all have a **nucleus** and other kinds of structures called **organelles** within the cell. This type of cell is called a **eukaryotic cell.** Because the Archaea and Eubacteria are extremely small and difficult to see, in this exercise we will look only at Eubacteria as examples of a prokaryotic cell. We will spend the majority of our time looking at the various classifications of the Domain Eucarya, which is divided into the following kingdoms: Protista (algae and protozoa), Fungi, Plantae, and Animalia.

Robert Hooke was the first person to use the word *cell* in reference to the units that make up organisms. He examined cork under a microscope and saw the cell walls of these plant cells. Hooke recounts his important observation.

> I took a good clear piece of cork, and with a Penknife sharpen's as keen as a Razor, I cut a piece of it off . . . then examining it very diligently with a Microscope . . . I could exceeding plainly perceive it to be all perforated and porous, much like a Honey-Comb in these particulars . . . in that these pores, or cells, were not very deep, but consisted of a great many little Boxes. . . . For, as to the first, since our Microscope informs us that the substance of Cork is altogether fill'd with Air, and that Air is perfectly enclosed in little Boxes or Cells distinct from one another.[1]

Today, we recognize that Hooke saw only the cell walls of plant cells. However, he did recognize that living material, is made of many similar subunits, which he called cells. We continue to use his terminology today.

[1]Source: Robert Hooke (1635–1703), quoted in Gabriel and Fogel, *Great Experiments in Biology,* 1955, Prentice Hall, Inc.

Preview

In this exercise, you will look at a number of different types of cells, describe their three-dimensional shape, and identify some of their structures.

During this lab exercise, you will

1. prepare a temporary wet mount of sections of onion membrane, view the specimen through a microscope, identify common structures, and make a three-dimensional drawing of a typical onion cell.
2. make a temporary wet mount of an *Elodea* leaf and view its cellular structure through a microscope, identify common structures, and make a three-dimensional drawing of a typical cell.
3. make a temporary wet mount of *Spirogyra, Euglena,* and *Paramecium,* view the cells through a microscope, identify common structures, and make a three-dimensional drawing of a typical cell.
4. observe cheek epithelial cells through a microscope, identify common structures, and make a three-dimensional drawing of a typical cell.
5. observe slides of fungi, soil bacteria, and *Anabaena* and note their structures and characteristics.

Procedure

Kingdom Plantae

We will begin with plant cells, because they are relatively large and have several organelles that can be easily identified. Plants have many different kinds of cells organized into complex structures, such as leaves, fruits, and stems. We will look at two examples of plant cells: onion and *Elodea.*

Onion

An onion is composed of overlapping layers, which form rings when the onion is sliced. Cut a small piece of an onion ring approximately 1 cm × 1 cm. On the *concave* surface of the piece of onion is a thin membrane, which consists of many onion cells attached to one another. This membrane is one layer of cells thick. Peel this membrane from the rest of the piece of onion. Be careful to not wrinkle it and place it in a drop of water on a slide. Place a coverslip over the entire preparation and examine under a microscope. It should look something like figure 6.1. Begin viewing the onion cells with low power and proceed to high power to see detail.

You should be able to see the following structures that are typical of plant cells: (1) cell wall, (2) nucleus, (3) one or more nucleoli in the nucleus, (4) a large central vacuole, and (5) cytoplasm. The **cell wall** is found on the outside of the cell and provides a "box" within which the rest of the cell is found. The cell walls of plant cells are composed of a complex carbohydrate known as *cell*ulose. The **nucleus** will appear as a round or egg-shaped structure inside the cell and the small structures seen inside the nucleus are the **nucleoli** (singular, **nucleolus**). The vacuole and cytoplasm will probably be the most difficult to recognize. The **cytoplasm** will appear as a granular material near the cell wall. This will be a very thin layer. There is an outer boundary to the cytoplasm, known as the **cell membrane,** which is located inside the cell wall and outside the cytoplasm. However, it is very thin and in plant cells it is pressed up against the cell wall, making it difficult to see. (The term **plasma membrane** is often used to refer to this outer membrane of the cell. Thus, the terms *cell membrane* and *plasma membrane* mean the same thing.) Often in unstained cells, which are still alive, the cytoplasm may be seen to flow. Look closely at the granules or specks in the cytoplasm to see if they are moving. These granules are objects or cell structures too small to be seen clearly with the light microscope. The **vacuole** is a large, water-filled space in the center of the cell. Because it does not have any particles in it, it appears to be empty but it is not.

Figure 6.1 Onion cells.

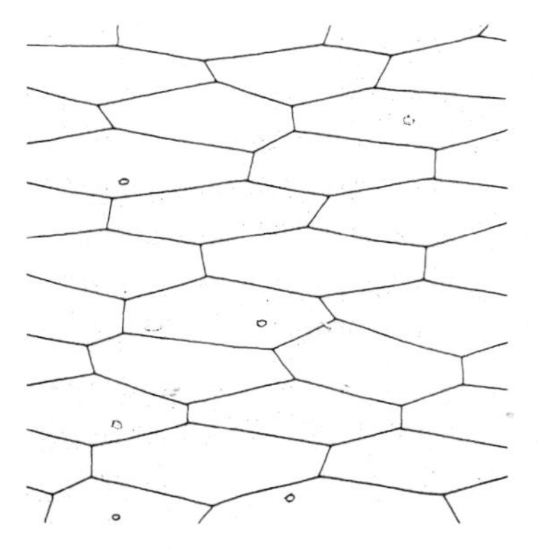

These cells are not flat but resemble a structure similar to a shoebox. The box itself represents the cell wall, the space in the center represents the vacuole, and all the other structures (nucleus and cytoplasm) are squeezed in between the cell wall and the vacuole.

Staining

After you have examined the wet mount of the living onion cells, you can stain the cells to make some of the structures easier to see. Biologists often use **stains** that bind to various macromolecules and structures in the cell to make the structures more visible. Living cells may be stained to show cilia, flagella, the nucleus, or other cell organelles. Some stains destroy cells immediately, whereas others, called "vital stains," kill cells more slowly. The organisms absorb these stains and continue to carry on their life functions for some time.

1. *Use caution when using all stains. Many will stain your hands and clothing.* Use Lugol's solution (composed of iodine and water) to stain your onion tissue, as demonstrated in figure 6.2. Lugol's solution stains carbohydrates, such as starches and glycogen.

Figure 6.2 Adding stain to cells.

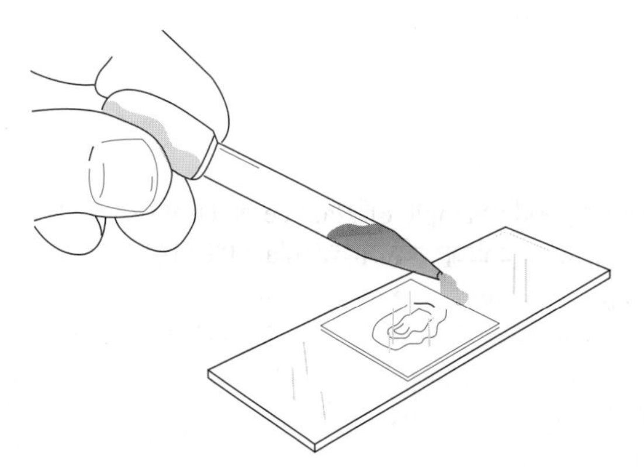

(a) Add one or two drops of stain to edge of coverslip.

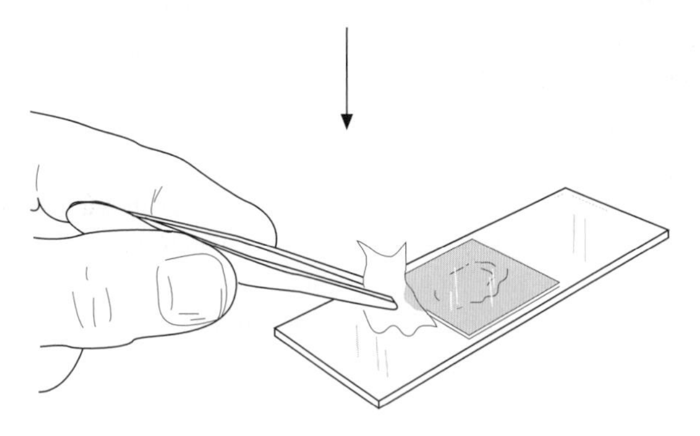

(b) Draw the stain under by touching lens paper to the opposite side of coverslip.

Microscopic Drawings

The average student is not gifted with a "photographic mind." Many observations, including those done using a microscope, must therefore be recorded for later study and review. That means drawings must be made. Make drawings in pencil, so that you can make modifications easily. Biological drawings should be simple but accurate representations of your observations. Make your drawings large enough that you can show clear details. You are not expected to be an artist; such drawings are for your benefit. Drawings should be labeled with a title and the total magnification used. As you make sketches, you will look more closely at the cells and this will help you remember what you have seen. If you have seen an object well enough to reproduce it accurately in a drawing, then you have seen it well.

1. Sketch the three-dimensional shape of an onion cell in the outline in the space provided. Draw the cell structures in their proper relationships to one another as you viewed them, and label the following: cell wall, vacuole, cytoplasm, nucleus, nucleolus, and the position of the cell membrane.

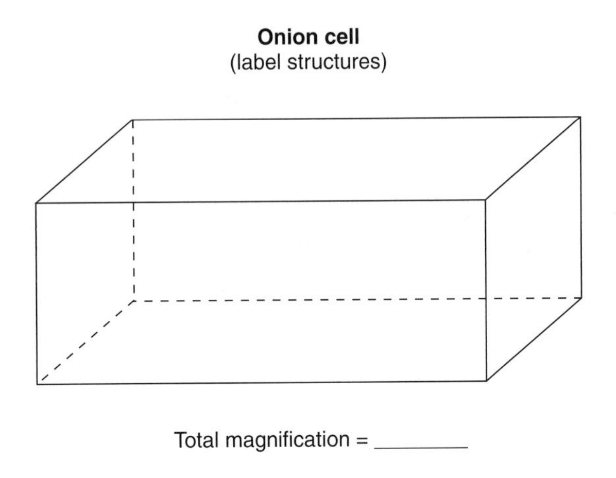

Elodea

The aquatic plant *Elodea* provides another good example of plant cells. Use forceps to pluck a young leaf from the tip of a sprig of *Elodea* and place it on a slide with a drop of water and a coverslip. Examine the leaf under the microscope. Begin with low power and switch to high power to see detail. The leaf is two layers of cells thick. Use the fine adjustment knob to focus up and down with your microscope, so that you can see the two layers. Examine the cells under high power. You should be able to see the following structures: (1) **cell wall,** (2) **vacuole,** and many small, green (3) **chloroplasts** in the (4) **cytoplasm.** There is also a nucleus but it is difficult to see among all the chloroplasts. You will also be able to see the cell membrane later. If you scan your slide, you should be able to see some cells in which the chloroplasts are moving along the inside of the cell wall. It is actually the cytoplasm that is in motion; therefore, this phenomenon is known as **cytoplasmic streaming.**

Draw and label the three-dimensional structure of an *Elodea* cell in the space provided. Label the cell wall, cytoplasm, cell membrane, chloroplasts, and vacuole.

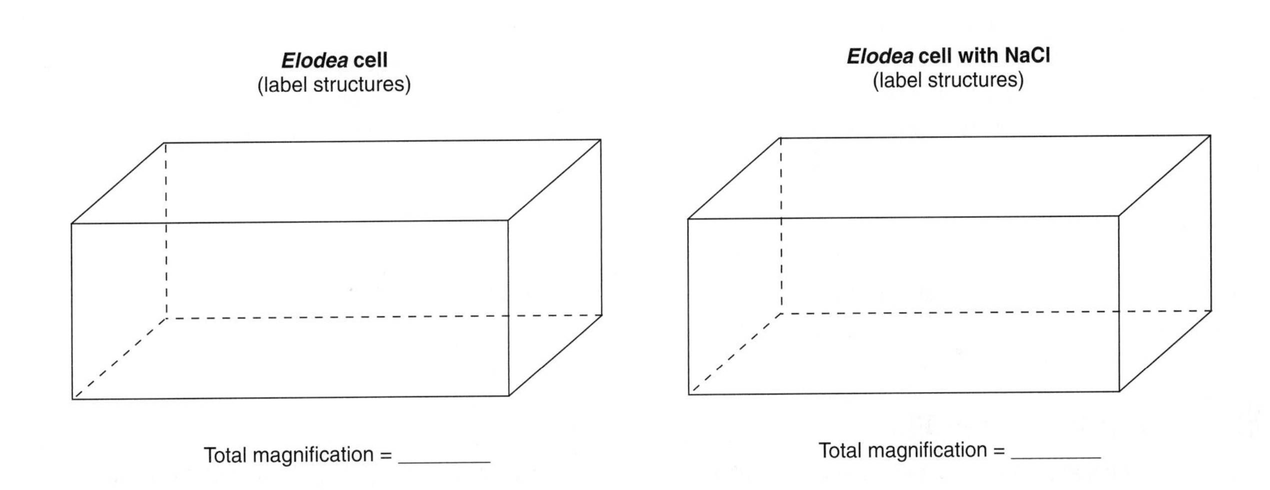

After you have examined the living cells of the *Elodea* leaf gently remove the coverslip, add a drop of 5% salt solution to the slide and replace the coverslip. The salt solution will cause water to leave the large central vacuole and the cell will shrink and pull away from the cell wall. This will allow you to see the **cell membrane,** which is on the outside of the cytoplasm. In the space provided above, draw and label an *Elodea* cell after salt water is added.

Kingdom Protista

The kingdom **Protista** contains many different kinds of organisms in which each cell functions as a separate unit. The many kinds of Protista are lumped together in one kingdom for convenience and are subdivided into two major types of organisms, **algae** and **protozoa.** Algae are either single cells or groups of similar cells that have cell walls and are capable of photosynthesis. Protozoa lack cell walls and with a few exceptions are not capable of photosynthesis.

Spirogyra

The freshwater organism *Spirogyra* is a good example of an alga. *Spirogyra* is composed of cells that are attached to one another end-to-end to form long, hairlike strands. Use an eyedropper to obtain a few of these strands from the culture provided and prepare a slide for examination. In many ways, a *Spirogyra* cell will appear similar to those of plants. It has a (1) **cell wall,** (2) **cytoplasm,** (3) a large **vacuole** between the strands of cytoplasm, and (4) one or two spiral-shaped **chloroplasts.** On the chloroplast, you will be able to see dots. These are (5) **pyrenoids,** which are places in the chloroplast where starch is manufactured. The cell also has a centrally located nucleus, which is suspended in the center of the cell by strands of cytoplasm, but this is difficult to see without special stains. The cell also has a different shape from that of the plants you looked at previously. These cells are cylindrical, rather than boxlike.

Draw and label a *Spirogyra* cell in the following space.

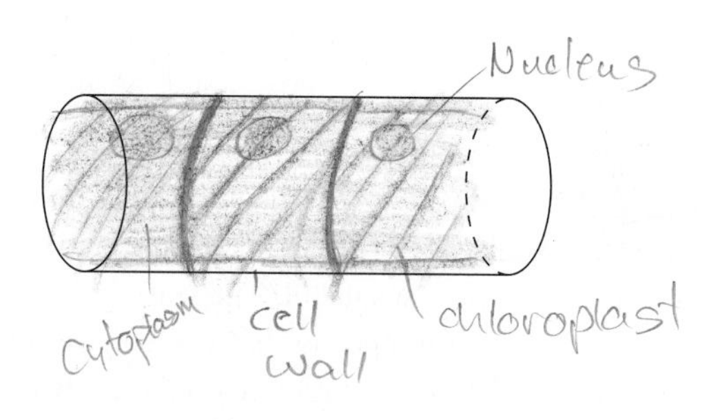

Euglena

Another member of the kingdom Protista is *Euglena.* Obtain a drop of the *Euglena* culture and place it on a slide with a coverslip. Examine it under the microscope. You should be able to see some organisms swimming around. *Euglena* has a single (1) **flagellum** at the anterior end, which whips about and pulls the cell through the water. You will be able to see that it also has a red (2) **eyespot** at the base of the flagellum. Within the cytoplasm of the cell, you will be able to see several green (3) **chloroplasts.** The eyespot allows the *Euglena* to swim toward a source of light and, therefore, position itself so that its chloroplasts receive sunlight. *Euglena* lacks a cell wall. Its outer covering is a flexible (4) **cell membrane,** so it is able to bend as it swims. It has a nucleus but this is often difficult to see without special staining. Because it has chloroplasts, it is able to carry on photosynthesis; however, it is also able to "eat" by taking up organic molecules from its surroundings. Because it has chloroplasts, some people classify it with the algae. Because it swims, lacks a cell wall, and eats, some people prefer to classify it with the protozoa.

Paramecium

Paramecium is a large protozoan, which is just visible to the naked eye. Obtain a drop of culture medium containing *Paramecium.* The paramecia will have been fed yeast cells that were stained with the dye congo red. You may need to place the organisms in a special syruplike, methyl cellulose solution to slow them down, so that you can see them. Your instructor will provide the solution if needed. On their surface protruding through the (1) **cell membrane** are hundreds of tiny, hairlike (2) **cilia,** which they use for movement. On the surface, you should be able to see a funnel-like structure through which the *Paramecium* feeds. Inside the organism, you will see a number of spherical (3) **food vacuoles** containing yeast. Food vacuoles that were recently formed as the organism fed on the yeast will be red. Older food vacuoles in which digestion has begun will turn blue. Although *Paramecium* does not have a cell wall, it does have a stiff outer layer of its cytoplasm. At each end of the cell, you should be able to see a (4) **contractile vacuole,** which periodically fills with water and collapses, expelling water from the cell. When the contractile vacuoles are empty, they will appear star-shaped. They become large, spherical, clear areas as they fill with water. *Paramecium* has a large **macronucleus** and one or more smaller **micronuclei** but these are often difficult to see without staining.

Label the structures in the cells of *Euglena* and *Paramecium*.

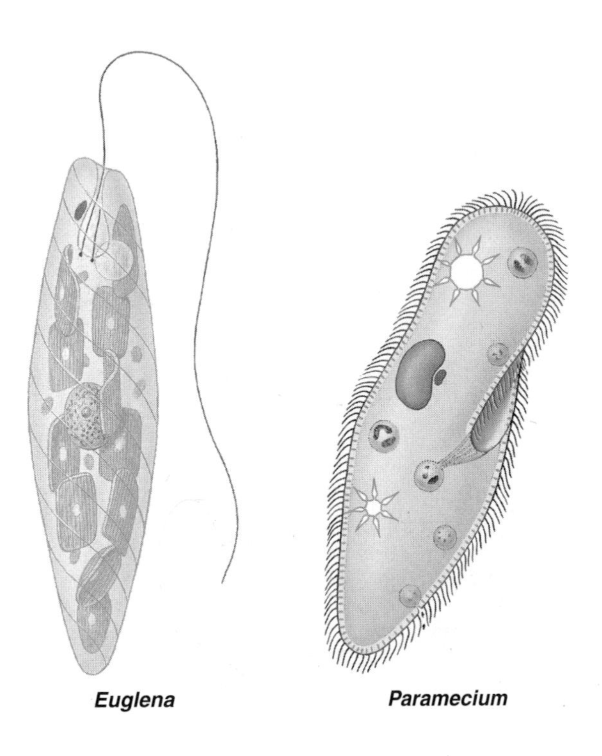

Kingdom Animalia

Animal cells are often difficult to study because they are small. They are also easily destroyed when making a slide, because they do not have a protective cell wall.

Cheek Epithelial Cells

> **Special Note**
>
> Whenever human tissue is used in the lab, you must follow special precautions. Therefore, the toothpick and the slides and coverslips you use must go in the special disposal container provided.

One kind of animal cell that is relatively easy to study is the cheek epithelial cells from the inside of your mouth. Take a toothpick and scrape the inside of your cheek (figure 6.3).

Smear the material from the toothpick onto a slide and add a drop of methylene blue stain and a coverslip. Use low power to locate some cells; then examine them under high power. You should be able to see flattened cells that have an irregular outline. (They will look like blue fried eggs.) The outside surface of the cell is the (1) **cell membrane.** You should also be able to see a football-shaped (2) **nucleus** in the (3) **cytoplasm** of the cell. On the surface of the cell, you will be able to see a large number of tiny dots. These are (4) **bacteria.**

Figure 6.3 Obtaining cheek epithelial cells.

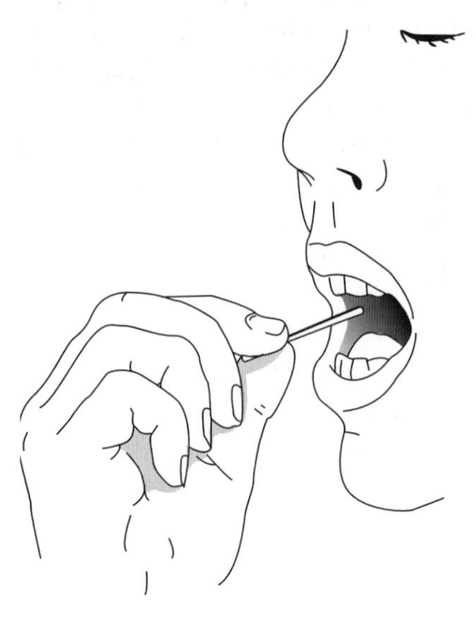

Gently scrape the inside of your cheek with the broad end of a toothpick.

Draw a cheek epithelial cell in the following space.

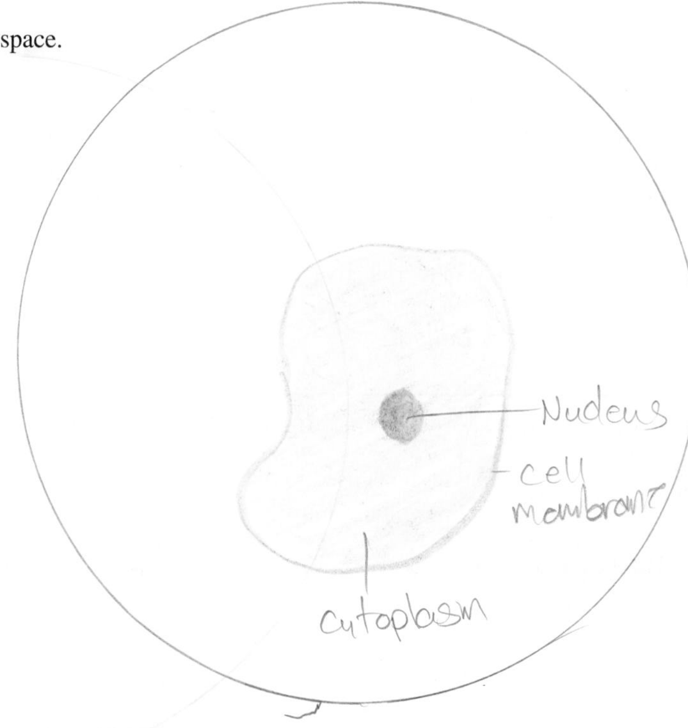

Kingdom Fungi

At the cellular level fungi are composed of, cylindrical filaments known as (1) **hyphae** (singular, **hypha**). These hyphae may form masses with no particular shape or may be organized to form very specific shapes such as mushrooms. Some fungi have the hyphae divided into individual cells by crosswalls. Other fungi do not have crosswalls separating the hyphae into individual sections. Fungi also differ from other organisms in that they have cellwalls of chitin rather than cellulose and the cellular units of many fungi have two nuclei rather than one.

We will examine the cells of the common mushroom sold in grocery stores, *Agaricus bisporus.* Obtain a mushroom and use a sharp knife or razorblade to cut the stalk of the mushroom lengthwise. From the freshly cut surface of the stalk make an extremely thin lengthwise slice. It should be as thin as tissue paper. Place the slice on a microscope slide. Add a drop of methylene blue stain and a coverslip. Examine the slide with a microscope. Look around on the slide to find the area that is the thinnest.

You will see a large number of filaments that are cylindrical in shape. Nearly all of them will be oriented so that you will see them from the side. But you should also be able to see places where the hyphae form a branching pattern. Thus, the stalk of the mushroom consists of a network of hyphae. If you examine the slide carefully, you will see that each hypha is divided into cellular units by crosswalls. These crosswalls have openings that allow cytoplasm to flow from one compartment to another within the hypha. You should be able to distinguish the following structures: (1) a **cell wall** on the outside of the hypha, (2) **crosswalls** that divide the hypha into sections, (3) and large, clear **vacuoles**, within the (4) **cytoplasm** of the cells. Although each section of a hypha has 2 **nuclei** present, they are very small and difficult to see without extremely high magnification.

You may also want to make a slide of the black material on the underside of the cap. Simply place some of this material on the slide with a drop of water and cover with a coverslip. You will probably see a large number of egg-shaped structures. These are reproductive structures called **spores**. The large number of spores produced and their small size makes them ideal mechanisms for distributing the mold to new sources of food.

Draw and label a fungal cell in the following space.

Domain Eubacteria

Most Eubacteria are extremely tiny and difficult to see. Because they lack a nucleus and most other kinds of organelles, it is difficult to see anything other than the general shape and size of the cells. You have already seen some bacteria on your cheek epithelial cells. Now, you will view a mixture of bacteria cultured from soil and an example of cyanobacteria (blue-green algae), which is easily collected and viewed.

Soil Bacteria

Obtain a drop of the culture of soil bacteria, make a wet mount slide, and examine it under low power. What you see will be a mixture of many different kinds of organisms: protozoa, worms, algae, etc. Use high power to look for the tiny bacteria. Some of the largest soil bacteria will be corkscrew-shaped and will be swimming. These organisms do not have a nucleus and lack the other cellular structures typical of eukaryotic organisms. Therefore, it will be impossible to see any characteristics other than the general shape of the cells.

Cyanobacteria

Obtain a drop of the culture of *Anabaena.* It consists of strings of cells attached end-to-end, like beads. These cells have a (1) **cell wall.** Often, you will be able to see some larger cells that are specialized to withstand harsh environmental conditions. These specialized cells are called (2) **heterocysts.** They typically form when the algae begins to dry up from lack of water or when there is a significant change in the temperature. Few other structures are identifiable.

Draw and label examples of the bacteria and *Anabaena* in the following space.

Diatom

Penicillium

Euglena

Mixed Protozoa

Mixed Bacteria

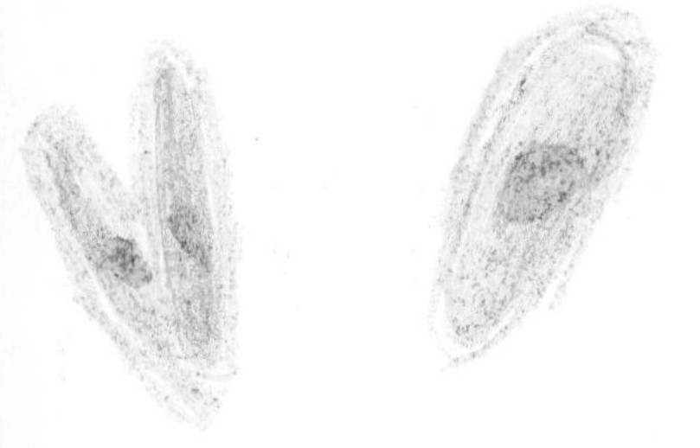

6 Survey of Cell Types: Structure and Function

Name ______________________________ Lab section ____________

Your instructor may collect these end-of-exercise questions. If so, please fill in your name and lab section.

End-of-Exercise Questions

1. List two structural differences between prokaryotic cells and eukaryotic cells.

Eukaryotic	Prokaryotic
complex organelles	simple organelles
No nucleus	No nucleus
Small ex Bacteria	Large

Animal

2. List two structural differences between plant and animal cells. How are these structural differences related to the ways the cells function?

Plants	Animal
- Cell wall	- Cell membrane
- Chloroplast	- no chloroplasts

3. In what ways do fungi resemble plant cells? In what ways are they different from plant cells?

Cell wall

Heterotrophic
Cell wall made of kutin

4. Describe three ways in which algal cells and plant cells are similar.

- Photosynthesis is in both
- Chloroplast
- Cell wall

5. Why are algae and protozoa placed in the same kingdom, Protista?

unicellular

single celled

6. Describe the size and location of the vacuole in an onion cell. What does the vacuole contain?

LABORATORY

7

Enzymes

+ Safety Box

- Hot waterbaths may be hazardous.
- Certain reagents used in this laboratory are acidic or basic. Care should be taken not to get them on clothing or skin.

Objectives

Be able to do the following:

1. Define these terms in writing.

substrate	enzyme	end product
enzyme-substrate complex	turnover number	competitive inhibitor
activation energy	attachment site	active site

2. Describe enzyme specificity.
3. Describe the effect of enzymes on chemical reactions.
4. Describe how temperature, concentration of the enzyme, concentration of the substrate, pH, and inhibitors change the turnover number.

Introduction

Every living organism carries out a large number of chemical reactions. It is essential for the life of the organism that these reactions occur at an extremely rapid rate and at a safe temperature. All organisms contain **enzymes,** which are protein molecules that speed up the rate of chemical reactions without increasing the temperature. Enzymes are named for the kind of reactions in which they are involved. The name also includes the suffix *ase* at the end. Thus, an enzyme that breaks down protein can be called a protein*ase*. Tyrosinase, the enzyme used in this exercise, normally alters the structure of the amino acid tyrosine.

All chemical reactions require an initial input of energy to get them started. This is called **activation energy.** Enzymes do not start a reaction; they merely speed up the reaction already in progress by reducing the need for large amounts of activation energy. Life on Earth would not be possible without this increased rate of reaction. Each reaction in a cell requires a specific enzyme to allow the reaction to proceed at the proper rate. Because there are hundreds of different reactions necessary in the life of the cell, hundreds of different enzymes are present in the cell. For an enzyme to work in a reaction, it must fit with its **substrate** (the molecule that will be altered). Each type of enzyme has a specific physical shape that fits the physical shape of its substrate. Each enzyme has a particular **attachment site,** where it binds with its substrate. In addition the enzyme has an **active site** that is involved in changing the substrate. In order for an enzyme to work, it must have a particular shape that produces the correct three-dimensional shape for the attachment and active sites.

When an enzyme reacts with a substrate, the two molecules physically combine to form an **enzyme-substrate complex.** The substrate binds to the attachment site of the enzyme. As a result of activities at the active site, the substrate is changed into the new **end product,** but the enzyme is not changed by the reaction. The number of times one molecule of enzyme can react with a substrate in a period of time is known as the **turnover number** (for example, 500,000 per second). This means that one enzyme molecule reacts with 500,000 substrate molecules in a second.

If a molecule has a shape that is almost the same as that of the normal substrate, this non-substrate molecule might bind to the enzyme. When the enzyme is attached to this molecule, the enzyme is not free to combine with the normal substrate molecule. The non-substrate molecule attached to the enzyme is called a *competitive inhibitor.* An **inhibitor** slows down the normal turnover number of an enzyme because it does not allow the normal substrate to have access to the enzyme. A general equation for an enzymatic reaction is shown in figure 7.1.

Figure 7.1 Model of enzyme action.

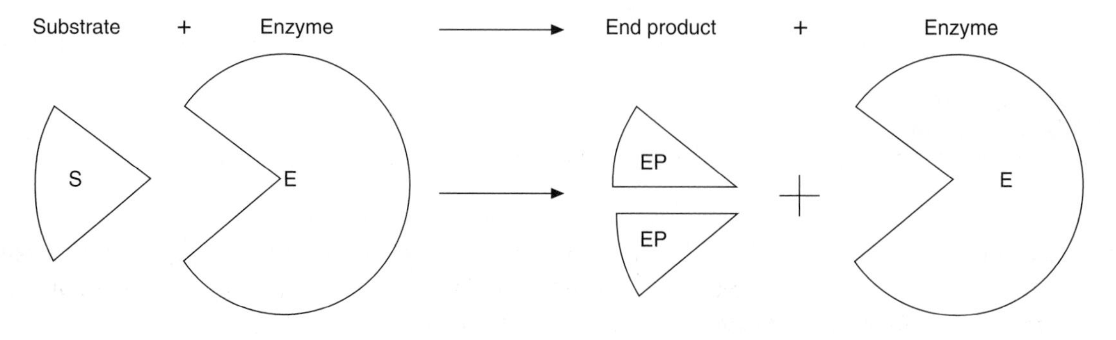

The specific reaction for today's exercise is

Substrate	+	Oxygen	+	**Enzyme** - - - - - - - - →	**End product**	+	Water	+	Enzyme
Pyrocatechol (colorless)	+	O_2	+	Tyrosinase - - - - - - →	Hydroxyquinone (yellow-brown)	+	H_2O	+	Tyrosinase
OH, OH	+	O_2	+	Tyrosinase - - - - - - →	HO, O, O	+	H_2O	+	Tyrosinase

Preview

In this exercise, you are asked to determine how various factors influence the turnover number of an enzyme. The substrate is a clear, colorless molecule known as *pyrocatechol.* The enzyme is one naturally found in potatoes called *tyrosinase.* (You will use ground-up potatoes as an inexpensive source of the enzyme.) A reaction between this enzyme and the substrate results in the formation of an end product having a yellowish-brown color. The appearance of this color indicates that a reaction between the enzyme and the substrate has occurred. The degree of color change corresponds to the amount of end product produced. When performing this experiment, use the following scale to rank the degree of color change:

No Change	Light Yellow	Medium Yellow	Golden Brown
O	+	++	+++

Note: Potatoes are an inexpensive source of the enzyme tyrosinase. The enzyme is obtained by grinding a potato in water and straining the blended potato through a strainer. You will have best results if you do the following:

1. Grind up the potato just before you need it.
2. Use a chilled potato from a refrigerator and ice cold water to blend the potato.
3. After filtering the ground-up potato, store the liquid containing the enzyme in a small container in a beaker of ice.

You may see a pink color develop after a few minutes. This color change is not related to the reaction we are using in this exercise. Ignore any pink color that develops and record only the intensity of the yellow color that develops. Your instructor may use a pure solution of the enzyme tyrosinase. If this is used, no pink color appears.

During this lab exercise, you will observe the

1. normal reaction of the enzyme, value of a control in an experiment, and enzyme specificity.
2. influence of temperature on the turnover number.
3. influence of concentration of the enzyme on the turnover number.
4. influence of concentration of the substrate on the turnover number.
5. influence of pH on the turnover number.
6. influence of inhibitors on the turnover number.

When everyone has completed this exercise, your instructor may wish to discuss the results obtained by the class.

Procedure

Throughout this exercise, it is important to clearly label all test tubes so that you can identify them.

Control

This portion of the exercise demonstrates the normal reaction that occurs between the enzyme (tyrosinase) and the substrate (pyrocatechol) used in today's exercise. You can also determine the value of a control and whether an enzyme is substrate specific. **Note:** *All glassware must be clean for this experiment to work properly.*

1. Put 10 mL of distilled water into a test tube and add 10 drops of enzyme. To this mixture add 10 drops of substrate. Mix the contents of the tube by holding the tube in one hand near the top and gently tapping the base of the tube with your other hand. Do not cover the opening of the test tube with your thumb. Begin timing the reaction as soon as you have added the substrate to the tube.
2. Put 10 mL of distilled water into a second test tube. Add 10 drops of substrate and mix. This tube contains no enzyme.
3. Put 10 mL of distilled water into a third test tube. Add 10 drops of enzyme and 10 drops of sucrose (the wrong substrate). Mix this tube.
4. Observe each tube after 5 minutes and note any changes in color. Record the color in table 7.1. The color change in the first test tube indicates what normally occurs in this particular enzyme-substrate reaction. A control, a basis of comparison for a reaction, is an essential part of any experiment. This first test tube serves as a control during the remainder of this exercise. You can compare experimental tubes with this one to determine whether a reaction has occurred.
   a. What is the purpose of the second tube containing only water and substrate?

   b. What can you learn from the third tube containing water, enzyme, and sucrose?

Table 7.1 **Enzyme Reactions**

Tube	Contents of Tube	Original Color	Color After 5 Minutes	Rank
a	10-mL water 10 drops of enzyme 10 drops of substrate (pyrocatechol)	Clear	Medium Yellow	++
b	10-mL water 10 drops of substrate (pyrocatechol)	Clear	No color	0
c	10-mL water 10 drops of enzyme 10 drops of sucrose	Clear	No color	0

Temperature

In this portion of the exercise, you will examine the effect of different temperatures on the activity of the enzyme. You need to allow the water in the test tubes to come to a specific temperature before you add the substrate and enzyme. After these are mixed with the water, they need to stay at that specific temperature for 5 minutes to see if the temperature influences the enzyme's activity.

1. Take five test tubes and label them *a, b, c, d,* and *e*. Fill each with 10 mL of distilled water. Add 10 drops of enzyme to each tube.

2. Place tube *a* in an ice water mixture in a waterbath for 5 minutes to allow it to cool. Similarly, place tubes *b* through *e* in water baths of 20°, 40°, 60°, and 100°C, respectively. These tubes are to remain in their respective waterbaths until they reach the designated temperature.
3. After allowing the tubes to come to their appropriate temperature, add 10 drops of substrate to each tube. Mix the contents and immediately return the tubes to their appropriate waterbaths for an additional 5 minutes. Begin timing the reaction as soon as you have added the substrate.
4. After 5 minutes, remove the test tubes from the waterbaths and observe the color of the tubes. Record the color intensity of each tube in table 7.2 and rank them from darkest to lightest. Graph your results on the grid provided.

Table 7.2 Temperature

Tube	Contents of Tube	Original Color	Color After 5 Minutes	Rank
a 0°C	5 10-mL water 10 drops enzyme 10 drops of substrate (pyrocatechol)	Clear	10:26 Light Yellow	+
b 20°C	5 10-mL water 10 drops enzyme 10 drops of substrate (pyrocatechol)	Clear	10:29 Medium Yellow	++
c 40°C	5 10-mL water 10 drops enzyme 10 drops of substrate (pyrocatechol)	Clear	10:32 Light Yellow	+
d 60°C	5 10-mL water 10 drops enzyme 10 drops of substrate (pyrocatechol)	Clear		
e 100°C	10-mL water 10 drops enzyme 10 drops of substrate (pyrocatechol)	Clear	10:34 No color	0

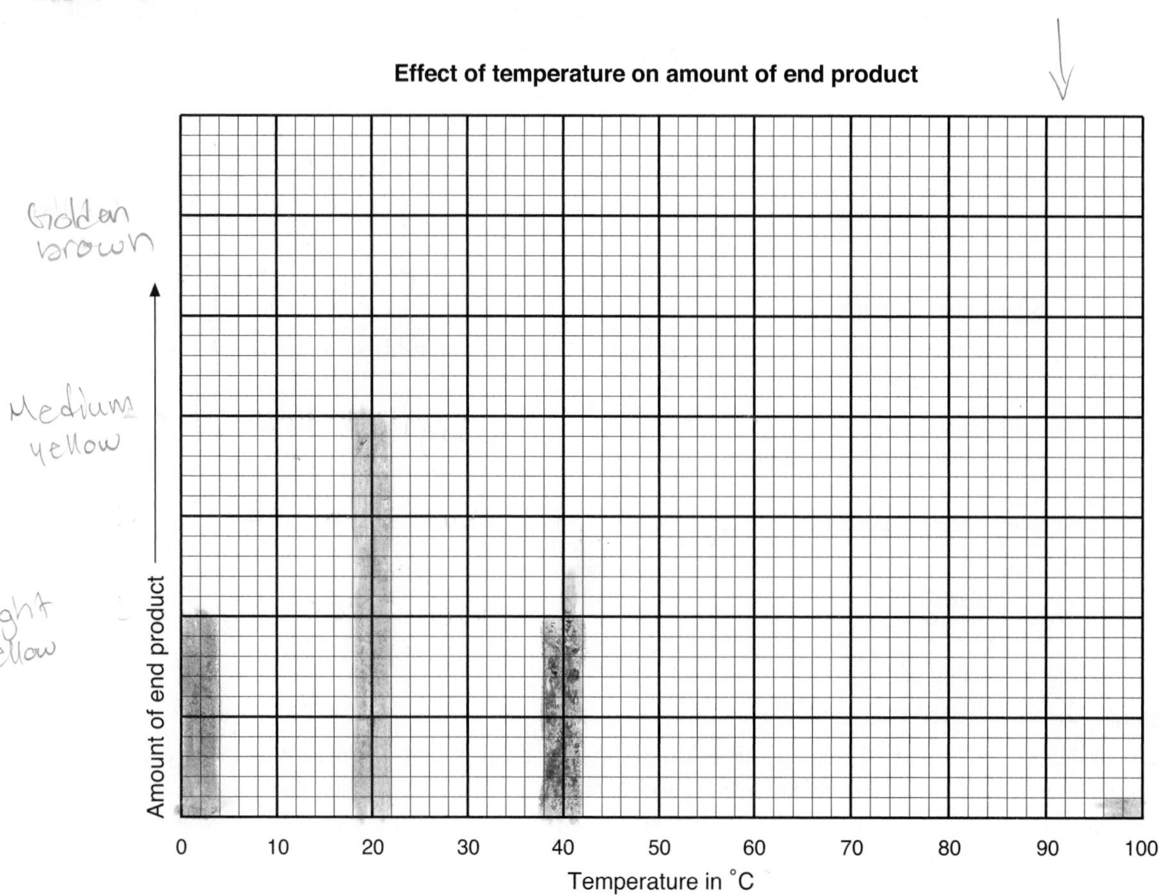

Concentration of Enzyme

In this experiment, you will examine the effect of altering the number of enzyme molecules on the number of product molecules produced.

1. Take three test tubes and label them *a*, *b*, and *c*. Fill each with 10 mL of distilled water. Place them in a test tube rack.
2. Add 3 drops of enzyme to tube *a*, 9 drops of enzyme to tube *b*, and 27 drops of enzyme to tube *c*.
3. Add 10 drops of substrate to each of these tubes. Mix thoroughly and observe after 5 minutes. Begin timing the reaction as soon as you have added the substrate.
4. Record the color intensity of each tube in table 7.3 and rank them from darkest to lightest. Graph your results on the grid provided.

Table 7.3 Concentration of Enzyme

Tube	Contents of Tube	Original Color	Color After 5 Minutes	Rank
a	10 mL water 3 drops of enzyme 10 drops of substrate (pyrocatechol)	Clear	No color	o
b	10 mL water 9 drops of enzyme 10 drops of substrate (pyrocatechol)	Clear	Light yellow	+
c	10 mL water 27 drops of enzyme 10 drops of substrate (pyrocatechol)	Clear	medium yellow	++

Effect of enzyme concentration on amount of end product

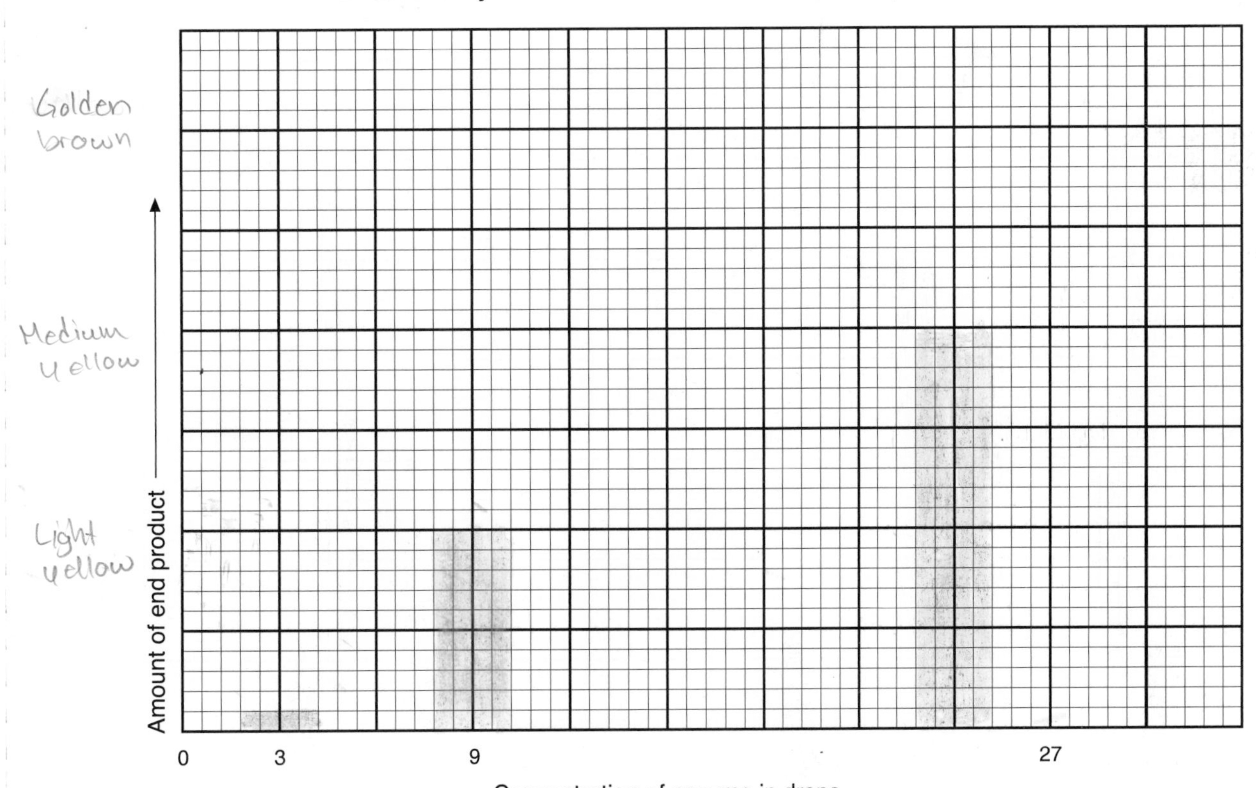

Concentration of Substrate

In this experiment, you will examine the effect of altering the number of substrate molecules on the number of product molecules produced.

1. Take three test tubes and label them *a, b,* and *c*. Fill each with 10 mL of distilled water. Place them in a test tube rack.
2. Add 3 drops of substrate to tube *a,* 9 drops of substrate to tube *b,* and 27 drops of substrate to tube *c*.
3. Add 10 drops of enzyme to each of these tubes. Mix thoroughly and observe after 5 minutes. Begin timing the reaction as soon as you have added the enzyme.
4. Record the color intensity of each tube in table 7.4 and rank them from darkest to lightest. Graph your results on the grid provided.

Table 7.4 Concentration of Substrate

Tube	Contents of Tube	Original Color	Color After 5 Minutes	Rank
a	10-mL water 10 drops of enzyme 3 drops of substrate (pyrocatechol)	Clear	Light yellow	+
b	10-mL water 10 drops of enzyme 9 drops of substrate (pyrocatechol)	Clear	Medium yellow	++
c	10-mL water 10 drops of enzyme 27 drops of substrate (pyrocatechol)	Clear	Medium yellow	++

Effect of substrate concentration on amount of end product

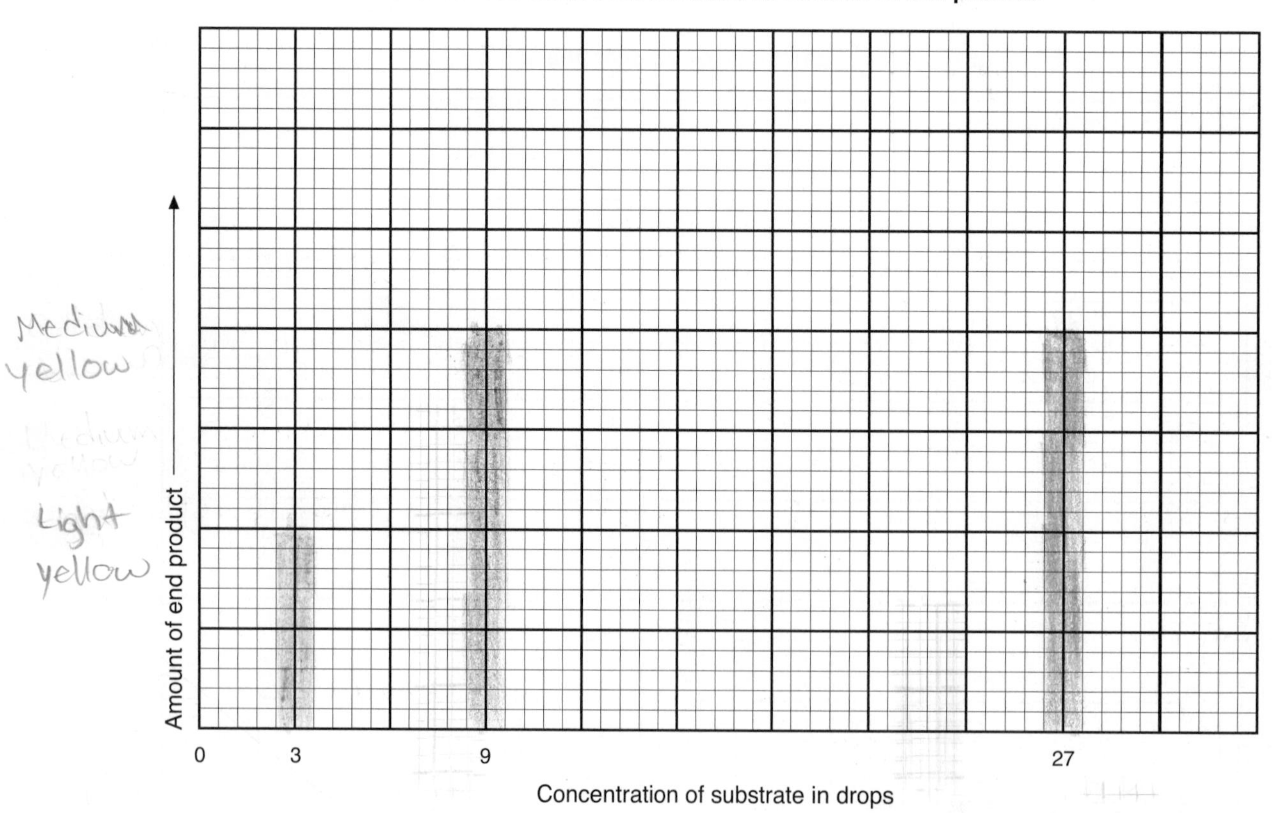

pH

pH is measure of the number of hydrogen ions (H^+) present in a solution. Acids are substances that release hydrogen ions in solution. Solutions that have many hydrogen ions are acidic and have a low pH. Bases are substances that remove hydrogen ions from solution. Solutions with few hydrogen ions are basic and have a high pH. Solutions with a pH = 7 are called neutral solutions. In this experiment, you will assess the effect of different pH values on the effectiveness of enzymes.

1. Take five test tubes and label them *a, b, c, d,* and *e*. Fill tube *a* with 10 mL of water that has been adjusted to a pH of 3. Place the tube in a test tube rack.
2. Fill tube *b* with 10 mL of water that has been adjusted to a pH of 5. Place the tube in a test tube rack.
3. Fill tube *c* with 10 mL of water that has been adjusted to a pH of 7. Place the tube in a test tube rack.
4. Fill tube *d* with 10 mL of water that has been adjusted to a pH of 9. Place the tube in a test tube rack.
5. Fill tube *e* with 10 mL of water that has been adjusted to a pH of 11. Place the tube in a test tube rack.
6. Add 10 drops of enzyme to each tube.
7. Add 10 drops of substrate to each tube. Mix the tubes thoroughly and observe after 5 minutes. Begin timing the reaction as soon as you have added the substrate.
8. Record the color intensity of each tube in table 7.5 and rank them from darkest to lightest. Graph your results on the grid provided.

Table 7.5 Effect of pH

Tube	Contents of Tube	Original Color	Color After 5 Minutes	Rank
a	10-mL water adjusted to pH of 3 10 drops of enzyme 10 drops of substrate	Clear		
b	10-mL water adjusted to pH of 5 10 drops of enzyme 10 drops of substrate	Clear		
c	10-mL water adjusted to pH of 7 10 drops of enzyme 10 drops of substrate	Clear		
d	10-mL water adjusted to pH of 9 10 drops of enzyme 10 drops of substrate	Clear		
e	10-mL water adjusted to pH of 11 10 drops of enzyme 10 drops of substrate	Clear		

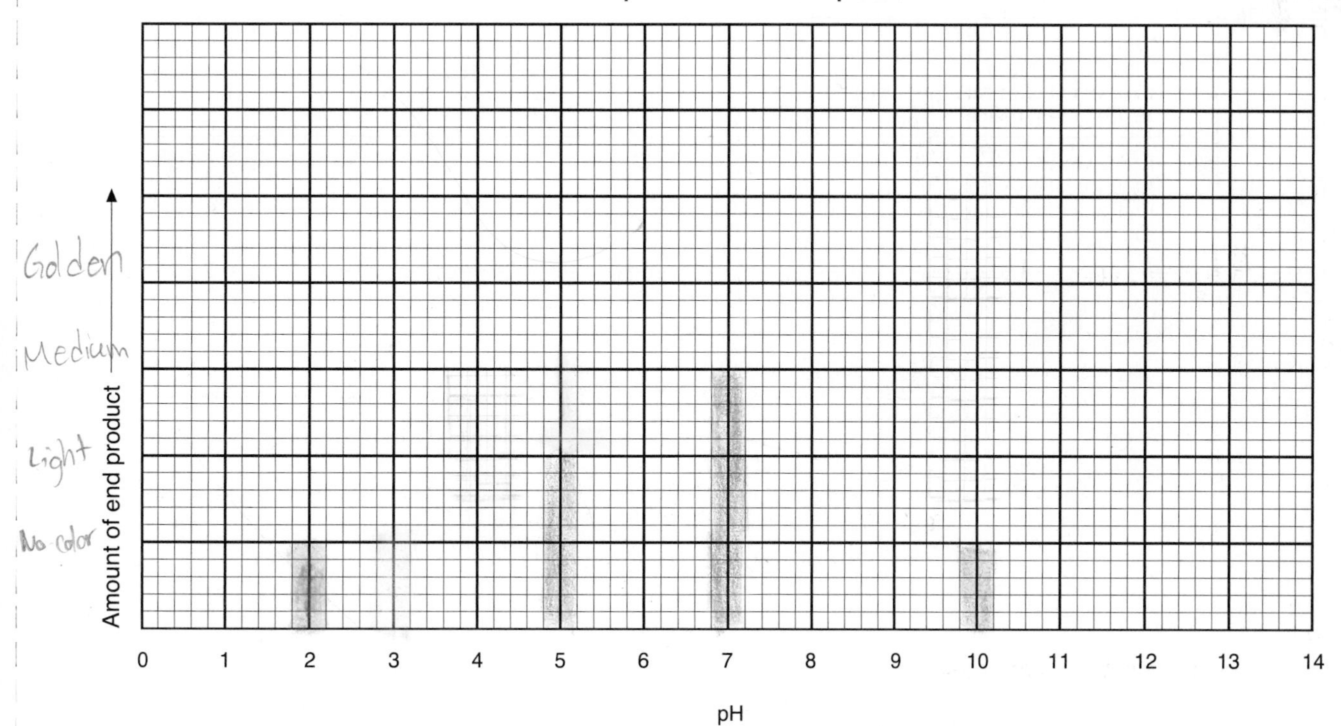

Inhibitors

Inhibitors interfere with the ability of enzymes to interact with enzymes. In this experiment, you will identify one of two molecules as being an inhibitor. (The following steps must be performed in the order presented or results will vary.)

1. Take six test tubes and label them *a* through *f*. Fill each tube with 10 mL of distilled water. Place them in a test tube rack.
2. Add 1 drop of phenylthiourea to tube *a*, 10 drops of phenylthiourea to tube *b*, and 20 drops of phenylthiourea to tube *c*.
3. Add 1 drop of tyrosine to tube *d*, 10 drops of tyrosine to tube *e*, and 20 drops of tyrosine to tube *f*.
4. Add 10 drops of enzyme to each of the six tubes.
5. Add 10 drops of substrate to each of the six tubes. Mix the tubes thoroughly and observe after 5 minutes. Begin timing the reaction as soon as you have added the substrate.
6. Record the color intensity of each tube in table 7.6 and rank them from darkest to lightest.

Table 7.6 **Inhibitors**

Tube	Contents of Tube	Original Color	Color After 5 Minutes	Rank
Phenylthiourea				
a	10-mL water 10 drops of enzyme 10 drops of substrate (pyrocatechol) 1 drop of phenylthiourea	Clear		
b	10-mL water 10 drops of enzyme 10 drops of substrate (pyrocatechol) 10 drops of phenylthiourea	Clear		
c	10-mL water 10 drops of enzyme 10 drops of substrate (pyrocatechol) 20 drops of phenylthiourea	Clear		
Tyrosine				
d	10-mL water 10 drops of enzyme 10 drops of substrate (pyrocatechol) 1 drop of tyrosine	Clear		
e	10-mL water 10 drops of enzyme 10 drops of substrate (pyrocatechol) 10 drops of tyrosine	Clear		
f	10-mL water 10 drops of enzyme 10 drops of substrate (pyrocatechol) 20 drops of tyrosine	Clear		

7 Enzymes

Name ______________________________ Lab section ____________

Your instructor may collect these end-of-exercise questions. If so, please fill in your name and lab section.

End-of-Exercise Questions

1. Which of the two substances (phenylthiourea or tyrosine) was the inhibitor in the exercise you just completed? What was your evidence?

2. If you were to look at the three test tubes involved in the enzyme concentration experiment after 24 hours, how would the colors of each compare? Explain.

   They would all be the same thats because enzymes recycle

3. Both high and low temperatures reduced the amount of color (product) produced. However, the cause of the reduction is different in the two cases. Explain what the differences are.

   At extremely high temp. they denatured
   At low temp it reacts slowly

4. Because an enzyme-substrate complex, involving the substrate and the attachment site of the enzyme, must be formed for an end product to be produced, how might an inhibitor reduce the effectiveness of an enzyme?

   Outcompete's the substrate

5. What is a turnover number?

6. How did you estimate turnover number in this series of experiments?

7. Why does increasing the amount of substrate increase the amount of color produced in the test tubes?

8. At what pH was the enzyme most effective?

pH 7

9. At what pHs did the enzyme work but not as effectively?

5 & 10

10. Explain why changing the pH alters the effectiveness of the enzyme.

change pH- you move the enzume out of its optimal range

LABORATORY

8 Photosynthesis and Respiration

+ Safety Box

- Certain reagents used in this exercise may be caustic. Be careful not to get them on your clothing or skin.
- Follow your instructor's directions for the disposal of waste solutions.

Objectives

Be able to do the following:

1. Define these terms in writing.

   chlorophyll | aerobic respiration | parts per million (ppm)

   P/R ratio | photosynthesis

2. List the raw materials needed, products released, and conditions necessary for an organism to engage in photosynthesis.
3. List the raw materials needed, products released, and conditions necessary for an organism to engage in aerobic respiration.
4. Describe in writing the interdependency of the two processes of photosynthesis and respiration.
5. Realize that the processes of photosynthesis and aerobic respiration are not mutually exclusive but can occur simultaneously in the same organism.

Introduction

Photosynthesis is a metabolic process that combines carbon dioxide (CO_2) and water (H_2O) to form sugar ($C_6H_{12}O_6$) and oxygen (O_2). The process takes place within the chloroplasts of plants and algae. The chloroplasts contain the green pigment **chlorophyll** and the enzymes necessary for photosynthesis. The chlorophyll traps light, which serves as the energy source that allows the process to take place. A simplified equation for photosynthesis is

$$\underbrace{6\,H_2O + 6\,CO_2}_{\text{inorganic raw materials}} \xrightarrow[\text{chloroplasts}]{\text{light}} \underbrace{C_6H_{12}O_6 + 6\,O_2}_{\text{end products}}$$

All organisms require energy to sustain themselves. Nearly all organisms, including plants and animals, carry on **aerobic respiration,** in which sugar and oxygen react to form carbon dioxide, water, and a source of energy known as ATP (adenosine triphosphate). Mitochondria are cellular structures that contain the enzymes necessary for the many individual steps of aerobic respiration. A simplified equation for aerobic respiration is:

$$\underbrace{C_6H_{12}O_6 + 6\,O_2}_{\text{food and oxygen}} \xrightarrow[\text{mitochondria}]{\text{enzymes}} \underbrace{6\,H_2O + 6\,CO_2 + \text{energy (ATP)}}_{\text{end products}}$$

Look closely at the balanced equations for photosynthesis and respiration and notice that the end products of one reaction are the raw materials for the other. Only organisms containing chlorophyll can perform photosynthesis, whereas respiration can take place in virtually every organism.

Many plants perform photosynthesis and respiration simultaneously. The **P/R ratio** (photosynthesis/respiration ratio) compares the rate of photosynthesis to the rate of respiration. Knowing this ratio can help explain what happens in a plant at different times in its life. For instance, the P/R ratio is different for a corn plant during spring, summer, and fall. It is also different for day and night. Animals, on the other hand, can engage only in respiration.

Because animals are incapable of converting inorganic raw materials into organic molecules, they must obtain energy-rich organic molecules by eating plants or other animals. They also need oxygen to allow them to release energy from organic molecules. Thus, animals are dependent on plants for the two end products of photosynthesis, organic molecules (glucose) and oxygen.

Water has many gases dissolved in it, including oxygen and carbon dioxide. Aquatic organisms use these gases when they carry on photosynthesis and aerobic respiration and release gases into the water. This exercise allows you to make measurements of the rates of photosynthesis and respiration by measuring the amount of oxygen and carbon dioxide present in the water in which the organisms live.

This laboratory activity gives you an opportunity to set up an experiment with proper controls, quantitatively test water samples for oxygen and carbon dioxide content, and collect and analyze data. You may also learn that experimental work sometimes yields results that are difficult to interpret.

Preview

During this exercise, you will allow organisms to engage in their normal biochemical processes. Evidence that the organisms have carried on photosynthesis or respiration will be revealed by sampling the oxygen and carbon dioxide content of the water in which they live. The tests for measuring the oxygen and carbon dioxide content appear on page 98. Follow these test procedures carefully, because you will be measuring very small quantities—**parts per million (ppm)**—of oxygen and carbon dioxide. If your water sample contains 8 ppm of oxygen, it means that there are eight oxygen molecules dissolved in every 1 million molecules of your sample.

During this lab exercise, your group will

1. determine dissolved oxygen and dissolved carbon dioxide concentration in aged tap water. (The purpose for doing this is to get baseline data as well as to give you practice with these complex tests.)
2. set up controls of aged water and three experimental situations: plants in light, plants in darkness, and fish.
3. determine dissolved oxygen and dissolved carbon dioxide concentrations of the controls and the three experimental situations at the end of an hour.
4. use the data collected to answer questions.

Procedure

Initial Trial

Fill a large beaker from the container labeled *aged water.* The aged water is simply tap water that has been sitting open to the atmosphere overnight. Therefore, the aged water has the same concentrations of carbon dioxide and oxygen dissolved throughout the container. In addition, it is equilibrated to room temperature. Because everyone in class will use this aged

Table 8.1 **Dissolved Oxygen and Carbon Dioxide Results**

	Oxygen Test	Carbon Dioxide Test
Test Tube	**Dissolved O_2 (ppm) (Number of Drops of Sodium Thiosulfate)**	**Dissolved CO_2 (ppm) (Number of Drops of NaOH)**
Initial aged water		
Control in light		
Control in dark		
Plant in light		
Plant in dark		
Goldfish		

water to set up his or her experiment, everyone will start with water that contains the same amount of carbon dioxide and oxygen and has the same temperature. Test this aged water for dissolved O_2 and CO_2. The directions for the dissolved oxygen test and the dissolved carbon dioxide test are found on page 98. Record the results in the proper column of table 8.1.

Controls

Fill four large test tubes with aged water; cork the tubes in such a way that no air is trapped. Label these tubes *Controls* and place two of them in a test tube rack marked *Light.* Place the other two control tubes in a test tube rack in the dark. Your instructor will designate this location. At the end of the hour, you will use the water from one of each pair of tubes to test for oxygen content and the water in the other tube to test for carbon dioxide content.

Experimental Tubes

Plant in Light

Fill two other large tubes with aged water. Place several healthy, green sprigs of *Elodea* or another water plant in each tube. There should be plants from the top to the bottom of the test tube, but the plants should not be jammed together in a clump. Cork the tubes without trapping air. Label these tubes *Plant in Light* and place them in the test tube rack in front of a fluorescent light source for 1 hour. It is best to use fluorescent lights because incandescent lights tend to heat up the water in the tubes and change the amount of gases that can remain dissolved in the water. See table 8.2.

Plant in Darkness

Fill two more large tubes with aged water and *Elodea* or another water plant; cork tubes. Label these tubes *Plant in Dark* and place them in the designated dark area for 1 hour.

Animal

Fill two other large tubes with aged water and place a goldfish in each test tube. Cork the tube without trapping air bubbles. Label these tubes *Goldfish* and place them with the plant in the light for 1 hour.

Analysis of Results

Often, a visual presentation of data helps one see patterns or trends and makes interpretation easier. Use the data from table 8.1 to construct a bar graph in the space provided. Use two different colors or shades to distinguish between oxygen and carbon dioxide.

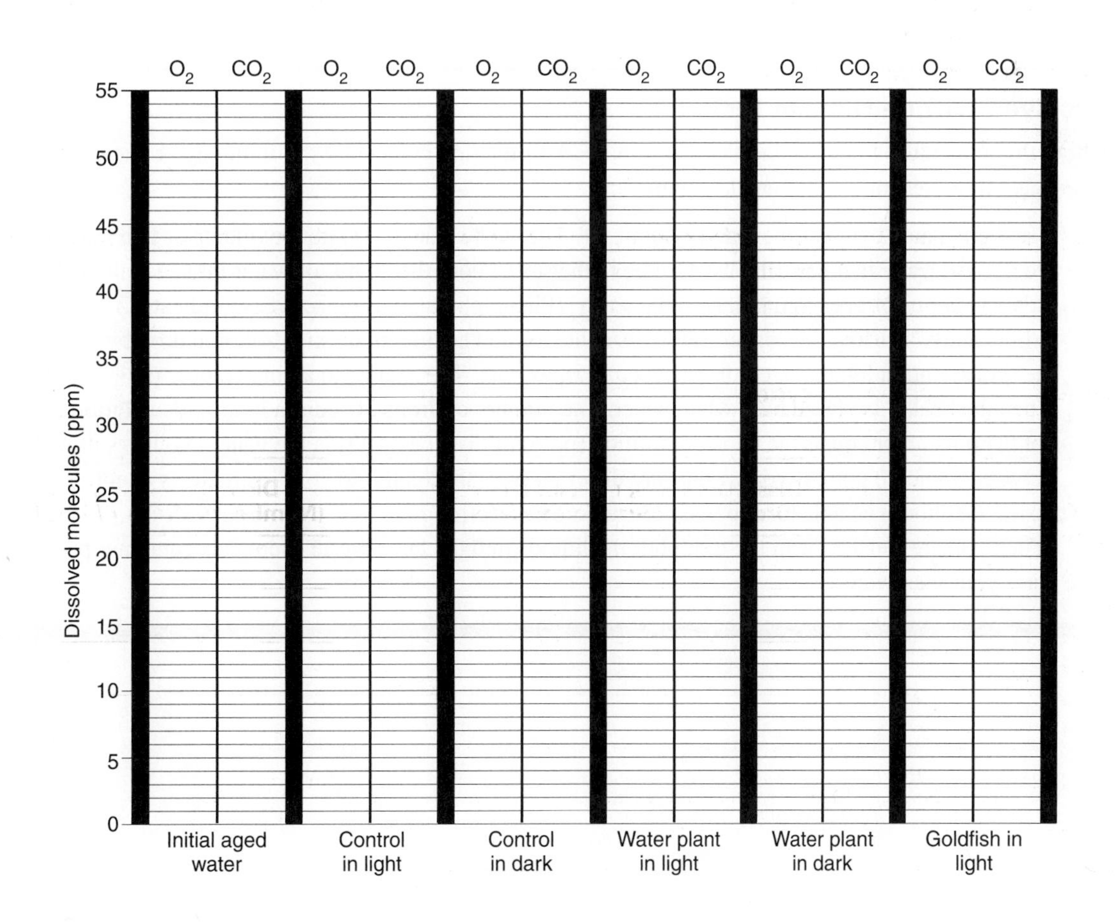

Dissolved Oxygen Test

Follow this procedure to determine the dissolved oxygen content of the initial aged water, control containers, and three sets of experimental containers.

1. Carefully fill (overflow) the small glass-stoppered bottle with the water to be tested. (Do not put plants or fish into this bottle. Do not use water that has been used for any other test.) Insert the glass stopper and pour off any excess water trapped on the outside of the bottle around the stopper.
2. Remove the stopper. Cut open chemical packets *a* and *b* and carefully empty the contents of both into the bottle. Because the quantity of chemicals in the packets is premeasured, it is important that you not spill any of the contents and that you get all the contents into the bottle. Replace the stopper carefully (do not trap any air bubbles) and shake for 30 seconds or until all granules are dissolved. A brown precipitate will form. (If granules still remain after 30 seconds of shaking, proceed to step 3.)
3. Set the bottle on the lab table. After the brown precipitate has settled out about halfway, empy the contents of the large packet (*c*) into the bottle. Stopper and shake again until the precipitate has completely dissolved. If you have brown particles in the water, vigorous shaking should get rid of them. You should now have a clear, yellowish liquid.
4. Fill the small measuring test tube with the yellow liquid and pour one full tube of the yellow liquid into a clean glass beaker. It is a good idea to rinse any glassware before you use it to be sure that it is clean.
5. While gently swirling the beaker to mix its contents, add sodium thiosulfate solution drop by drop (count the drops) until the yellow color turns clear, like water. The color change is best seen when the beaker is placed over white paper. The number of drops of sodium thiosulfate used equals the parts per million (ppm) of dissolved oxygen. Your result should be between 1 and 20 ppm. If your result is different from this, call it to your instructor's attention. Your instructor should be able to help you determine what is wrong.
6. When you are finished with the test, pour the liquids into the large waste chemicals container provided and rinse your glassware.

Dissolved Carbon Dioxide Test

Follow this procedure to determine the dissolved carbon dioxide content of the initial aged water, control containers, and three sets of experimental containers.

1. Use a graduated cylinder to measure 30 mL of water to be tested. (Do not put plants or fish into the bottle. Do not use water that has been used for any other test.) Pour the 30 mL of water into a small beaker.
2. Add five drops of phenolphthalein to the water. (Phenolphthalein is an acid/base indicator.)
3. While gently swirling the beaker to mix its contents, add sodium hydroxide solution drop by drop (count the drops) until a light pink color appears *and stays pink*. The change to a pink color is easiest to detect when the beaker is placed over white paper. The number of drops of sodium hydroxide solution used to get the pink color equals the parts per million (ppm) of dissolved CO_2 in the water. Your result will be between 1 and 20 ppm. If your results are different from this, call it to your instructor's attention. Your instructor should be able to help you determine what is wrong.
4. When you are finished with the test, pour the liquids into the large waste chemicals container provided and rinse your glassware.

Table 8.2 Effect of Temperature on Water Solubility of O_2 and CO_2

Temperature °C	O_2 Solubility (mg/kg or ppm)	CO_2 Solubility (mg/kg or ppm)
0	14.62	3,347
5	12.77	2,782
10	11.29	2,319
15	10.08	1,979
20	9.09	1,689
25	8.26	1,430
30	7.56	1,250
35	6.95	1,106
40	6.41	970

8 Photosynthesis and Respiration

Name ______________________________ Lab section ______________

Your instructor may collect these end-of-exercise questions. If so, please fill in your name and lab section.

End-of-Exercise Questions

Use the equations for photosynthesis and aerobic respiration to help you think about the questions. Remember, reactants decrease in concentration as products increase in concentration.

Photosynthesis:

$$6\,CO_2 + 6\,H_2O \xrightarrow[\text{chloroplasts}]{\text{light}} C_6H_{12}O_6 + 6\,O_2$$

Aerobic respiration:

$$C_6H_{12}O_6 + 6\,O_2 \xrightarrow[\text{mitochondria}]{\text{enzymes}} 6\,CO_2 + 6\,H_2O + \text{energy (ATP)}$$

1. How does the P/R ratio change from summer to winter for a plant growing in Canada?

2. What does it mean if the P/R ratio is 2:1? Which of the tubes (1. plant in light, 2. plant in dark, 3. goldfish) had a P/R ratio similar to 2:1?

   What does it mean if the P/R ratio is 1:2? Which of the tubes (1. plant in light, 2. plant in dark, 3. goldfish) had a P/R ratio similar to 1:2?

3. Why was there less CO_2 in the *Plant in Light* tube at the end of the hour than in the control?

4. If you could have measured the number of individual water molecules, would the number of water molecules in the tube have differed at the beginning and at the end of the hour for the *Plant in Light?* Why?

5. Why was there less oxygen in the *Plant in Dark* tube at the end of the hour?

6. Why was there less oxygen in the *Goldfish* tube at the end of the hour?

7. How do you think the results would have differed if the *Goldfish* tube had been placed in the dark?

8. Compare the processes occurring in the plants in the dark with the processes occurring in the goldfish in the light or in the dark.

9. If a fish had been placed in the tube with the plants in the light, what results would you expect? Why?

10. If a fish had been placed in the tube with the plants in the dark, what results would you expect? Why?

11. If you have $10,000 and you lend your friend a nickel, how many parts per million of your money have you given to your friend?

LABORATORY

9 The Chemistry and Ecology of Yogurt Production

+ Safety Box

- When using open flames and hot plates, be careful of fire, particularly if you have loose clothing or long hair. Hot waterbaths may also be hazardous.
- When the fermentation is complete and the yogurt has been formed, have your instructor check it before you sample the product. Should it appear unsuitable because of contamination, discard it in an appropriate manner.
- Use hot pads when removing the beaker from the heat source. Be careful as the milk approaches the boiling point that it does not boil over and burn.

Objectives

Be able to do the following:

1. Define these terms in writing.

   anaerobic respiration
   fermentation
   substrate
   lactose
   end product
   lactic acid
   enzyme

2. List the raw materials (substrate), end products, and conditions necessary for bacteria to engage in anaerobic respiration.
3. List two factors that inhibit the growth of bacteria in milk.
4. Recognize that the change in viscosity (thickness) of the milk is a result of a change in pH caused by the bacteria converting sugar to lactic acid.

Introduction

Yogurt is the product of the *fermentation* of milk sugar (lactose) by two species of bacteria, *Lactobacillus bulgaricus* and *Streptococcus thermophilus.* **Fermentation** is a metabolic process in which organic molecules are broken down to simpler compounds by organisms *without the use of oxygen.* It is an example of **anaerobic respiration.** Organisms that carry on fermentation obtain energy in the form of ATP molecules. In the process of either aerobic or anaerobic respiration, each step in the process is controlled by an enzyme that converts a **substrate**, the material acted upon by the enzyme, into an **end product.** The end product of the first reaction becomes the substrate for the next reaction in the series. The **enzymes** in the bacteria we are using today convert **lactose** into **lactic acid.** Lactic acid and ATP are the final end products of this example of fermentation. A simplied version of fermentation is

$$\text{glucose} \xrightarrow{\text{enzymes}} \text{2 lactic acid} + \text{2 ATP}$$

Figure 9.1 provides a more detailed description of what happens during fermentation.

Figure 9.1 Biochemical pathway for conversion of lactose to lactic acid.

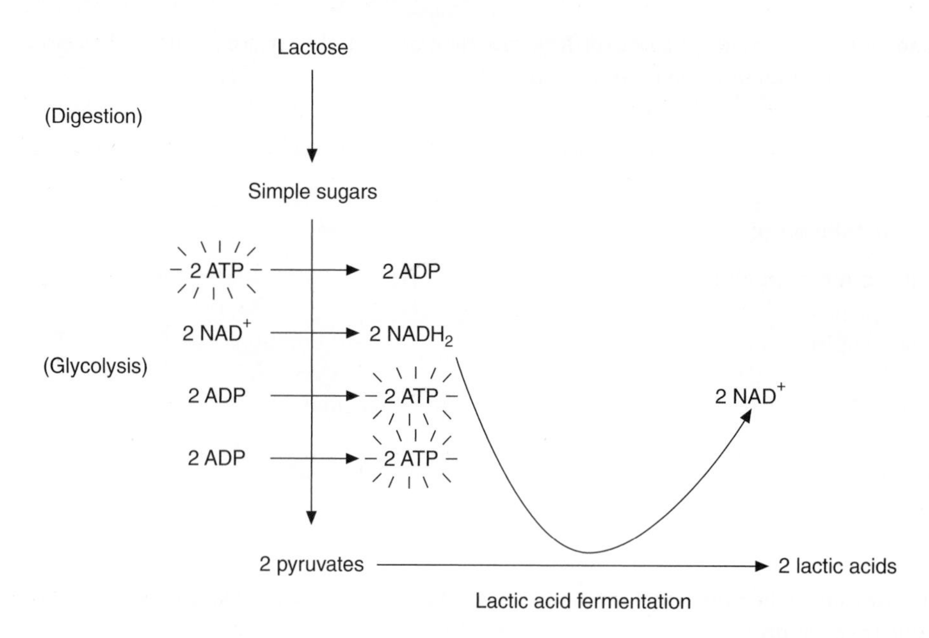

Because acids have a sour taste, the result of the activity of these bacteria is a sour-tasting dairy product. The production of the acid is responsible for the change in the consistency of the milk. The greatly lowered pH causes the milk protein to coagulate and become thick and viscous. *Streptococcus thermophilus* also produces some other compounds that are important in determining the final flavor of the yogurt.

To provide a consistently palatable product, commercial producers of yogurt must have a thorough understanding of the biology of the bacteria they use. They must be aware that changes in the environment alter the activities of the bacteria. Fresh milk normally contains a mixture of different species of bacteria, which eventually ferment, or sour, the milk. Because these bacteria produce unwanted flavors and compete with the desired bacteria, it is desirable to reduce the natural bacterial population as much as possible. This is usually done by heating the milk for a specified period of time.

Substances that retard the growth of the desired species of bacteria—inhibitors—are another problem for the yogurt producer. These inhibitors are of two varieties. One type is produced naturally by the cow (it makes sense that cows would have evolved a method of inhibiting fermentation while the milk was still in the cow; otherwise, cows might give yogurt). The second kind of inhibitor is a much greater problem and results from the treatment of cows with antibiotics. If cows are being treated with penicillin or another antibiotic that slows down the growth of disease-causing bacteria, this antibiotic can get into the milk and prevent the desired bacteria from doing their job of making yogurt.

Preview

With one carton of milk per student, you will

1. prepare and inoculate the container of milk with a bacterial culture that will produce yogurt.
2. incubate the mixture at the proper temperature.

Procedure

1. Empty a ½-pint container of milk into a clean beaker and place on a hot plate.
2. Add to the milk 11 g of powdered milk. Stir and heat to 96°C. Do not allow the milk to boil. Stir constantly to avoid burning. *As you approach 96°C, stir constantly or the milk will boil over.*
3. As soon as the milk reaches 96°C, use hot pads to remove it from the hot plate. Allow the milk to cool to 46°C. Stirring occasionally will reduce the time needed to cool the milk.
4. Add a teaspoon of starter yogurt culture to your empty milk carton from the bulk culture provided.
5. Pour the cooled milk into the carton; close and staple shut. Label the carton with your name and the date.
6. Place your culture in the designated incubator (39°C) until it coagulates (about 6–8 hours). A high temperature of incubation results in a more sour yogurt; a lower temperature of incubation results in a more viscous product. The 39°C temperature is a compromise between these two temperatures. Your product should be sour and viscous, like yogurt.
7. Remove the carton from the incubator following coagulation, and cool to 10°C in a refrigerator.
8. When it is cooled (any convenient time within a week), note the odor and flavor. Compare this with a commercially prepared carton of yogurt.

Flavored Yogurts (Optional)

You may wish to add some flavoring to the yogurt. The flavorings are added after the incubation period to prevent their change in taste by the bacteria. Here are some suggestions for flavoring your yogurt:

cucumbers and black pepper
dry, fruit-flavored gelatins
honey
chocolate syrup
fresh or frozen fruit
fruit preserves
nuts
granola
cereals

Yogurt Cheese

Yogurt and cheese are both fermented milk products; however, cheeses have the curd (the solid portion) separated from the whey (the liquid portion). The curd of yogurt is also softer and more tart than most cheeses. To convert yogurt into a cheese, you merely need to allow the whey to separate from the curd, so that the curd can knead together into a firmer product. This is easily done by placing the yogurt into a bag of closely knit cheesecloth and suspending the bag over a container to collect the whey. If you suspend the yogurt and allow the whey to drain away for 24 hours, the cheese that is formed has the consistency of cream cheese. If it is allowed to drain for a longer period, the curd becomes more dense and can be sliced. The whey can be saved. It is sometimes used in place of milk as a beverage or on cereals or as an additive in other recipes.

9 The Chemistry and Ecology of Yogurt Production

Name ____ Lab section __________

Your instructor may collect these end-of-exercise questions. If so, please fill in your name and lab section.

End-of-Exercise Questions

1. Why was the milk heated prior to adding the starter culture?

2. What active ingredient(s) can be found in the yogurt starter culture?

3. What would happen if you added the yogurt culture to the heated milk before it had been allowed to cool?

4. What things might act as inhibitors in the yogurt-making process?

5. Now that you have made yogurt, what do you think is different about the process of making Swiss cheese, cheddar cheese, and yogurt cheese? (See directions for making yogurt cheese on page 105.)

6. What would happen if you added the yogurt culture to the cooled milk and immediately refrigerated it?

7. Which of the following terms apply to the process of making yogurt (circle them)?

aerobic	anaerobic
bacteria	yeast
carbon dioxide	methane
glycolysis	Krebs cycle
coenzyme	alcohol
lactic acid	fermentation
cellular respiration	

LABORATORY

DNA and RNA: Structure and Function

10

+ Safety Box

- No unusual hazards are associated with this laboratory experience. Please follow standard laboratory safety procedures.

Objectives

Be able to do the following:

1. Define these terms in writing.

nucleotide	codon	base pairing
ribose	replication	DNA
translation	polypeptide	RNA
anticodon	structural gene	double-stranded DNA
transcription	deoxyribose	

2. Construct the four different DNA nucleotides using models or symbols of the nucleotide parts.
3. Construct a portion of a DNA molecule using models or symbols of the nucleotide parts.
4. Describe in writing, or by the use of models, the process of DNA replication.
5. Explain how the process of DNA replication ensures that exact copies of genetic information are available for distribution to daughter cells.
6. Construct the four different RNA nucleotides using models or symbols of the nucleotide parts.
7. Describe in writing, or by the use of models, the process of transcription.
8. Describe the similarities in structure and function between DNA and RNA.
9. Describe the differences in structure and function between mRNA and tRNA.
10. Describe in writing, or by the use of models, the process of translation.

Introduction

It is to your advantage when doing this exercise to keep your textbook open to the chapter that describes nucleic acids and the processes of replication, transcription, and translation. Refer to the information and diagrams in your text as needed.

Both of the nucleic acids, **deoxyribonucleic acid (DNA)** and **ribonucleic acid (RNA),** are constructed of smaller subunits called **nucleotides** (see laboratory 4, "Structure of Some Organic Molecules"). The various kinds of nucleotides synthesized in the cell differ from one another in the kinds of sugar (**deoxyribose** or **ribose**) and the kind of nitrogen base (adenine, guanine, cytosine, thymine, or uracil) they contain. Cells maintain a supply, or pool, of these nucleotides for use in the processes of *DNA replication* and *RNA transcription.*

DNA replication is the series of chemical reactions that result in the formation of two identical double-stranded DNA molecules from one original molecule. This is an assembly process that uses the two strands of the existing DNA molecules as templates upon which new DNA nucleotides are aligned. **Transcription** involves the formation of a copy of the DNA code in the form of RNA molecules. The process of manufacturing the RNA is similar to replication, except that RNA nucleotides are matched to only one side of the DNA double helix. RNA molecules are single-stranded. There are three types of RNA produced: messenger RNA (mRNA), transfer RNA (tRNA), and ribosomal RNA (rRNA). Each of these RNA molecules has a special role to play in the functioning of a cell. The mRNA carries the message from the DNA in the nucleus to the cytoplasm of the cell. The tRNA carries amino acids used for the manufacture of proteins. The rRNA, along with some proteins, forms ribosomes, structures needed to assemble proteins. **Translation** involves the co-operation of all the kinds of RNA to bring about the synthesis of specific proteins. During translation, the structure of the ribosome promotes complementary **base pairing** between mRNA and tRNA, leading to the formation of peptide bonds between specific amino acids. Strings of amino acids are known as **polypeptides**. One or more polypeptides are combined to form essential proteins, such as enzymes, antibodies, hemoglobin, collagen found in connective tissue, or myosin found in muscle tissue.

This exercise will help you understand the details of replication, transcription, and translation. Figure 10.1 summarizes these central concepts of molecular genetics.

Figure 10.1 The central concept of molecular genetics.

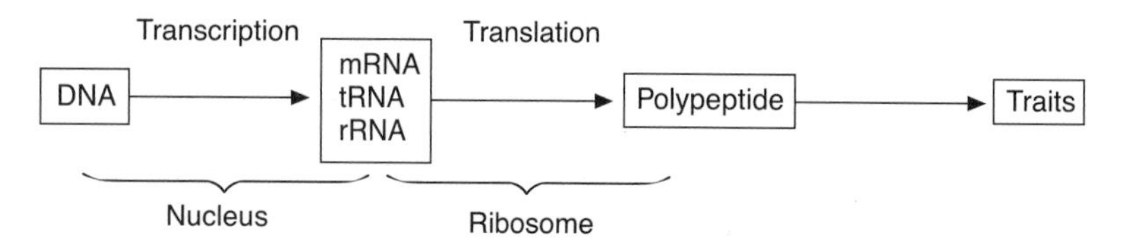

Preview

This exercise simulates the processes of DNA replication, RNA transcription, and translation. Remember, real DNA is a three-dimensional double helix. The flat plastic structures used in this exercise cannot be made to exactly duplicate the real molecules involved in these cell processes.

Work in pairs and be sure that you discuss with your partner what each stage of the exercise represents. If you are unsure at any time, check with the instructor.

During this lab exercise, you will

1. separate the individual pieces of the molecular model into groups.
2. use the individual model pieces to construct DNA nucleotides.
3. use the DNA nucleotides to construct a model of a single strand of DNA.
4. use the DNA nucleotides to base-pair with the nucleotides on the single strand, thus forming a double strand of DNA.
5. replicate the double strand of DNA.
6. use the individual model pieces to construct RNA nucleotides.
7. use the RNA nucleotides to construct tRNA molecules and attach the appropriate amino acid to these molecules.
8. transcribe DNA into mRNA.
9. translate and form a protein model.

Procedure (Carefully Follow the Directions Step-by-Step. Don't Take Shortcuts!)

Building Double-Stranded DNA

1. **Separate the pieces of the molecular model into groups**. Each pair of students should get a DNA/RNA model kit. Empty the plastic parts onto the table and separate the various parts into piles. These parts represent the types of molecules that are necessary to construct nucleotides. Table 10.1 lists the parts and the numbers of each that should be in your kit. If you find that you do not have the proper number, or if pieces are broken, obtain the parts you need from your instructor.

Table 10.1 DNA-RNA Model Kit

Part Name	Symbol	Number Needed	Number Present
Deoxyribose sugar	D	36	
Ribose sugar	R	18	
Phosphate	P	36	
Uracil	U	5	
Thymine	T	10	
Guanine	G	8	
Cytosine	C	8	
Adenine	A	10	
Hydroxyl group	OH	3	
Hydrogen atom	H	3	
Amino acids	Leu, His, Gly	1 of each type	

2. **Construct DNA nucleotides.** A DNA nucleotide is composed of a phosphate molecule bonded to a deoxyribose sugar molecule bonded to a nitrogen-containing base, such as adenine. Use the model pieces to construct DNA nucleotides. You will use these nucleotides as building blocks to construct a DNA molecule.

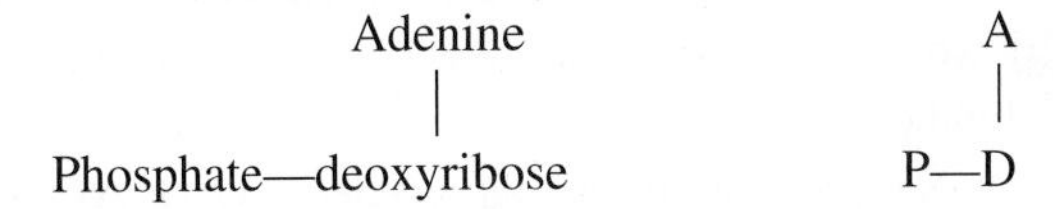

   a. Assemble 36 separate DNA nucleotides. Each nucleotide should look like one of the diagrams in figure 10.2.
   b. Assemble the following:

      10 separate nucleotides using adenine
      8 separate nucleotides using guanine
      8 separate nucleotides using cytosine
      10 separate nucleotides using thymine

   c. *Make sure that all the phosphates are on the left side of the sugar.* These 36 nucleotides make up the DNA nucleotide pool of your cell.

Figure 10.2 DNA nucleotides. Notice that each nucleotide differs in the type of nitrogen base it contains and not in the sugar or the phosphate.

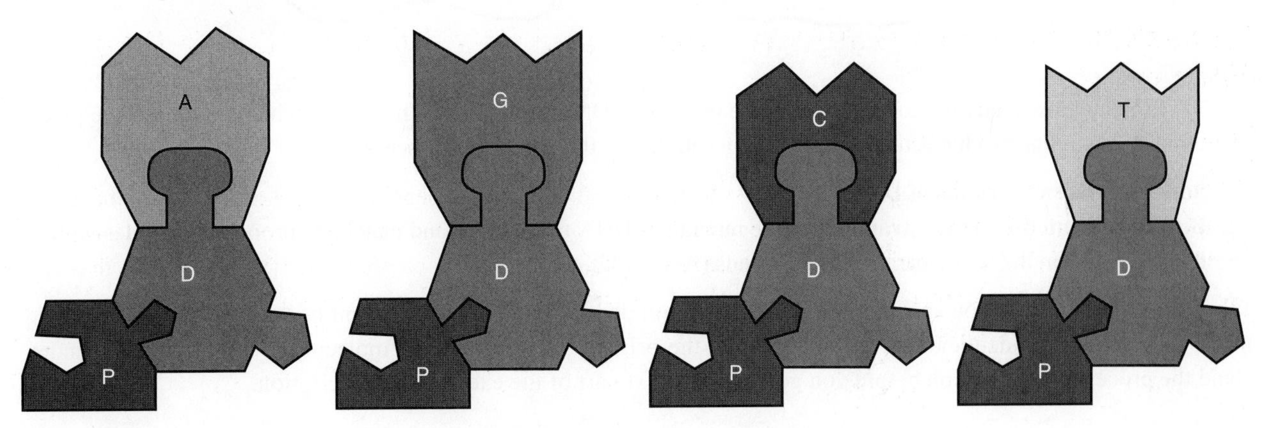

3. **Construct one strand of a DNA molecule**. Use the DNA nucleotides you have just assembled to construct a model of one side of a DNA helix using precisely the sequence given in figure 10.3. Attach the nucleotides end-to-end, with the deoxyribose to the right and the phosphates to the left. The phosphate of one nucleotide should link with the sugar of the next. *This sequence of nucleotides will represent the gene in our simulation.* Place a piece of masking tape along its length to help you identify it and to help hold it together.

Figure 10.3 Single DNA strand.

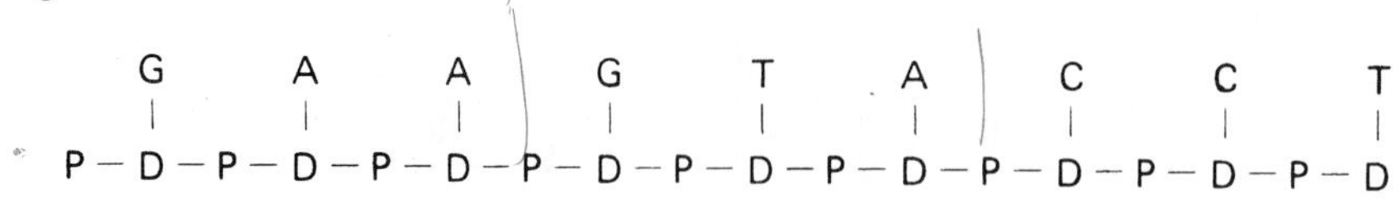

4. **Construct a DNA double helix**. Use complementary nucleotides from your pool of DNA nucleotides to base-pair with the nitrogenous bases on the single strand you just assembled. This will result in a double-stranded DNA molecule. To assemble the second strand properly, follow the rule that the nitrogen base adenine always base-pairs with the base thymine in DNA and that the nitrogen base guanine always base-pairs with the base cytosine. Slide the appropriate nucleotides into position, so that a ladderlike molecule is formed on the table. In the new strand that you are forming, link the sugar of one nucleotide to the phosphate of the next nucleotide. Because of the nature of the pieces in this kit, we will use this ladderlike molecule to represent the double-stranded DNA molecule, although in reality it is a spiral-shaped double helix.

Replication: Synthesis of DNA

When a cell divides, each of the two new daughter cells must receive the same DNA. To accomplish this, the cell usually has two sets of DNA molecules on chromosomes that may be separated into the two newly forming cells. The process of constructing copies of a double-stranded DNA molecule is called DNA replication. If the genetic information is not replicated and separated equally when cells divide, the new cells will not contain the genetic messages necessary to manufacture the proteins needed by the cell.

Replication of DNA is going on within your body at this moment. Whenever a wound is repaired, growth occurs, or old cells are replaced, DNA must first replicate. In other words, before any cell division occurs, the DNA molecules make copies of themselves so that each new cell receives the same genes. The diagrams in figure 10.4 show a cell in the process of division after the DNA has replicated.

Figure 10.4 Cell division in progress.

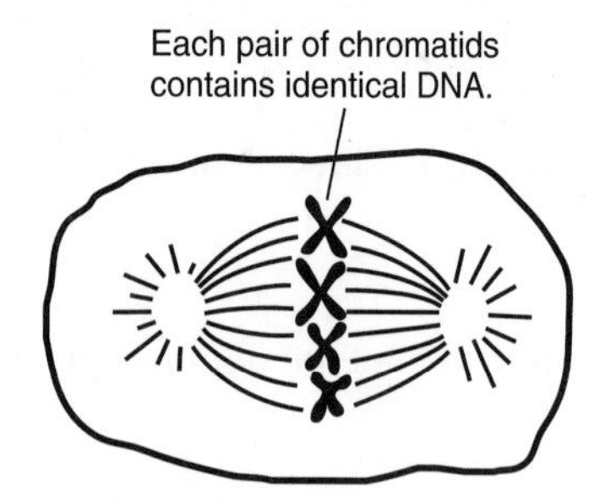

Chromatids beginning to separate and the identical DNA moves into two new daughter cells.

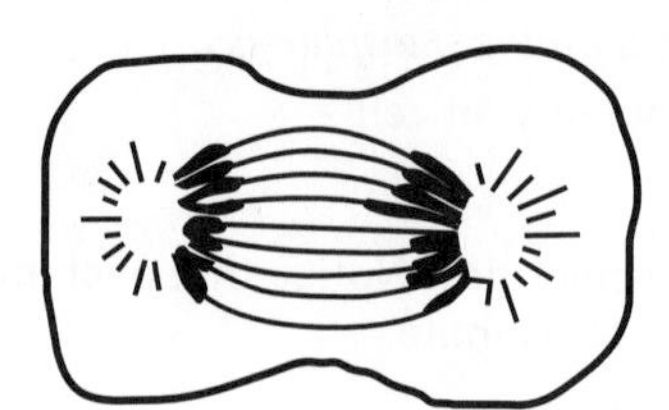

You are now ready to proceed through the process of DNA replication. The way in which you move the pieces of the model is similar to what actually happens in a cell during the replication process.

1. **Separate the two strands of DNA**
2. **Build new matching DNA strands**. Use the remaining DNA nucleotides and match the proper nucleotides with their partners on the two separated DNA strands (A-T, G-C).

When this is completed, you should have two double-stranded DNA molecules in front of you on the table. Furthermore, they should be identical to each other and to the original double-stranded molecule. Be sure that you understand the process of replication before you go on to the next part of the exercise—transcription.

Transcription: Synthesis of RNA

Transcription is the process of synthesizing RNA. Three forms of RNA are produced by transcription: messenger RNA (mRNA), transfer RNA (tRNA), and ribosomal RNA (rRNA). In this exercise, we examine the formation of mRNA only, but remember that transfer RNA and ribosomal RNA are formed in much the same way. During transcription, RNA nucleotides base-pair with DNA nucleotides on the gene side (template strand) of the DNA molecule.

To prepare for the production of ribonucleic acids during the process of transcription, we must first synthesize RNA nucleotides. Remember that RNA nucleotides have ribose sugar in place of deoxyribose sugar and the uracil base in place of thymine.

1. **Retain the copy of double-stranded DNA that has the original single strand of DNA with the tape on it.** Break down both strands of the other DNA molecule into its smallest subunits (phosphate, sugar, base). Some of these subunits are now needed to assemble RNA nucleotides. *(DNA is not disassembled to make RNA in cells, but we must do this with this model so that we have enough pieces to make the RNA we need.)*
2. **Construct RNA nucleotides** using ribose, phosphate, and the bases adenine, guanine, cytosine, and uracil. You need

   5 separate adenine nucleotides
   4 guanine nucleotides
   4 cytosine nucleotides
   5 separate uracil nucleotides

   Be sure that all the phosphates are on the left side of the sugar (figure 10.5). These 18 nucleotides represent the RNA pool in the cell.

Figure 10.5 Individual RNA molecules. Notice that each nucleotide differs in the type of nitrogen base it contains and not in the sugar or phosphate.

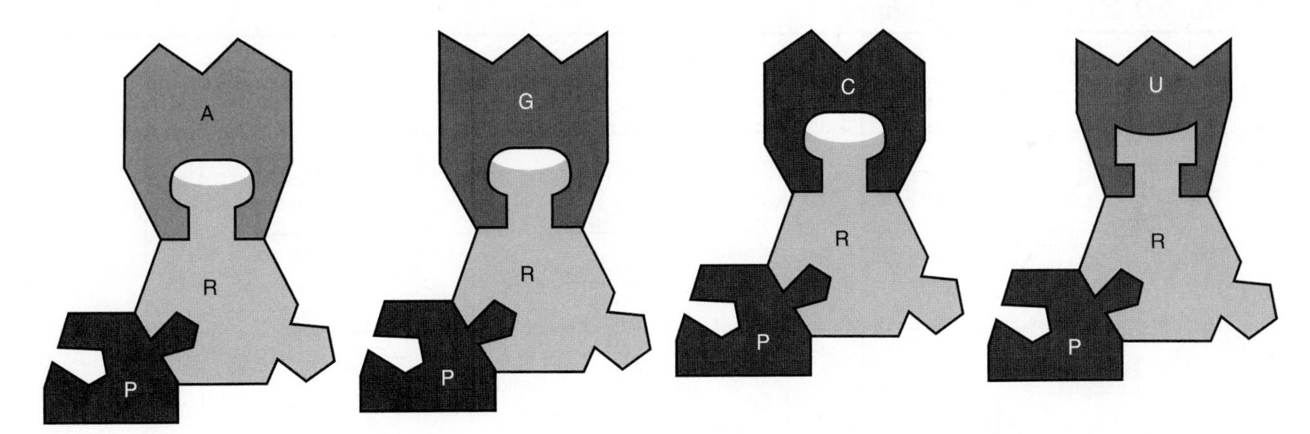

3. **Separate the DNA double helix into two strands.**
4. **Build an RNA molecule.** Use the RNA nucleotides you constructed and match the proper RNA nucleotide with the appropriate nitrogenous base on the gene side (template strand) of the DNA. Remember that it is the strand with the tape on it. This side is the genetic code; the other side does not carry genetic information but is important in the replication process.
5. **This newly constructed mRNA molecule should consist of nine nucleotides.** The order of the RNA nucleotides in the mRNA you constructed is predetermined by the order of nucleotides along the coding strand (gene) of the DNA molecule.
6. **Remove the mRNA molecule from the DNA strand and move the RNA to the side.** Now, put the two separated strands of DNA back together as they were. You have just simulated the process of transcription. The mRNA molecule that was formed has picked up the code from DNA. The DNA is intact and not damaged. The same DNA code can be used again for the transcription of additional mRNAs if necessary. In fact, it is common for more than one RNA transcript to be produced each time the cell expends energy to unwind and open up the DNA helix.

The Codon Dictionary

The information stored in nucleic acids is in the form of a linear arrangement of three nitogenous bases. Because both DNA and RNA are linear molecules, you can start at one end and proceed toward the other end, reading three bases at a time. In the mRNA each of these sequences of three bases is known as a **codon.** There are 64 possible codons, and each codon corresponds to one amino acid, with the exception of three codons that are stop signals. The codon dictionary in table 10.2 shows each of the codons and the specific amino acid each codes for. Note that most amino acids are coded for by more than one codon.

Table 10.2 Codon Dictionary

First Letter	Second Letter: U	Second Letter: C	Second Letter: A	Second Letter: G	Third Letter
U	UUU, UUC: Phe – Phenylalanine UUA, UUG: Leu – Leucine	UCU, UCC, UCA, UCG: Ser – Serine	UAU, UAC: Tyr – Tyrosine UAA, UAG: Stop	UGU, UGC: Cys – Cysteine UGA: Stop UGG: Trp – Tryptophan	U C A G
C	CUU, CUC, CUA, CUG: Leu – Leucine	CCU, CCC, CCA, CCG: Pro – Proline	CAU, CAC: His – Histidine CAA, CAG: Gln – Glutamine	CGU, CGC, CGA, CGG: Arg – Arginine	U C A G
A	AUU, AUC, AUA: Ile – Isoleucine AUG: Start / Met – Methionine	ACU, ACC, ACA, ACG: Thr – Threonine	AAU, AAC: Asn – Asparagine AAA, AAG: Lys – Lysine	AGU, AGC: Ser – Serine AGA, AGG: Arg – Arginine	U C A G
G	GUU, GUC, GUA, GUG: Val – Valine	GCU, GCC, GCA, GCG: Ala – Alanine	GAU, GAC: Asp – Aspartic Acid GAA, GAG: Glu – Glutamic Acid	GGU, GGC, GGA, GGG: Gly – Glycine	U C A G

Synthesis of tRNA and rRNA

Although we will not go through the process, the synthesis of transfer RNA (tRNA) and ribosomal RNA (rRNA) takes place in the same manner as mRNA. However, different genes are involved. In our model, we will construct only three of the many possible tRNAs. You will make the tRNA molecules directly from the remaining plastic parts to save time. Also keep in mind that, because the plastic pieces are not flexible, these tRNA models are different from the real thing. Our models lack a phosphate at one end and are much shorter than actual tRNA molecules. Actually a transfer RNA molecule is composed of a single strand of about 100 nucleotides that folds back on itself. One end of the tRNA has a coding portion, called an **anticodon,** which matches a codon on mRNA at one end and a **binding site** for a specific amino acid at its other end.

1. **Assemble the three tRNA models as they appear in figure 10.6.**

Figure 10.6 tRNA nucleotides. Each tRNA has a different anticodon.

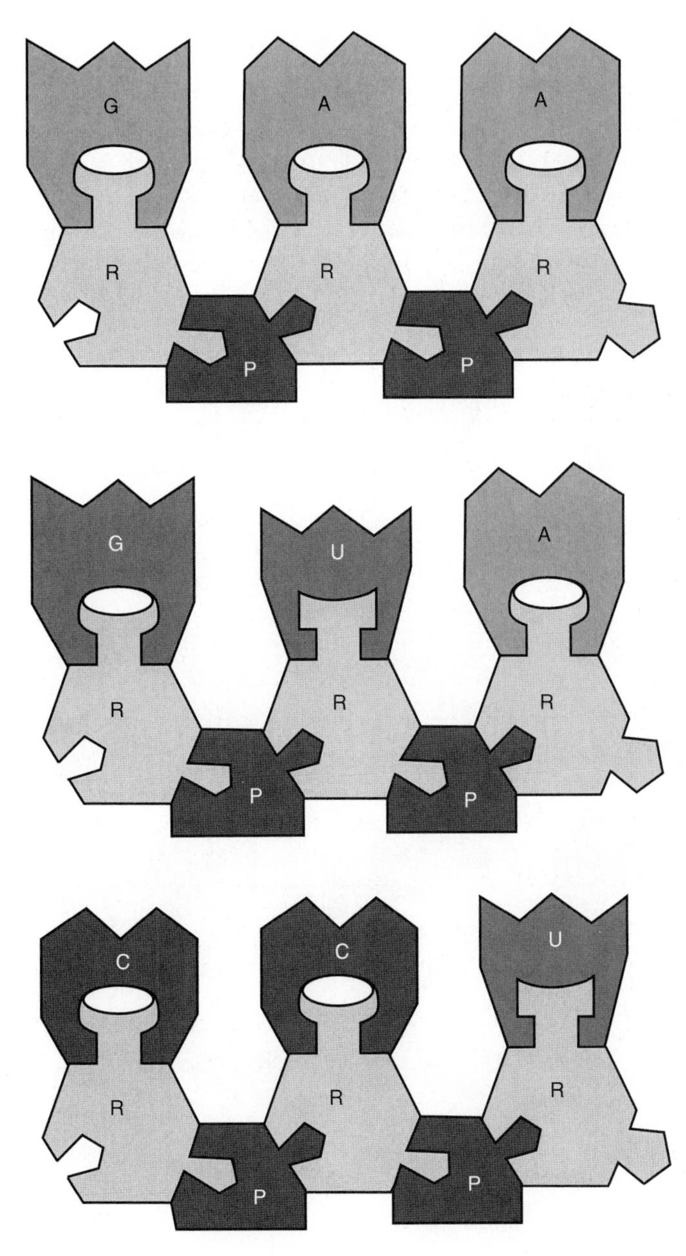

2. **Attach the appropriate amino acid to each of the three tRNA molecules (figure 10.7)**. Make sure that the H and OH units are attached to the amino acids.

Figure 10.7 tRNA nucleotides and attached amino acid. Each tRNA has a different sequence of bases and a different type of amino acid attached.

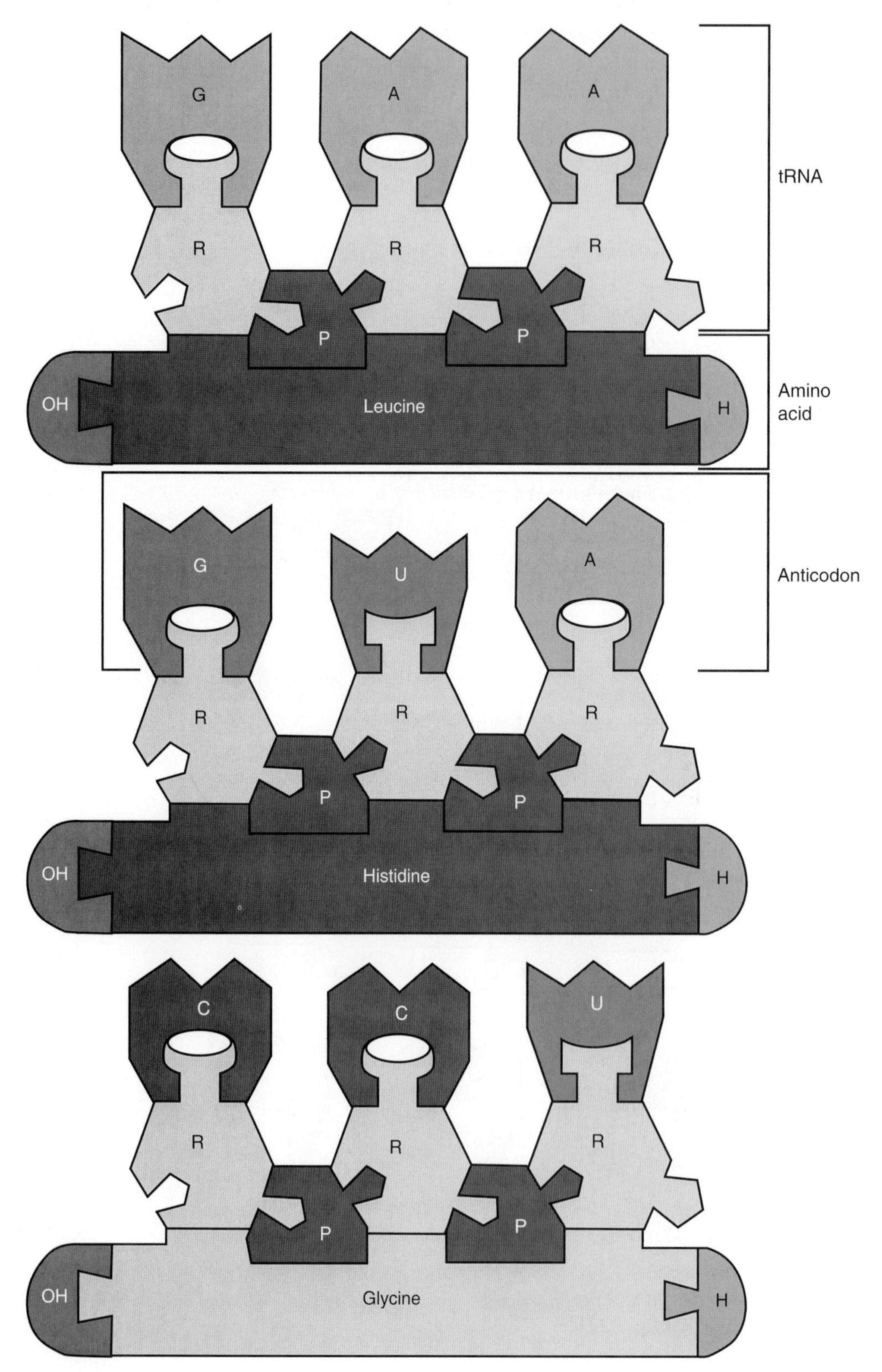

There are several different rRNA molecules that become associated with certain proteins to form ribosomes, which are necessary for protein synthesis. We will use the box the model came in to simulate the ribosomes.

Translation: Synthesis of Proteins

During translation, the genetic message, which is coded by the order of bases in the DNA, is translated into the structure of a protein by determining the order of amino acids in the protein. Messenger RNA carries the message from the DNA in the nucleus to the ribosomes in the cytoplasm. Transfer RNA picks up amino acids in the cytoplasm and carries them to particular places on the mRNA called codons. Ribosomal RNA is part of the structure of the ribosome. The mRNA and tRNA come together at the ribosome during the synthesis of proteins from individual amino acids. In an actual cell, the ribosomes are composed of proteins and rRNA. We will not manufacture a model of the ribosome but will use the box the model came in to simulate the ribosome as it moves down the mRNA.

For translation to occur in the cytoplasm of the cell, all three forms of RNA are needed. You will use the mRNA just constructed as well as three molecules of tRNA.

Simulate the process of translation as follows:

1. **Place the model's box on the first set of three nucleotides on the mRNA molecule** (CUU) to simulate the presence of a ribosome. Pair the appropriate tRNA to the first three nucleotides (codon) on the mRNA. Because the three bases on the tRNA pair with the three bases (codon) on the mRNA, the bases on the tRNA are often referred to as the **anticodon.** The process should occur as shown in figure 10.8.
2. **Next, attach the appropriate tRNA to the second mRNA codon** by matching the mRNA codon to its complementary tRNA anticodon. Note that the two amino acids are brought together so that the H end of one overlaps the OH of the other.
3. **The two amino acids that have been brought adjacent to one another become chemically bonded to each other**. Bonds between amino acids are called peptide bonds. To simulate the formation of a peptide bond, remove the H and OH from the adjacent amino acids and bond the two amino acids together. Link the H and OH to form a molecule of water. Because a water molecule is removed as the two amino acids are joined, it is often called a dehydration synthesis reaction.
4. Once this bonding has occurred,
   a. **the ribosome (box) shifts to the next codon on the mRNA** and the anticodon of the next tRNA can be aligned with the codon on the mRNA.
   b. **the first tRNA (GAA) is removed from the site** and can bond with another leucine present in the cell cytoplasm.
   c. **the original leucine molecule stays as part of the growing polypeptide chain**.
5. **When the third tRNA (CUU) binds to its mRNA codon, a second dehydration synthesis takes place**, forming another peptide bond.

The completion of this process should result in the synthesis of a short **polypeptide** composed of the three amino acids leucine, histidine, and glycine, in that order. (see Figure 10.9) You should also be able to identify three separated tRNA molecules ready to function in the transfer of more amino acids, two water molecules from dehydration synthesis reactions, and two peptide bonds holding the amino acids together.

Transfer RNA molecules may be used over and over again as amino acid–carrying molecules. A messenger RNA molecule can be used several times to produce the same polypeptide, but then it is broken down into its parts, which can be used to make other RNAs. All polypeptides, no matter how long they are, are constructed in this manner. Each gene is responsible for the synthesis of a particular polypeptide that differs from other polypeptides in the sequence of the amino acids it contains and the length of the sequence. Each piece of DNA that contains information for the building of a particular sequence of amino acids is called a **structural gene.**

Table 10.3 shows the standard flow of genetic information from the DNA gene sequence in the nucleus to mRNA, which travels to the ribosome, where mRNA meets tRNAs, which carry the amino acids. The specific complementary **base pairing** allows one to work forward or backward along this informational transfer scheme.

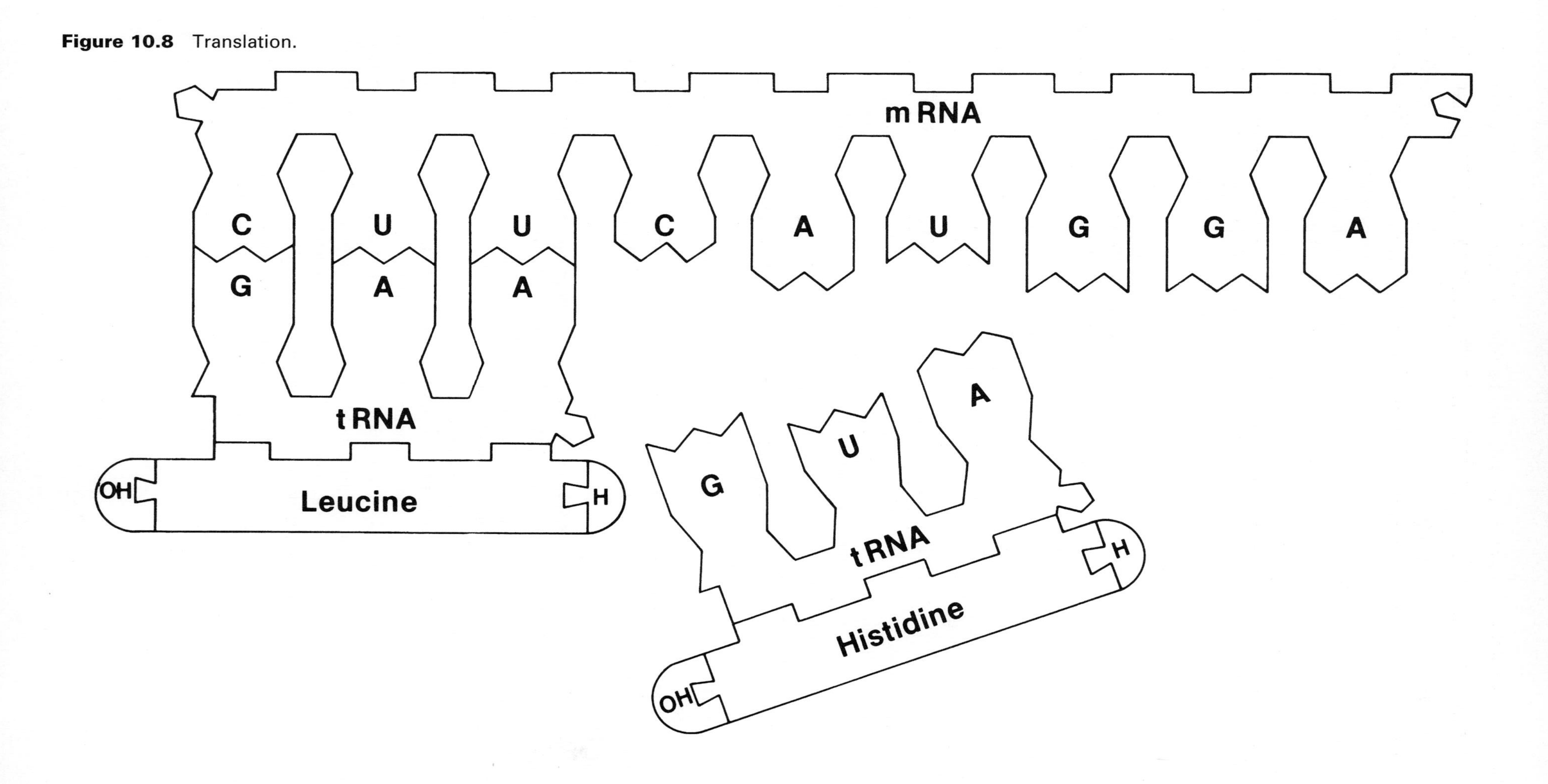

Figure 10.8 Translation.

Figure 10.9 A Polypeptide.

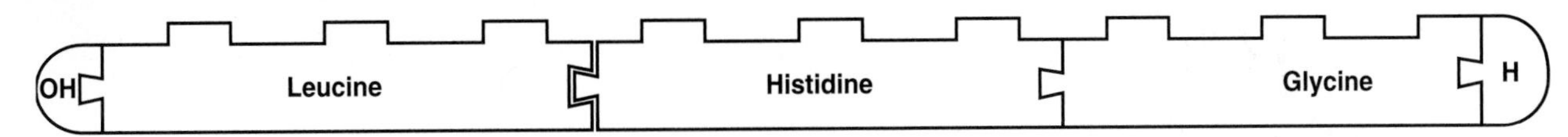

Table 10.3 Complementary Base Pairing

DNA Sequence		mRNA Codon	tRNA Anticodon	Carrying Amino Acid
C	G	C	G	
T	A	U	A	Leucine
T	A	U	A	
C	G	C	G	
A	T	A	U	Histidine
T	A	U	A	
G	C	G	C	
G	C	G	C	Glycine
A	T	A	U	
Complement	Gene			

10 DNA and RNA: Structure and Function

Name ______________________ Lab section __________

Your instructor may collect these end-of-exercise questions. If so, please fill in your name and lab section.

End-of-Exercise Questions

Directions: Use the complete codon dictionary (table 10.2) to answer these questions.

1. Use the base sequence for mRNA to complete the columns on the following table (table 10.4). (Use the mRNA sequence shown in the table. Do not use the mRNA base pair sequence from the exercise you just completed.) Remember that complementary base pairing is the key. Refer to table 10.3 if you have problems.

Table 10.4 **Complementary Base Pairing**

DNA Complement	DNA Gene	mRNA	tRNA	Amino Acid
		U		
		U		
		U		
		A		
		U		
		C		
		U		
		G		
		U		

2. Use one word to describe the relationship between the gene sequence and the mRNA sequence in table 10.4.

   - Complimentary

3. Describe in a sentence the relationship between the gene sequence and the tRNA sequence in table 10.4.

   - Complintory except U substitutes T

4. What will the sequence of mRNA nucleotides be if the following represents the bases in a DNA molecule of a structural gene?

DNA gene	A	A	T	G	G	T	C	C	A	C	C	G	C	T	G
mRNA	U	U	A	C	C	A	G	G	U	G	G	C	G	A	C

5. If a structural gene contains 300 DNA nucleotides, how many amino acids will be used in the protein synthesis process? 100

6. If a protein has 150 amino acids, how many DNA nucleotides will make up the structural gene?

450

7. A protein has the following amino acid sequence. Construct a DNA nucleotide sequence of the structural gene.

Phenylalanine — Glycine — Glycine — Alanine — Proline —Valine— Asparagine — Alanine

Amino Acid	Phenylalanine	Glycine	Glycine	Alanine	Proline	Valine	Asparagine	Alanine
mRNA codon								
DNA sequence								

8. Compare your DNA sequence from question 7 to that prepared by the others in the lab. Are there any variations? Did such variations occur in your answer to question 4? What is the reason for the difference? Are there any advantages to these variations? What are they?

9. Fill in table 10.5.

Table 10.5 Nucleotide Components and Function

Nucleic Acid Type	DNA	mRNA	tRNA
Name of sugar present in nucleotides	deoxyribose	ribose	ribose
Name of bases present in nucleotides	A,C,G,T,	ACGU	ACGU
Number of phosphates present in one nucleotide	1	1	1
Function of each of the kinds of nucleic acid	Info. storage	Delivers message	Delivers amino acids
Relative size and number of strands in each of the nucleic acids	2	1	1
Where each of these nucleic acids can be found in a cell	Nucleus	cyto-plasm	cyto-plasm.

LABORATORY

Mitosis: Cell Division

11

Safety Box

- A microscope is very expensive to replace; therefore, be particularly careful when handling one. Use both hands to carry it around the room. Hold it upright, so that the ocular lens does not slip out. Use only clean, dry lens paper to remove dust from the glass lenses. Do not use wet paper, paper towels, or other materials that may scratch the lenses.
- Microscope slides prepared by professionals are expensive to replace. They generally need to be viewed using a high-power lens system. Remember that the distance between the high-power lens and the surface of the coverslip is extremely narrow; therefore, adjust the high-power focus by using the fine adjustment knob only.

Objectives

Be able to do the following:

1. Define these terms in writing.

chromosome	spindle	anaphase
chromatid	cytokinesis	telophase
centromere	cell plate	interphase
nucleolus	prophase	mitosis
poles	metaphase	daughter nuclei

2. Write an explanation of why both nuclear and cytoplasmic division are important.
3. Locate and identify the principal stages of mitosis on slides of plant and animal tissue.
4. Diagram and label or describe in writing the principal stages of plant and animal mitosis.
5. Describe in writing how cytokinesis differs between plant and animal cells.

Introduction

All large organisms are composed of many cells. Growth of many-celled organisms involves an increase in the number of cells followed by an increase in size of the new cells. This is the basic mechanism by which a body grows and wounds are repaired. Cell division involves two major events: the distribution of identical copies of the genetic information from the parent cell to two daughter cells followed by cytoplasmic division. As a result of cell division, all the cells of a multicellular organism have the same genetic information. The parent cell divides by mitosis, producing two daughter cells. These two identical daughter cells contain exactly the same genetic material as the parent cell. Mitosis assures the production of identical sets of genetic information in the daughter cells.

Mitosis is an orderly series of events that results in the equal distribution of the **chromosomes** that carry the genetic information to the two new cells. The process flows from one stage to the next without interruption. Traditionally, mitosis has been artificially divided into phases: **prophase, metaphase, anaphase,** and **telophase.** It is a convenience to be able to classify parts of the process in this way, but it can be misleading if we forget that there is really no interruption in the events. It might be helpful to consider these four phases as being pictures taken with a camera. In such pictures, the motion is frozen, so that we can examine details that may interest us. Similarly, when we look at a plant or animal cell that has been killed and stained while in the process of mitosis, we are looking at a cell that was at a particular point in the mitotic process.

It is also important to recognize that mitosis is a relatively brief period in the life of a cell. A typical mitosis will take 2 to 4 hours to occur. The majority of the life of a cell is spent in a nondividing condition known as **interphase.** During interphase, the cell is participating in many important activities, including growth and the replication of DNA. Many cells differentiate into specialized cells that do not divide. They remain permanently in interphase and they synthesize molecules, move, or perform other activities typical of the cell. Figure 11.1 shows the cell cycle and the part mitosis plays in it. Figure 11.2 shows the events of mitosis.

Before a cell can divide, it must duplicate its DNA and chromosomes. This happens during interphase. The G_1 phase of interphase involves preparations for DNA synthesis. During the S phase, DNA is replicated. Thus, following the S phase, each chromosome consists of two parallel parts. Each part is known as a chromatid. During the G_2 phase, the cell goes through additional preparations for the division of the chromosomes during mitosis.

Figure 11.1 Stages of the cell cycle.

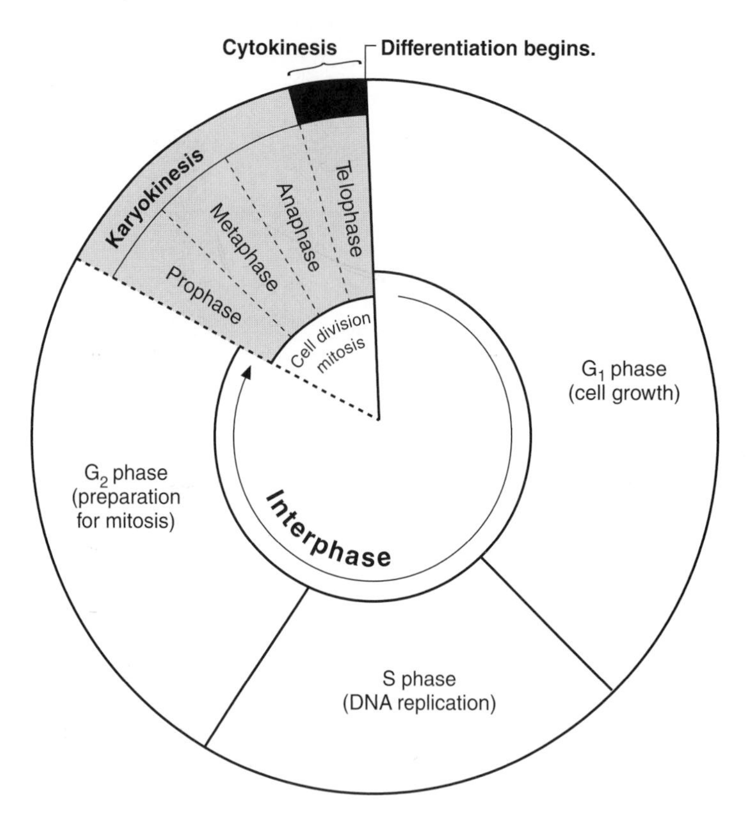

Figure 11.2 Parent cell dividing into two identical daughter cells.

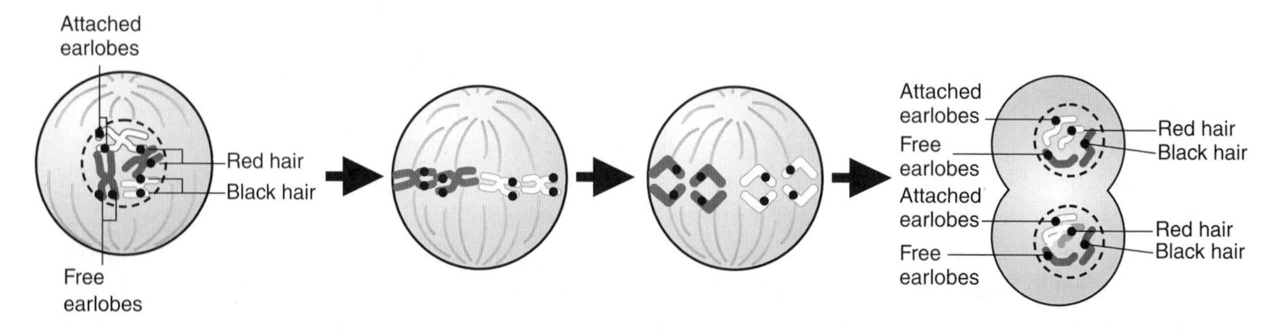

Preview

Each student should work with an assigned microscope and slides. Share your examples of each stage with other students. Quiz one another.

During this lab exercise, you will

1. locate the interphase stage preceding mitosis in a plant cell and an animal cell.
2. locate each of the following stages in a plant cell:
   a. prophase
   b. metaphase
   c. anaphase
   d. telophase
3. locate each of the following stages of mitosis in animal cells:
   a. prophase
   b. metaphase
   c. polar view of metaphase
   d. anaphase
   e. telophase
4. select one stage of mitosis in either plant or animal cells to be used as part of a trial quiz.
5. take the trial quiz.
6. sketch the condition and arrangement of the chromosomes of a cell as it proceeds through prophase, metaphase, anaphase, and telophase.

Procedure

1. Obtain a slide of *Allium* (onion) root tip or other plant material provided. Refer to figure 11.3 to orient yourself to the plant tissues you will study. The cells of the onion root tip were killed and stained. Then, the root tip was sliced lengthwise into thin strips, in which you will be able to see cells that were in different stages of the cell cycle when they were killed and stained.

Figure 11.3 *Allium* (onion) root tip.

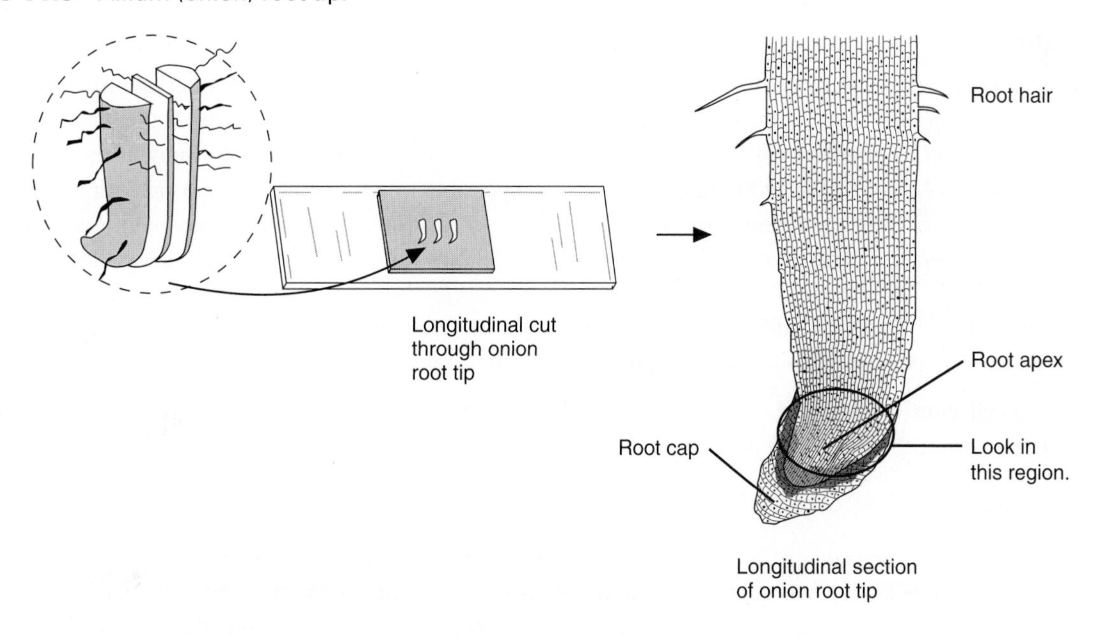

2. Locate the tip of a root on low power and focus. Then, switch to high power. It is important that you have clear focus using high power before you attempt to identify the various stages of mitosis.
3. You are now ready to scan your slide to locate the stages of mitosis in plant tissue. It is easiest to proceed in a step-by-step fashion, reading about the key events in the division of a "typical" cell in the paragraphs that follow before looking for that cell type. As you read, compare the description of the events with the photographs.

Interphase

Interphase is an active metabolic stage, during which the cell performs its normal functions. It carries on metabolism, grows, and replicates DNA during this stage. However, an interphase cell is not dividing. The nucleus contains chromosomes, but they are in a tangled mass of threads, which presents a uniform appearance of tiny dots. Complete chromosomes cannot be seen. The nuclear membrane is present and one or two **nucleoli** are visible in the nucleus. During this stage, the DNA replicates, but these molecules are much too small to be visible. Replication is necessary if the cell is going to divide at some time in the future. Figure 11.4 shows a chromosome before and after replication.

Figure 11.4 Chromosome before replication and after replication, late interphase.

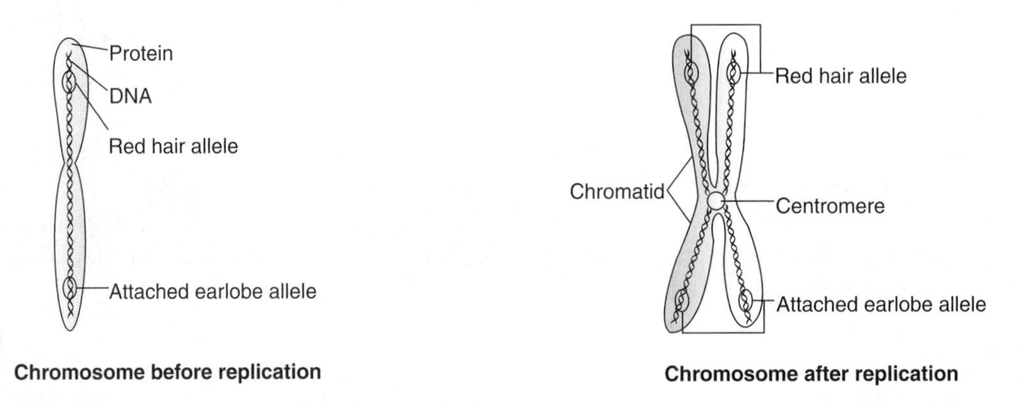

1. Locate a cell that you think is in the interphase stage and place the cell at the end of your pointer. Call the instructor over to check it. Let your neighbors look through your microscope while you examine the cell that they have selected as an example of interphase. When you are sure you can identify an interphase cell, find a prophase cell.

Prophase

During prophase, the cell prepares itself for division. One of the preparations is the shortening and thickening of the chromosomes into structures that can be seen. Their first appearance is as a mass of fine, tangled threads. The chromosomes shorten and thicken as a result of coiling. They gradually become more easily seen as individuals. Although the chromosomes and DNA were replicated in interphase, it is not possible to see that the chromosomes are double structures until late prophase. In late prophase, you should be able to see that the chromosomes consist of two **chromatids** joined at a point called the **centromere.** Meanwhile, a series of tiny tubules forms a structure called the **spindle** (figure 11.5).

At the same time, the nucleoli disappear. Eventually, at the very end of the prophase stage, the nuclear membrane disintegrates and the chromosomes are lying free in the cytoplasm.

2. Look for a cell showing prophase. Again, place the cell at the end of the pointer in your microscope and share your slide with your neighbors. If you have any doubts, ask your instructor to check the slide.

Metaphase

During metaphase, the chromosomes become attached by their centromeres to spindle fibers. The chromosomes appear to be tugged by the spindle fibers so that they form a flat sheet of chromosomes on a plane that passes through the equator of the cell. Viewed from the side, they appear to form a line; viewed from the **pole,** they can be seen to form a ring or disk. (Onion slides will not show polar views, because all the cells have a lengthwise orientation in the root tip.)

3. Look for a cell showing metaphase. Again, place the cell at the end of the pointer in your microscope and share your slide with your neighbors. If you have any doubts, ask your instructor to check the slide.

Figure 11.5 Onion root tip mitosis. (© The McGraw-Hill Companies, Inc./Kingsley Stern, Photographer)

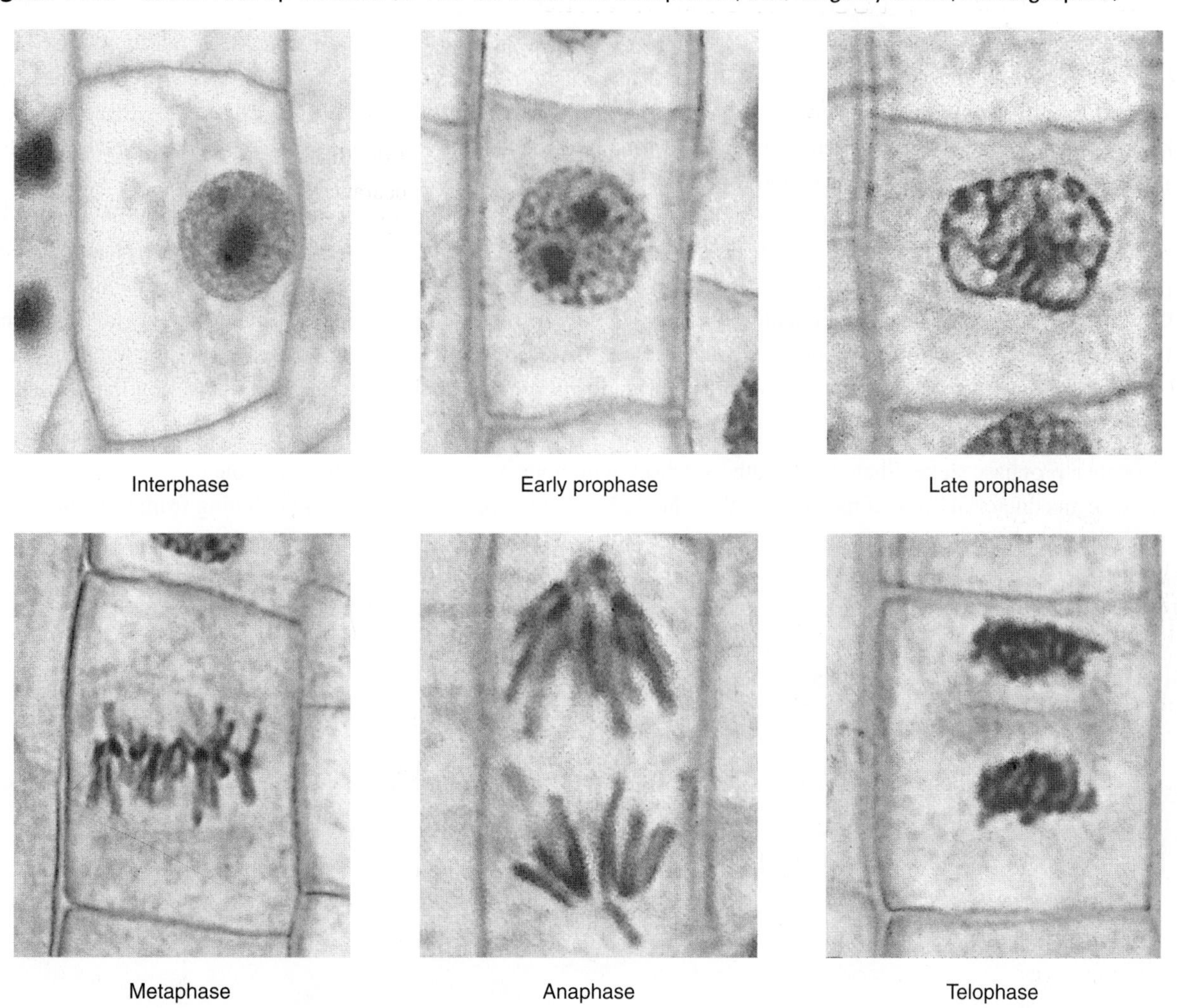

Anaphase

It is during anaphase that the equal distribution of genetic information is accomplished. Each centromere divides, and the two identical chromatids of each chromosome are pulled to opposite sides (poles) of the cell by the spindle fibers. When the two chromatids of a chromosome separate, each is called a *daughter chromosome.*

4. Look for a cell showing anaphase. Again, place the cell at the end of the pointer in your microscope and share your slide with your neighbors. If you have any doubts, ask your instructor to check the slide.

Telophase

A number of changes occur during the telophase stage of division. The chromosomes form two groups at opposite ends of the cell and the individual chromosomes become difficult to see as they uncoil. The spindle apparatus breaks down and a nuclear membrane forms around each group of chromosomes. These are now called **daughter nuclei.** Nucleoli also re-form within the nucleus. As cell division is completed, **cytokinesis** (division of the cytoplasm) occurs by the formation of a **cell plate** between the two daughter nuclei. If you find evidence of the plate being formed, you can be sure that the cell is in telophase.

5. Look for a cell showing telophase. Again, place the cell at the end of the pointer in your microscope and share your slide with your neighbors. If you have any doubts, ask your instructor to check the slide.

Whitefish Mitosis look don't draw

Obtain slides of whitefish blastula cells showing mitosis. A blastula stage is an embryonic stage that is shaped like a ball of cells. Because embryos grow rapidly, they must be made of cells that divide frequently. The blastula has been sliced into thin disks and stained to help you see the chromosomes and other structures. The slide will have several disks on it and you will probably need to look at all of them to see the various stages of the cell cycle. Locate a section of a whitefish blastula under low power and focus; then switch to high power. Refer to figure 11.6.

It is often difficult to clearly distinguish interphase cells in the whitefish blastula slide preparation. This is because the cells divide so rapidly that interphase is a very short period of time; thus, few cells are in interphase. Sections like the ones you are viewing can be sliced at any angle through a given cell. Therefore, some of the cells will be sliced so that you will see the cell from the equator, whereas others will show the cell from the pole. You may also find cells that do not appear to have any chromosomes because the knife removed all the chromosomes and left an empty cell. Or the slice of cell may have included only part of the chromosomes. If you look for the spindle fibers, you may be able to figure out the orientation more readily.

1. Locate a prophase stage. Share it with others. Check with your instructor if you have trouble.
2. Locate an equatorial view of metaphase. You should be able to see the spindle fibers radiating from the poles and the chromosomes lined up at the equatorial plane.
3. Locate a polar view of the metaphase. Note that this view does not show either of the poles or any of the spindle fibers. The chromosomes are grouped in a ring or disk but are not enclosed by a nuclear membrane. Figure 11.6 is especially helpful in identifying this view.
4. Locate a cell in the anaphase stage.
5. Locate a telophase cell.

Figure 11.6 Whitefish mitosis. (© The McGraw-Hill Companies, Inc./Kingsley Stern, Photographer)

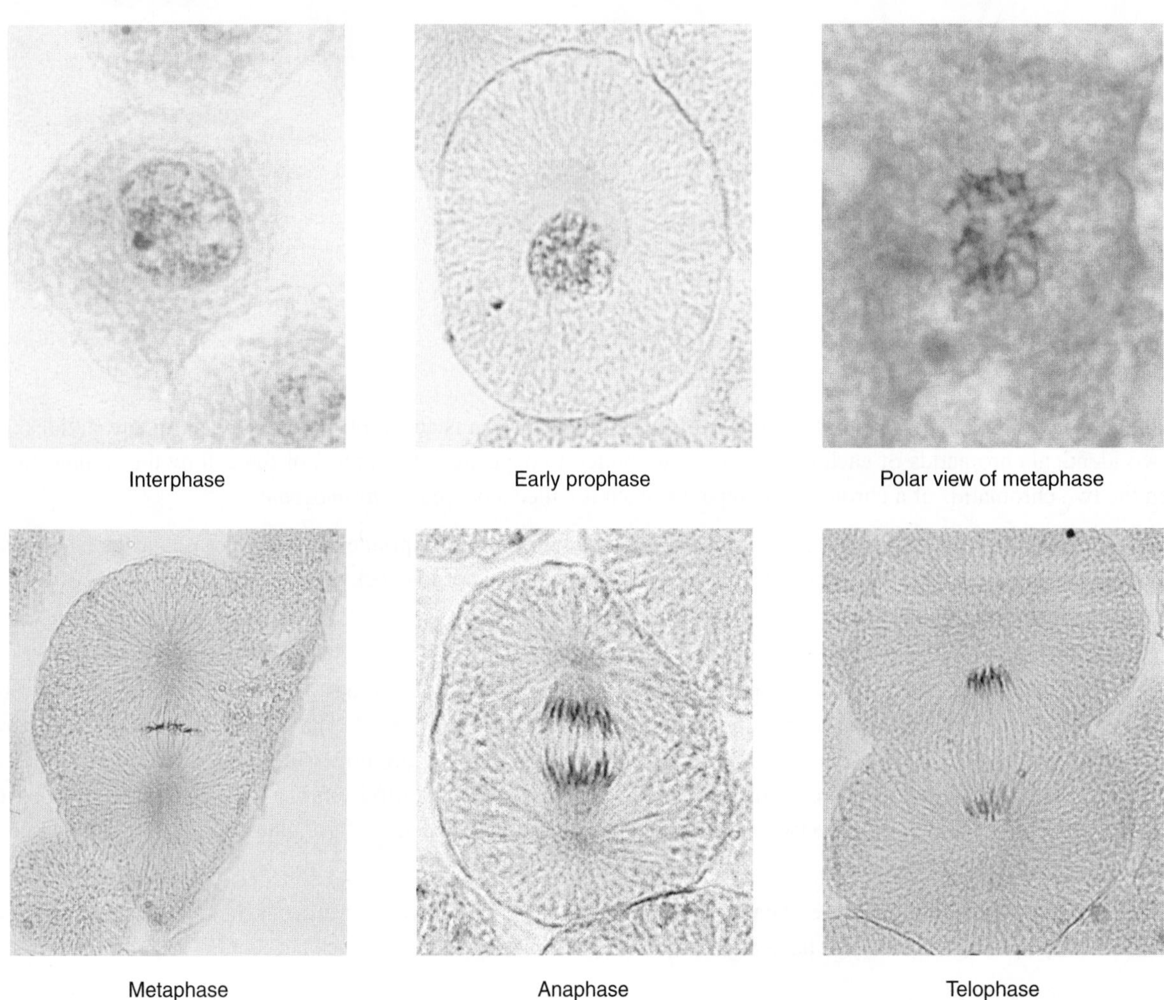

Cooperative Trial Quiz

Choose one stage of either plant or animal mitosis and put your pointer on it. Pick a cell that you think is a good example of a particular stage of mitosis. Write the name of this stage on a slip of paper. On the opposite side of the slip of paper, write the number you are assigned by your instructor. Put the slip by your scope, so that the number side is up and the side with the name of the stage is down. Use the Mitosis Quiz Answer Sheet found at the end of this lab. Visit each student's scope and identify the stage at the end of the pointer. Your instructor will also take the trial quiz and make a key. If you have made any errors, be sure to study the cells you identified incorrectly, so that you do not make the same mistake again.

DNA-Mitosis Relationship

The chromosomes found in cells are made up of DNA and protein molecules. Both chromatids of a chromosome contain identical DNA molecules. Different chromosomes have different sequences of DNA. Therefore, you should be able to follow DNA molecules as you proceed through the cell cycle. The first drawing on page 130 represents a cell that was just formed from mitosis and is entering the G_1 stage of interphase. The nucleus of the cell contains four single-stranded chromosomes. Each of the chromosomes has a sketch of DNA superimposed on it. In the cell outlines that follow the interphase G_1 stage, sketch a series of views of this cell as it proceeds through interphase and enters mitosis (prophase, metaphase, anaphase, and telophase). Show what happens to the chromosomes and the DNA that is a part of each chromosome.

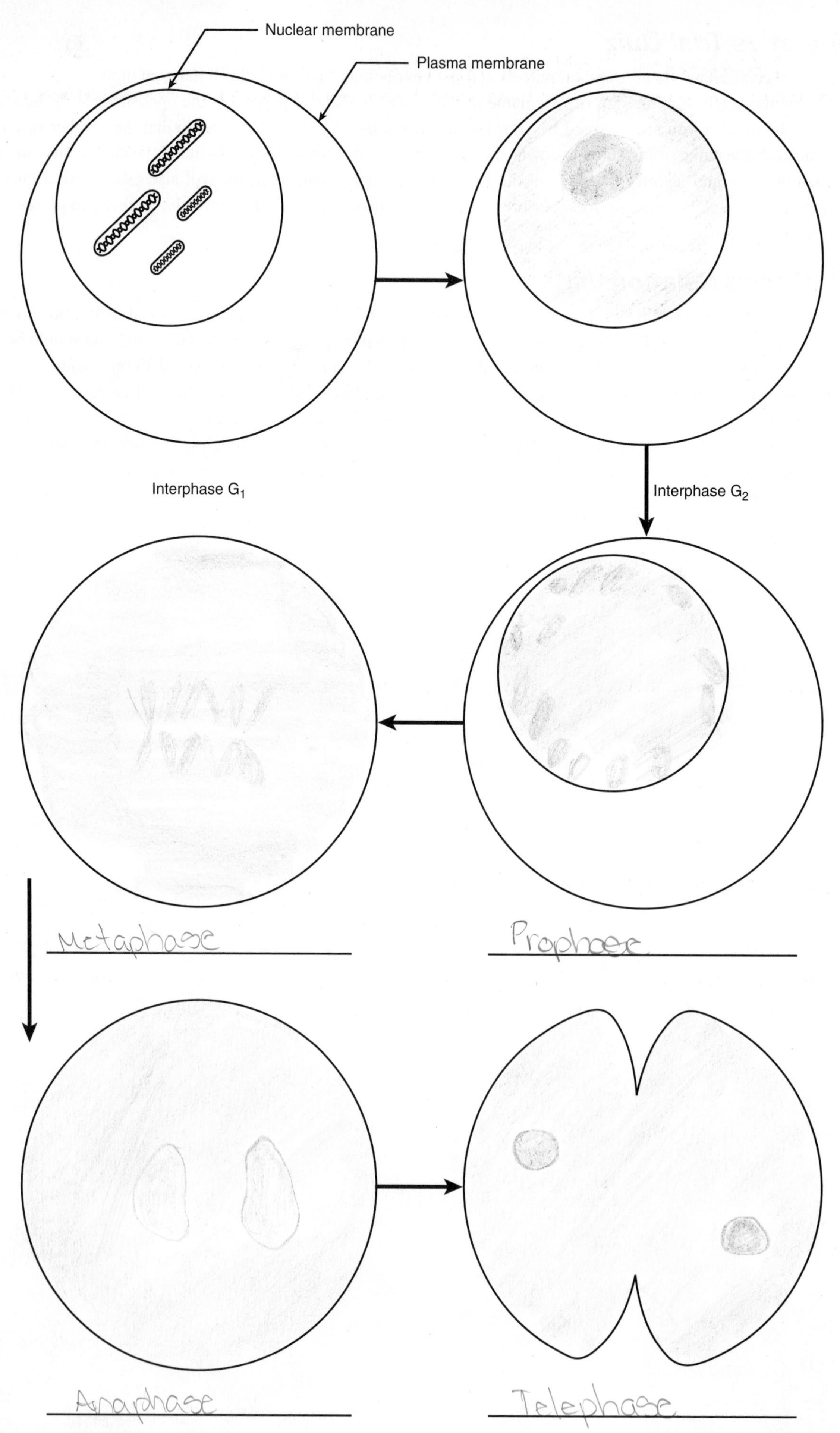
Nuclear membrane
Plasma membrane
Interphase G_1
Interphase G_2
Metaphase
Prophase
Anaphase
Telephase

11 Mitosis: Cell Division

Name ______________________ Lab section ____________

Your instructor may collect these end-of-exercise questions. If so, please fill in your name and lab section.

All

End-of-Exercise Questions

1. During which part of the cell cycle does DNA replication occur? Interphase
2. When do chromosomes first become visible in mitosis? Late prophase
3. What is the difference between the terms *chromosome* and *chromatid?*

   Chromose is the whole structure
   Chromatid is only half of the structure

4. What is a spindle? What is its function?

   micro tubules pull the chromosomes to opposite ends
   stage in which mitosis occurs

5. During what stages of mitosis are chromosomes composed of two chromatids?

   Metaphase, prophase, telophase

6. During what stages of mitosis are chromosomes single structures composed of one chromatid?

   Anaphase, telephase

7. How does cytokinesis differ in plant and animal cells?

   -Cell plate for plants, cleavage furrows for animals

8. Most cells spend the longest amount of time in Interphase. What evidence do you have to support this statement? Most cells are in interphase.

9. Why was it difficult to find interphase cells in the whitefish blastula slide?

   because they replicate quickly

10. Why are you more likely to see a polar view in animal cells than in plant cells?

    animal cells are round

Mitosis Quiz Answer Sheet

1. ______________________________

2. ______________________________

3. ______________________________

4. ______________________________

5. ______________________________

6. ______________________________

7. ______________________________

8. ______________________________

9. ______________________________

10. ______________________________

11. ______________________________

12. ______________________________

13. ______________________________

14. ______________________________

15. ______________________________

16. ______________________________

17. ______________________________

18. ______________________________

19. ______________________________

20. ______________________________

21. ______________________________

22. ______________________________

23. ______________________________

24. ______________________________

25. ______________________________

LABORATORY 12

Meiosis

+ Safety Box

- No unusual hazards are associated with this laboratory experience. Please follow standard laboratory safety procedures.

Objectives

Be able to do the following:

1. Define these terms in writing.

diploid	meiosis	gametes
homologous chromosomes	independent assortment	fertilization
synapsis	ovary	zygote
crossing-over	egg	phenotype
alleles	testis	dominant
locus	sperm	recessive
segregation	haploid	genotype
gene	homozygous	heterozygous

2. Explain why the generalized life cycle of an organism that reproduces sexually must contain a mechanism to reduce the chromosome number from diploid to haploid.
3. Diagram and/or describe the principal stages of meiosis I and II.
4. Predict the possible gametes a parent could produce when given the genotype of that parent.
5. Describe the process of crossing-over of a pair of homologous chromosomes.
6. Explain how crossing-over in meiosis I generates variety in the gametes.
7. Describe the process of independent assortment in meiosis I.
8. Explain how independent assortment generates variety in the gametes.
9. Explain how meiosis results in the segregation of alleles from one another.
10. Describe how sexual reproduction generates variety in the offspring.

Introduction

Nearly all organisms have life cycles in which sexual reproduction occurs. Sexual reproduction involves the joining of sex cells (**gametes**) from two different parent organisms to produce a new individual. The joining of sex cells is called **fertilization** and the new cell formed from their union is called a **zygote.** Because two gametes join to form one zygote during fertilization, the zygote must have twice as many chromosomes as either of the gametes that formed it. The zygote is said to have the **diploid** (two sets) number of chromosomes and the gametes the **haploid** (one set) number of chromosomes. Often the diploid cells are designated as *2n* because they have two sets of chromosomes; the haploid

Figure 12.1 The cells of this adult penguin have, for our purpose, eight chromosomes in their nuclei. In preparation for sexual reproduction, the number of chromosomes must be reduced by half, so that fertilization will result in the original number of eight chromosomes in the new individual. The offspring will grow and produce new cells by mitosis.

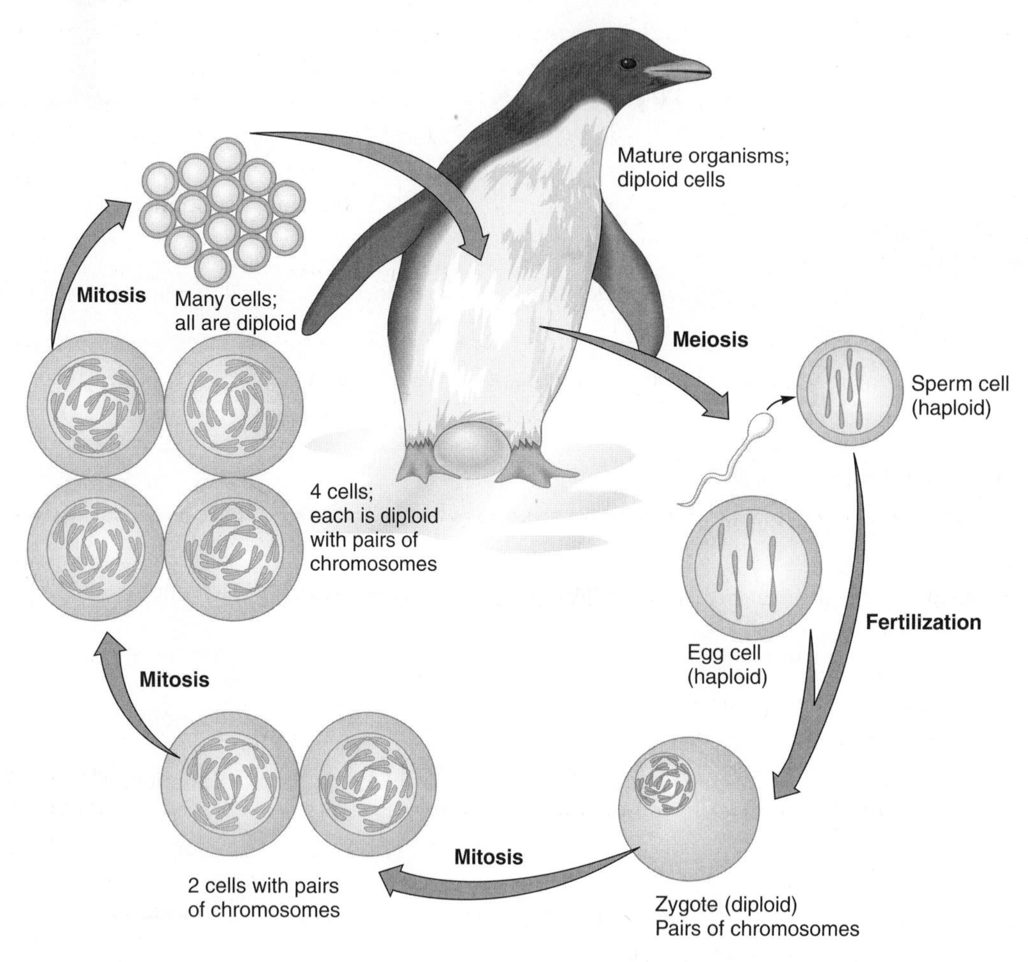

cells are designated as *1n* or *n* because they have only one set of chromosomes. If a sexually reproducing species is to retain a fixed number of chromosomes, the parents must have a mechanism to produce gametes with half the number of chromosomes typical for the organism (see figure 12.1).

Meiosis is a kind of cell division in which the cells produced (gametes) have half the number of chromosomes typical for the species. Meiosis occurs only in the gonads (testes and ovaries) of sexually reproducing organisms. The process of meiosis involves two divisions, with the first division (meiosis I) immediately followed by the second (meiosis II). During the first division of meiosis (meiosis I), the chromosomes join in homologous pairs. **Homologous chromosomes** are the same length and carry genetic information for the same characteristics. One chromosome of each pair originates from each of the parents of the individual in which meiosis is occurring. The pairing of homologous chromosomes is known as **synapsis.** Following synapsis, the pairs of homologous chromosomes line up at the equator of the cell. They then separate, and one member of each pair moves to each pole of the cell. When the two nuclei reorganize and cytokinesis occurs, the resultant two cells have one-half as many chromosomes as the original cell. Reduction from the diploid number *(2n)* of chromosomes to the haploid number *(n)* of chromosomes has been accomplished.

Each of the two cells produced as a result of meiosis I goes through a second division. This second division (meiosis II) results in the formation of four cells. During this division, individual chromosomes line up on the equators of the cells. The centromeres divide and one of each of the chromatids (daughter chromosomes) moves to each pole of the cell.

In males, the four cells that result from these meiotic divisions mature into sperm cells. Each cell has the haploid number of chromosomes, and each daughter chromosome (chromatid) is a single structure that carries a single

DNA molecule. Each cell contains a complete set of genetic information for all the characteristics of an organism but the combination of characteristics in each gamete is different from that in others. In females, only one of the four cells matures into an egg. The other three cells die. However, the same processes that generate variety in the kinds of sperm produced also are involved in producing eggs so that eggs are as different from one another as are sperm.

There are several other terms that need to be clarified, so that you can more easily follow the exercise. A **gene** is a piece of DNA that directs the expression of a particular characteristic. An **allele** is a gene for which there is an alternative expression. Scientists generally use the word *gene* when talking about characteristics in general and use the word *allele* when talking about specific examples of genes for which there are several alternatives. For example, we can talk about hair color *genes,* but when we want to talk specifically about the different possible hair colors we talk about the several *alleles* for hair color: red, blond, brown, and black. Some alleles are **dominant** to others and mask the presence of other alleles. The alleles that are masked are called **recessive** alleles. Each diploid organism has two alleles for each characteristic; one was received from each parent. The alleles may be identical (two alleles for red hair) or they may be different (one allele for red and one allele for black). When the two alleles are identical, the organism is said to be **homozygous.** When the two alleles are different, the organism is said to be **heterozygous.** The **genotype** of an organism is a *listing* of the two alleles for each trait that it possesses. The **phenotype** of an organism is a *description* of the way a characteristic is displayed in the structure, behavior, or physiology of the organism.

Preview

During this lab exercise, you will

1. determine the phenotype of a parent from the alleles on models of chromosomes.
2. manipulate the chromosome models to simulate synapsis.
3. follow the chromatids as they cross over.
4. simulate metaphase and anaphase stages of meiosis I.
5. manipulate the chromosome models as they proceed through meiosis II and produce four gametes.
6. unite one of your four gametes with one of the four gametes from another group of students to form a zygote.
7. compare the phenotype of your zygote with the phenotypes of other zygotes and with the parental phenotype.

Procedure

To understand the mechanism involved in the production of gametes, we will manipulate models of chromosomes. Each chromosome is composed of two identical chromatids.

1. Work in pairs. Obtain four model chromosomes that have already completed DNA replication and are, therefore, composed of two identical chromatids. The two chromatids of a duplicated chromosome are called sister chromatids. These four chromosomes should include a short pair of chromosomes with two alleles and terminal centromeres and a longer pair of chromosomes with four alleles. The centromere of this long pair of chromosomes is more centrally located. The two chromosomes of each pair should be different colors.
2. Place the four chromosomes randomly on a large sheet of paper (newsprint) to represent their presence in the cytoplasm of a cell.
3. Note the genes on the short pair of chromosomes. At one point **(locus)** on each chromosome there is information about insulin production. At a different locus is information about hair color. Because these chromosomes are the same size, have their centromeres in the same relative position, and have genes for the same characteristic at equivalent points along their lengths, they are called **homologous chromosomes.** The various alternative expressions of genes on homologous chromosomes (dark hair or light hair, insulin production or no insulin production) are known as **alleles.**
4. Compare your four chromosomes with the diagrams in figure 12.2 to make sure that you start with a proper set of information. Make sure that you have the correct alleles on the correct chromosomes and that you have the correct number of beads present between alleles. If your chromosomes are not as shown in figure 12.2, make corrections before proceeding. Notice that, on the figure and on your chromosome models, some of the genetic information is in capital letters and some is in lowercase letters. Those alleles that are **dominant** (those that always express themselves) are in capital letters. **Recessive** alleles (those expressed only when no dominant allele is present) are in lowercase letters.
5. Let's suppose that the models of the two pairs of homologous chromosome represent the complete set of genetic information of an organism. (In reality, humans have 46 chromosomes and thousands of genes.) This cell was

Figure 12.2 Homologous chromosomes.

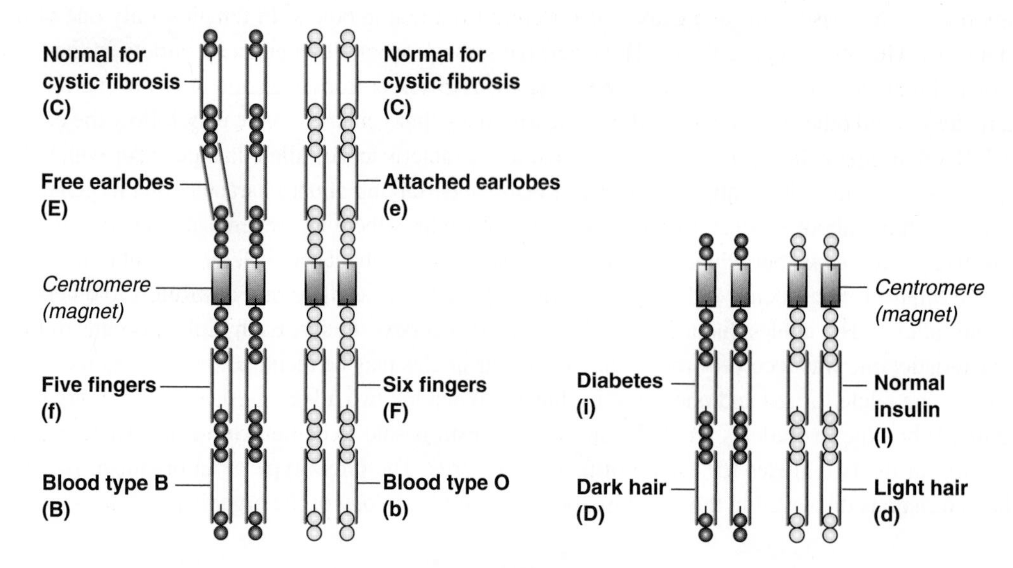

formed by the uniting of a sperm and an egg. One set of chromosomes is from the father and one set from the mother. Let the chromosomes with the darker beads represent the chromosomes donated by the father and those with the lighter beads represent the chromosomes donated by the mother. The cell you are dealing with has two sets of genetic information *(2n)* with one allele for each characteristic from the father and the other from the mother. In table 12.1, list the genotype and phenotype of this organism. Remember, the genotype is a listing of the two alleles for each characteristic and the phenotype is a description of the characteristics displayed in the structure, behavior, or physiology of the organism.

Table 12.1 List of Characteristics

Trait	Genotype of Organism		Phenotype of Organism
	Allele from Mother	*Allele from Father*	
Cystic fibrosis	C	C	*This person would be normal—would not have cystic fibrosis.*
Ear shape			
Finger number			
Blood type			
Insulin production ability			
Hair color			

Prophase I (Synapsis and Crossing-Over)

Because meiosis consists of two divisions, the various stages of the process are labeled I or II to indicate whether they are occurring in meiosis I or meiosis II. During the prophase I stage of meiosis, the nucleus prepares for a division. The chromosomes become visible, the nuclear membrane begins to disintegrate, the nucleolus disappears, and the spindle apparatus develops. In addition, during prophase I homologous chromosomes pair with each other along their entire length. While they are paired like this, they are in **synapsis.**

1. Put the two members of each homologous pair of chromosomes near each other to simulate their synapsis. While in the synapsed condition, equivalent pieces of homologous chromatids will be broken off and exchanged. This **crossing-over** can occur several times along the length of homologous chromosomes.

2. Simulate crossing-over in your models by detaching, exchanging, and reattaching exactly equal parts of chromatids between the two members of a homologous pair of chromosomes.

   (The pop beads will allow you to break a chromatid in a variety of places but for each crossover you will need to make the break at the same place on equivalent chromatids.)

   Each chromatid acts as an independent unit in this process. Therefore, you should not make exactly the same changes to both chromatids of a chromosome. Make a minimum of two crossovers for each pair of homologous chromosomes. You will now have chromosomes that contain alleles that were originally on the other member of the homologous pair and each chromosome will contain both colors of beads. These are the chromosomes you will use for the remainder of the exercise.
3. As a result of the process of crossing-over, a new combination of alleles has been formed. Look at figure 12.2 and note the differences between the arrangement of alleles in the original chromosomes and the chromosomes you have just crossed over.
4. Compare your chromosome models with the models of other students. Are their models like yours? Do they need to be? Did they make crossovers at the same places you did? Did they make the same number of crossovers?

If you don't understand the process of crossing-over, ask your instructor for help before going on.

Metaphase I (Alignment of Chromosome Pairs)

The chromosome pairs, still in synapsis, line up at the equatorial plane.

1. Move your model chromosomes on your piece of paper to show this arrangement. When the chromosomes are in this position, the cell is in metaphase I.
2. How does the arrangement of chromosomes in metaphase I of *meiosis* differ from metaphase in *mitosis?* In the following space, sketch the arrangement of four chromosomes in metaphase I of meiosis and the same four chromosomes in metaphase of mitosis.

Anaphase I (Independent Assortment and Segregation)

During this stage, the separation of homologous chromosomes occurs. This separation is called *segregation.* **Segregation** is the separating of *homologous chromosomes* to opposite poles of the cell. Because the chromosomes carry genes, the genes are also separated (segregated) into two sets during anaphase I.

1. Segregate the two members of one pair of homologous chromosomes by moving their centromeres to opposite ends of the paper (cell). Similarly segregate the other pair. Segregation of one pair of chromosomes is independent of the segregation of the other pair. The way a pair of homologous chromosomes happens to be aligned at the equator determines the direction in which each moves. The fact that the alignment and segregation of one pair of chromosomes is independent of the alignment and segregation of other pairs of chromosomes is called **independent assortment.** Because each set of homologous chromosomes is carrying genes, the genes on non-homologous chromosomes are segregated independently of one another.

2. How do the chromosomes at the end of *anaphase I of meiosis* differ from the chromosomes at the end of the *anaphase stage of mitosis?* In the following boxes sketch the arrangement of four chromosomes in anaphase I of meiosis and the same four chromosomes in anaphase of mitosis.

Meiosis End of Anaphase I	**Mitosis** End of Anaphase

Telophase I and Cytokinesis

During telophase I, the nuclear membrane reforms and the chromosomes unwind and become difficult to see. As this process is taking place, cytokinesis occurs and results in the formation of two daughter cells.

1. Show this division by tearing your paper (cell) into two equal parts. Each part will represent half the original cytoplasm. Each of these daughter cells has one chromosome from each of the two homologous pairs and is haploid. Each has half as many chromosomes as the original cell, but both have a complete set of genetic information. Separate the two sets of chromosomes by enough distance to remind you that they are in different cells.
2. Compare the sets of genetic data in each of the two cells by listing the alleles present on the chromosomes in table 12.2.

Table 12.2 **Comparison of Genetic Data of Cells Resulting from Meiosis I**

Genetic Data, Daughter Cell—1		Genetic Data, Daughter Cell—2	
Longer Chromosome	Shorter Chromosome	Longer Chromosome	Shorter Chromosome

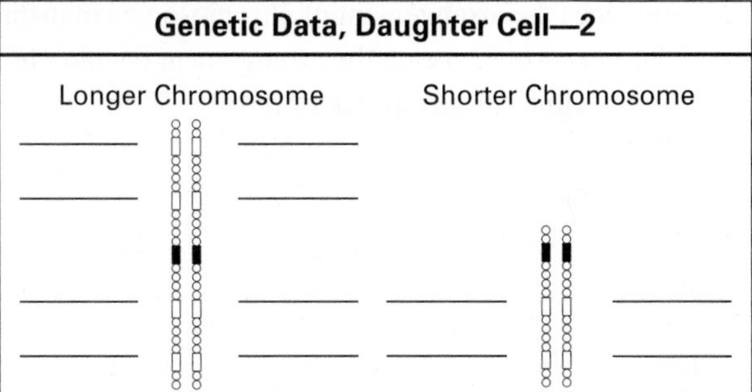

3. Compare your daughter cells to those formed by other students. Do all groups of students have the same combination of alleles in each cell? ________________
4. Is it necessary that they be the same? __________________
5. List two processes that occurred during meiosis that contributed to these differences.

Prophase II (Cells Prepare for Second Division)

No replication of chromosomes or DNA occurs after telophase I, and each of the two cells enters prophase II. The activities that occur in prophase II are the same as those that happen in prophase of mitosis. The chromosomes become visible, spindle fibers form, and the nuclear membrane breaks down.

1. List three ways that a prophase II cell differs from a prophase I cell.

Metaphase II (Alignment of Chromosomes at Equator)

During metaphase II, the chromosomes in each cell are lined up on the equatorial plane. They are not in homologous pairs; each chromosome, however, is still made up of two chromatids.

1. Move your chromosome models to the center of the half sheet of paper to represent this activity in both daughter cells.
2. The original diploid number of chromosomes in the cell you started with was four. How many *chromosomes* can be found in each daughter cell in metaphase II? _____________

Anaphase II (Chromatids Separate)

During anaphase II, the centromeres split and the chromatids (now referred to as daughter chromosomes) migrate toward opposite poles.

1. Separate your chromatids (daughter chromosomes) at the centromere and move them toward opposite poles of the cell (opposite ends of the paper). Do this in each of the two daughter cells.
2. List two ways in which anaphase I differs from anaphase II.

3. How does anaphase II of *meiosis* compare with anaphase of *mitosis?*

Telophase II (Haploid Gametes Formed)

During telophase II, the nuclear membrane reforms and the chromosomes unwind and become difficult to see. As this process is taking place, cytokinesis occurs in each cell. This results in the formation of four haploid cells called **gametes.**

1. Represent cytokinesis in each of the two daughter cells. Simulate this division of the cell by tearing the paper in half.
2. In addition to changes in the nucleus, what other processes would you expect to occur in a telophase cell?

3. List the alleles now found in each of the four gamete cells in table 12.3.

Table 12.3 Genes in Gametes After Telophase II

Gamete—1a	Gamete—2a

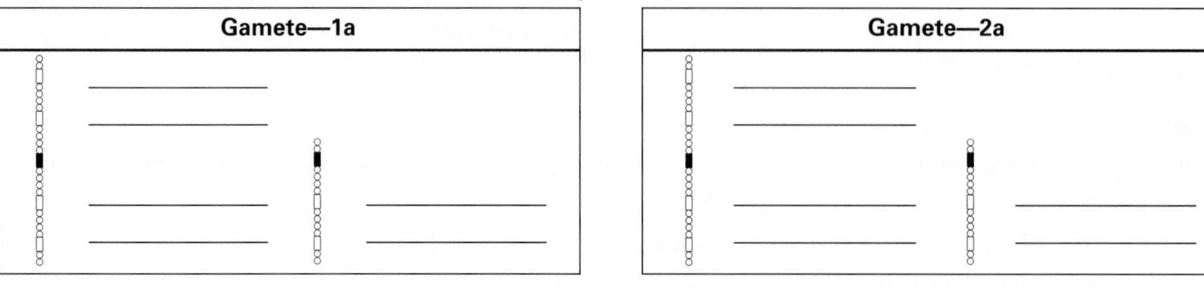

Gamete—1b	Gamete—2b

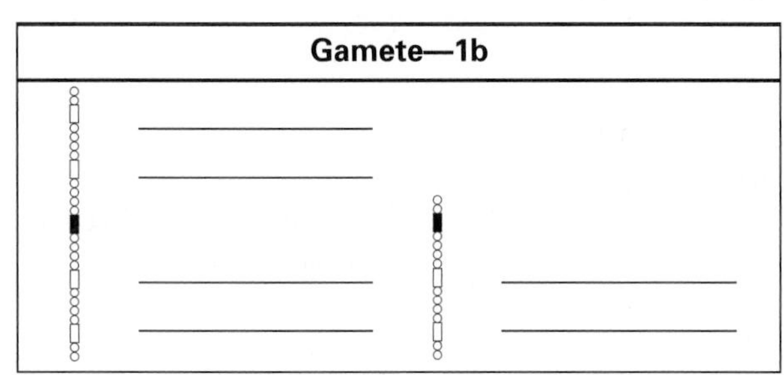

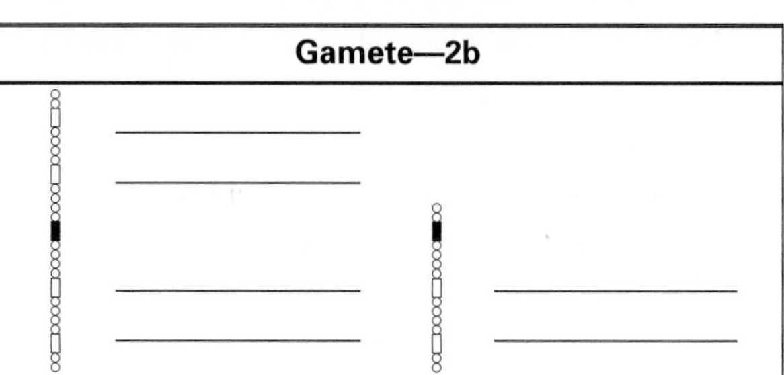

4. Note that the four cells you have formed are different from one another. At least two things have happened during meiosis to cause these four cells to be different. Look back over meiosis I and II and identify the processes that contribute to the differences in these four cells. List them in the space provided.

Fertilization (Joining of Gametes to Form a Zygote)

Your instructor will have designated each group in the class as being either a male or a female.

1. A male group randomly selects one of its four sperm cells and delivers it to a female group. The female group has previously selected at random one of its egg cells.
2. Unite these two gametes to represent **fertilization.** This fertilized egg cell is known as a **zygote.**
3. Record the genotype and phenotype of the offspring resulting from fertilization on table 12.4. Show the alleles it received from each parent and the phenotype that would be observable. Share your data with the other groups in class. Record the data from other groups in table 12.5.

Table 12.4 Genotypic and Phenotypic Characteristics of Offspring

Trait	Genotype of Offspring		Phenotype of Offspring
(Dominant traits are shown as capital letters, recessive as lower case.)	*Allele from Mother*	*Allele from Father*	
Cystic fibrosis	C	C	*This person would be normal—would not have cystic fibrosis.*
Ear shape			
Finger number			
Blood type			
Insulin production ability			
Hair color			

4. Compare the phenotype of your offspring with the phenotype of both parents.
5. Compare the phenotypes of all offspring produced in class (table 12.5).
6. Fertilization (the joining of two haploid gametes) results in a diploid zygote, which will develop into a new individual organism. What effect does being diploid rather than haploid have on determining what the phenotype will be?

At the end of this exercise, arrange the chromosome models exactly as they appear in figure 12.2. Have your instructor check the chromosomes before you leave the lab.

Table 12.5 Comparison of Offspring

Parent 1		Parent 2		Your Offspring		Offspring 2		Offspring 3		Offspring 4		Offspring 5		Offspring 6	
Geno-type	Pheno-type	Geno-type	Pheno-type	Geno-type	Pheno-type	Geno-type	Pheno-type	Geno-type	Pheno-type	Geno-type	Pheno-type	Geno-type	Pheno-type	Geno-type	Pheno-type
CC	Normal (No CF)	CC	Normal (No CF)												
Ee	Free earlobes	Ee	Free earlobes												
Ff	Six fingers	Ff	Six fingers												
BO	B blood type	BO	B blood type												
Ii	Normal insulin	Ii	Normal insulin												
Dd	Dark hair	Dd	Dark hair												

12 Meiosis

Name ______________________ Lab section __________

Your instructor may collect these end-of-exercise questions. If so, please fill in your name and lab section.

End-of-Exercise Questions

1. How many of the hypothetical offspring produced during this lab activity had the same phenotype? _______

   The same genotype? __________

2. How do the results of meiosis and mitosis differ in terms of chromosome numbers? Fill in the diagrams by assuming that each original cell represents a human cell with a diploid number of 46 chromosomes.

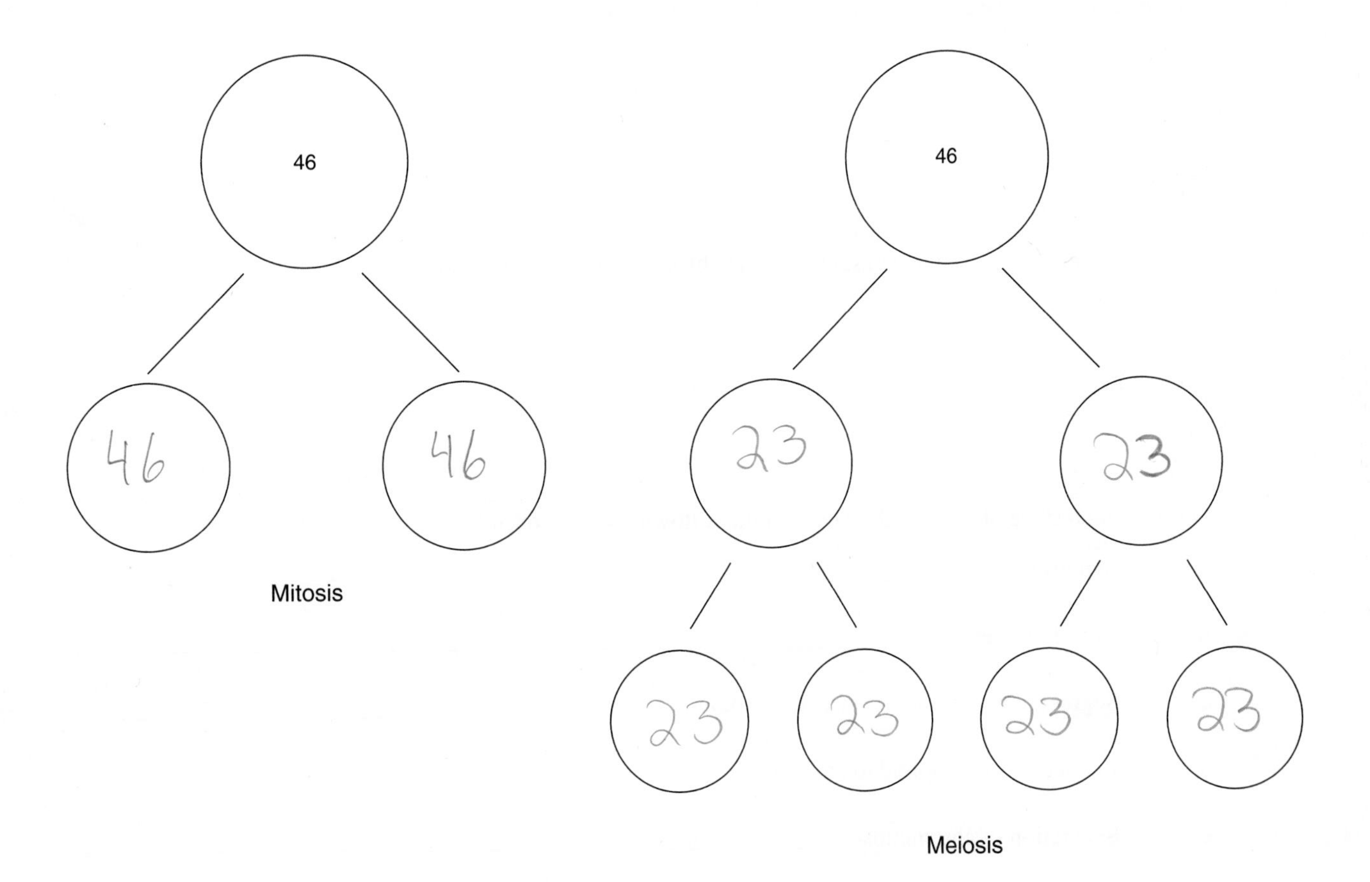

3. Many people use the terms *gene* and *allele* interchangeably. Explain how they are different.

4. Describe the main events happening in any cell undergoing metaphase?

Chromosome line up at equator

5. List the activities that occur during meiosis that contribute to the variety seen in the gametes produced.

Synapsis
Crossin over
Segregation
Independant assortment

6. Why is meiosis necessary in sexually reproducing organisms?

- Genetic diversity
- Cuts the chromosome number in half

7. Name the event of meiosis that ensures that the chromosome number is reduced.

8. During which stage of meiosis does each of the following events occur?

a. Synapsis ______________________________

b. Crossing-over ______________________________

c. Segregation of homologous chromosomes ______________________________

d. Reduction from diploid to haploid ______________________________

e. Separation of chromatids ______________________________

f. Independent assortment of homologous chromosomes ______________________________

9. Many people have the misconception that a dominant allele must be the most common in a population. Address this misconception by using the finger number trait as an example.

LABORATORY

13 DNA Extraction

+ Safety Box

- Avoid prolonged breathing of alcohol vapors.
- Be careful with hot water.

Objective

Be able to do the following:

1. Extract DNA from wheat germ.

Introduction

DNA is present in all cells. To extract DNA easily, you need a source that has a large amount of DNA. Wheat germ is the part of a wheat seed that will grow into the wheat plant when the seed germinates. It is a good, inexpensive source of cells that contain DNA. Because the wheat germ is dry, it needs to be moistened. The use of moderately hot water (60°C) performs two functions. It moistens the wheat germ, but more importantly, the high temperature of the water denatures enzymes that would break down the DNA. Since DNA is only denatured at higher temperatures (about 80°C), the use of 60°C water will not affect it.

In order to extract the DNA from cells, the DNA must be freed from the cells and dissolved in water. Cellular membranes are made of phospholipids, which have a water-soluble end and a fat-soluble end. The molecules in detergent interact with fat-soluble molecules and cause them to separate from one another and form into tiny spheres. Therefore, detergent can be used to cause the cellular membranes that surround the cell and the nucleus to break down. Thus, DNA is released into the watery solution surrounding the cells.

In order to extract DNA, it must be removed from the water. Because DNA is soluble in water but not soluble in alcohol, when DNA comes in contact with alcohol it will precipitate.

Preview

During this lab exercise, you will

1. Extract DNA from wheat germ.

Procedure

DNA Extraction

You will use wheat germ and other readily available compounds to extract DNA.

1. Place 1 g of raw (not roasted) wheat germ in a 50-mL test tube.
2. Add 20 mL of hot (50–60°C) tap water and mix constantly for 3 minutes.
3. Add 1 mL of clear dish detergent and mix ***gently*** every minute for 5 minutes. Mixing can be accomplished by using a glass stirring rod or by covering the top of the tube and gently inverting the tube. You do not want to produce foam. If any foam results at the end of the 5 minute mixing period, remove the foam with an eye dropper.
4. Tilt the test tube at an angle. ***SLOWLY*** pour about 15 mL of 90% isopropyl or ethyl alcohol down the side of the test tube, so that it forms a layer on top of the water/wheat germ/detergent solution. Because alcohol has a lower specific gravity than water, it will float on top. If you pour too rapidly, the alcohol will dissolve in water, all of the ingredients will mix together, and you will not be able to extract DNA. White, stringy, filmy DNA will begin to appear where the water and alcohol meet. You will usually see DNA precipitating from the solution at the boundary between the water and alcohol as soon as you pour in the alcohol.
5. Let the test tube sit for about 15 minutes. Then, use a toothpick, paperclip, or hooked stirring rod to remove the whitish DNA from the water-alcohol boundary.

Confirming the Presence of DNA

Your instructor may have you confirm that you have actually extracted DNA by using one of the following tests.

A test for purines
The use of a spectrophotometer
Electrophoresis

13 DNA Extraction

Name ______________________________ Lab section ____________

Your instructor may collect these end-of-exercise questions. If so, please fill in your name and lab section.

End-of-Exercise Questions

1. What is the function of each of the following substances used in this exercise?

   a. wheat germ - Provides DNA

   b. dish detergent - cause the membranes to break down in order for the DNA to come out

   c. alcohol Precipitate DNA

2. Why would it be difficult to extract DNA from the cells of fat tissue?

   have to break down the entire cell

3. Why should raw wheat germ be used instead of roasted wheat germ?

   Roasted wheat germ doesn't precipitate as well

4. Why does alcohol float on top of water? Alcohol has a lower specific gravity than water

LABORATORY

Genetics Problems 14

Safety Box

- No unusual hazards are associated with this laboratory experience. Please follow standard laboratory safety procedures.

Objectives

Be able to do the following:

1. Define these terms in writing.

alleles	homozygous	heterozygous
recessive	dominant	genotype
lack of dominance (incomplete)	Mendel's law of independent assortment	double-factor cross
multiple alleles	Punnett square	phenotype
sex-linked genes	carrier	single-factor cross
polygenic inheritance	probability	epistasis

2. Conduct a single-factor cross and predict the genotypes and phenotypes of the next generation.
3. Conduct a double-factor cross and predict the genotypes and phenotypes of the next generation.
4. Solve genetics problems that involve traits that are determined by recessive alleles, dominant alleles, lack of dominance, sex-linked genes, multiple alleles, and polygenic situations.

Introduction

One of the curiosities of life is the variety of offspring that can be produced as a result of sexual reproduction. The types of traits possible in an offspring have long been of interest to humans. Some people are interested for personal reasons—a new baby due in the family; some people are interested for business reasons—the desire to breed a new variety of plant. In either case, they want to know the probability of having a given type of offspring. Although offspring receive half their genes from each parent, because of chance combinations of genes, they may resemble one parent more than the other or may show characteristics that are not displayed by either parent.

To understand how characteristics are passed from one generation to the next, we need to know some basic information. Every individual produced by sexual reproduction has two genes for each characteristic. He or she receives one from each parent. However, there are alternative genes for the same characteristic, known as **alleles.** For example, there are alternative genes for eye color—the blue eye allele and the brown eye allele. Some alleles, called **dominant alleles,** are able to mask the presence of other **recessive** alleles. If an individual has two identical alleles for a characteristic (two blue eye alleles or two brown eye alleles), it is **homozygous.** If the two alleles are different from one another (one brown eye allele and one blue eye allele), the individual is **heterozygous.** Therefore, an individual may have some recessive alleles that do not express themselves but are still part of the individual's genetic catalog. All the genes that an individual has is its **genotype.** The observable characteristics displayed in the organism's structure, behavior, or physiology are known as the organism's **phenotype.**

To determine how characteristics are passed from one generation to the next, we must first know something about the parents. What do they look like? What alleles do they possess? We must know what alleles are possible and the probability of each allele appearing in the gametes produced by the parents. We must also know the ways these may combine during fertilization.

Preview

In this exercise, you are asked to determine how genes are passed from one generation to the next and determine the genotypes and phenotypes of parents and offspring.

During this lab exercise, you will

1. work a probability problem.
2. work single-factor inheritance problems.
3. work double-factor inheritance problems.
4. determine genotypes of parents and offspring.

Probability Versus Possibility

To solve heredity problems, you must have an understanding of probability. **Probability** is the chance that a particular desired outcome will happen divided by the total number of possible outcomes. Probability is often expressed as a percent or a fraction. *Probability* is not the same as *possibility.* It is possible to toss a coin and have it come up heads. But the probability of getting a head is more precise than just saying it is possible to get heads. The probability of getting heads is one out of two (1/2, or 0.5, or 50%) because a coin has two sides, only one of which is a head. The desired outcome (heads) is divided by the total number of outcomes (heads or tails). This probability can be expressed as a fraction:

$$\text{probability} = \frac{\text{the number of events that can produce a given outcome}}{\text{the total number of possible outcomes}}$$

What is the probability of cutting a deck of cards and getting the ace of hearts? The number of times that the ace of hearts can occur is 1. The total number of possible outcomes, the number of cards in the deck, is 52. Therefore, the probability of cutting an ace of hearts is 1/52.

What is the probability of cutting an ace? The total number of aces in the deck is 4, and the total number of cards is 52. Therefore, the probability of cutting an ace is 4/52, or 1/13.

It is also possible to determine the probability of two independent events occurring together. The probability of two or more events occurring simultaneously is the product of their individual probabilities. For example, if you throw a pair of dice, it is possible that both will be a 4. What is the probability that both will be a 4? The probability of one die being a 4 is 1/6. The probability of the other die being a 4 is also 1/6. Therefore, the probability of throwing two 4s is

$$1/6 \times 1/6 = 1/36$$

1/6 × 1/6 = 1/36 chance for two successive 4s

Probability Problem

Probability is a mathematical statement about how likely it is that something will occur. It is not certainty. To help you understand this concept, we will work with a deck of playing cards and determine the likelihood of getting each of the four suits.

Work in pairs. One student shuffles and cuts a standard deck of cards. The other student records the suit. Repeat this 100 times: shuffling, cutting, and recording. Record your information in table 14.1.

Table 14.1 Probability Problem Results

Suit	Actual	Expected	Difference
Hearts		25	
Clubs		25	
Diamonds		25	
Spades		25	

Your instructor will tell you how to record this information for the whole class. Record the actual number of times each suit is cut. Compare your results with the entire class. Why is there a difference between the results you got and the results of the whole class?

Single-Factor Inheritance Problems

Single-factor crosses are concerned with how a single genetic trait is passed from the parents to an offspring. Solving a heredity problem requires five basic steps.

Step 1: Assign a symbol for each allele.

Usually, a capital letter is used for a dominant allele and a lowercase letter is used for a recessive allele. For example, use the symbol E for free earlobes, which is dominant, and e for attached earlobes, which is recessive.

E = free earlobes
e = attached earlobes

Step 2: Determine the genotype of each parent and indicate a mating.

Suppose both parents are heterozygous; the male genotype is *Ee* and the female genotype is also *Ee*. The × between them is used to indicate a mating.

$$Ee \times Ee$$

Step 3: Determine all the possible kinds of gametes each parent can produce.

Remember that gametes are haploid; therefore, they can have only one allele instead of the two present in the diploid cell. Because the male has both the free earlobe allele and the attached earlobe allele, half his gametes contain the free earlobe allele and the other half contain the attached earlobe allele. Because the female has the same genotype, her gametes are the same as his.

For solving genetics problems, a *Punnett square* is used. A **Punnett square** is a box figure that allows you to determine the probability of obtaining each of the genotypes and phenotypes possible in the offspring resulting from a particular cross. Remember, because of the process of meiosis, each gamete receives only one allele for each characteristic listed. Therefore, the male gives either an E or an e; the female also gives either an E or an e. The possible gametes produced by the male parent are listed on the left side of the square; the female gametes are listed on the top. In our example, the Punnett square would show a single dominant allele and a single recessive allele from the male on the left side. The alleles from the female would appear on the top:

Female genotype
Ee
Possible gametes from female
E & *e*

Male genotype *Ee* — Possible male gametes } *E* & *e*

	E	*e*
E		
e		

<u>*Step 4:* Determine all the allele combinations that can result from the combining of gametes.</u>
To determine the possible combinations of alleles that could occur as a result of this mating, simply fill in each of the empty squares with the alleles that can be donated from each parent. Determine all the allele combinations that can result when these gametes unite:

	E	*e*
E	*EE*	*Ee*
e	*Ee*	*ee*

<u>*Step 5:* Determine the phenotype of each possible allele combination shown in the offspring.</u>
In this instance, three of the offspring (those with the genotypes *EE, Ee,* and *Ee*) have free earlobes, because the free-earlobe allele is dominant to the attached-earlobe allele. One offspring, *ee,* has attached earlobes. Therefore, the probability of having offspring with free earlobes is 3/4 and with attached earlobes is 1/4.

Additional Single-Factor Inheritance Problems (One Trait Followed from One Generation to the Next)

1. In humans, six fingers (*F*) is the **dominant** trait; five fingers (*f*) is the **recessive** trait. Assume both parents are **heterozygous** for six fingers.
   What are the phenotypes of the father and the mother?
   What is the genotype of each parent?
   What are the different gametes each parent can produce?
   What is the probability of having six-fingered children? Five-fingered children?

   a. Father's phenotype ____________; mother's phenotype ____________
   b. Father's genotype ____________; mother's genotype ____________
   c. Father's gametes ______ or ______ mother's gametes ______ or ______
   d. Probability of a six-fingered child ______
   e. Probability of a five-fingered child ______

2. If the father is heterozygous for six fingers and the mother has five fingers, what is the probability of their offspring having each of the following phenotypes?
   Six fingers ____________; five fingers ____________

3. In certain flowers, color is inherited by alleles that show **lack of dominance (incomplete dominance).** In such flowers, a cross between a **homozygous** red flower and a homozygous white flower always results in a pink flower. A cross is made between two pink flowers. Use F^w to represent the white allele and F^R to represent the red allele. What is the probability of each of the colors (red, pink, and white) appearing in the offspring?

4. Use the information given in the previous problem. A cross is made between a red flower and a pink flower. What is the expected probability for the various colors?

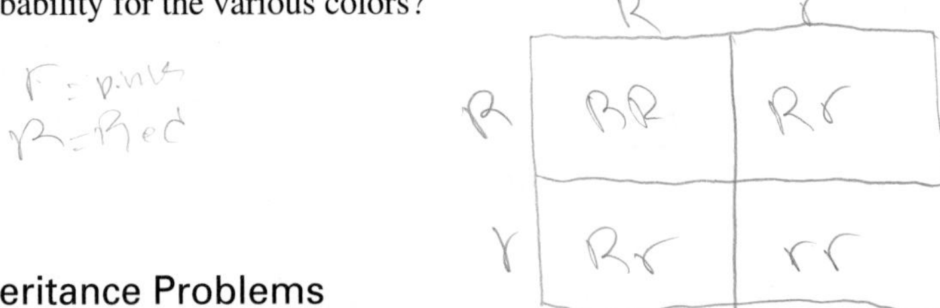

3:1 with red being the dominant one

Double-Factor Inheritance Problems

A **double-factor cross** is a genetics study in which two pairs of alleles are followed from the parental generation to the offspring. These problems are basically worked the same as a single-factor cross. The main differences are that in a double-factor cross you work with two different characteristics from each parent.

It is necessary to recognize that **independent assortment** occurs when two or more sets of alleles are involved. Mendel's law of independent assortment states that members of one allelic pair separate from each other independently of the members of other pairs of alleles. This happens during meiosis when the chromosomes which carry the alleles segregate. (Mendel's law of independent assortment applies only if the two pairs of alleles are located on different pairs of homologous chromosomes. This is an assumption we will use in double-factor crosses.)

In humans, the allele for free earlobes dominates the allele for attached earlobes. The allele for dark hair dominates the allele for light hair. If both parents are heterozygous for earlobe shape and hair color, what genotypes and phenotypes can their offspring have and what is the probability of each genotype and phenotype?

Step 1: Use the symbol *E* for free earlobes and *e* for attached earlobes. Use the symbol *D* for dark hair and *d* for light hair.

E = free earlobes
e = attached earlobes
D = dark hair
d = light hair

Step 2: Determine the genotype for each parent and show a mating.

In this example, the male genotype is *EeDd,* the female genotype is *EeDd,* and the × between them indicates a mating.

EeDd × *EeDd*

Step 3: Determine all the possible gametes each parent can produce and write the symbols for the alleles in a Punnett square.

Because there are two pairs of alleles in a double-factor cross, each gamete must contain one allele from each pair: one from the *E* pair (either *E* or *e*) and one from the *D* pair (either *D* or *d*). In this example, each parent can produce four different kinds of gametes. The four squares on the left indicate the gametes produced by the male; the four on the top indicate the gametes produced by the female. To determine the possible allele combinations in the gametes, select one allele from the "ear" pair of alleles and match it with one allele from the "color" pair of alleles. Then, match the same "ear" allele with other "color" allele. Then, select the second "ear" allele and match it with each of the "color" alleles. This may be done as follows:

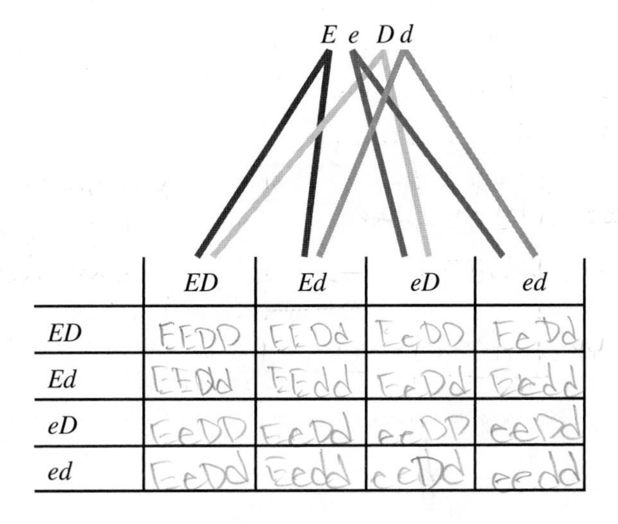

	ED	*Ed*	*eD*	*ed*
ED	EEDD	EEDd	EeDD	EeDd
Ed	EEDd	EEdd	EeDd	Eedd
eD	EeDD	EeDd	eeDD	eeDd
ed	EeDd	Eedd	eeDd	eedd

Step 4: Determine all the gene combinations that can result from the combining of gametes.
Fill in the Punnett square as follows:

	ED	*Ed*	*eD*	*ed*
ED	*EEDD* *	*EEDd* *	*EeDD* *	*EeDd* *
Ed	*EEDd* *	*EEdd* ^	*EeDd* *	*Eedd* ^
eD	*EeDD* *	*EeDd* *	*eeDD* "	*eeDd* "
ed	*EeDd* *	*Eedd* ^	*eeDd* "	*eedd* +

Step 5: Determine the phenotype of each possible allele combination shown in the offspring.
In this double-factor problem, there are 16 ways in which gametes could combine to produce offspring. There are four possible phenotypes in this cross. They are represented as

Genotypes	**Phenotype**	**Symbol**
EEDD, EEDd, EeDD, or *EeDe*	Free earlobes, dark hair	*
EEdd, Eedd	Free earlobes, light hair	^
eeDD, eeDd	Attached earlobes, dark hair	"
eedd	Attached earlobes, light hair	+

The probability of having a given phenotype is

9/16 free earlobes, dark hair
3/16 free earlobes, light hair
3/16 attached earlobes, dark hair
1/16 attached earlobes, light hair

BbTt

Additional Double-Factor Inheritance Problems (Two Traits Followed from One Generation to the Next)

5. In horses, black color (*B*) dominates chestnut color (*b*). Trotting gait (*T*) dominates pacing gait (*t*). A cross is made between a horse homozygous for both black color and pacing gait and a horse homozygous for both chestnut color and trotting gait. List the probable genotype and phenotype of offspring resulting from such a cross.

   Genotype BbTt; phenotype BbTt

6. Humans may have Rh^+ blood or Rh^- blood. A person who is Rh^+ (*R*) has a certain type of protein on the red blood cell. A person who is Rh^- (*r*) does not have this protein. In humans, Rh^+ dominates Rh^-. Normal insulin production (*I*) dominates abnormal insulin production (*i*) (diabetes). If both parents were heterozygous for both Rh^+ and normal insulin production, what phenotypes would they produce in their offspring? What would be the probabilities of producing each phenotype?

	RI	Ri	rI	ri
RI	RRII	RRIi	RrII	RrIi
Ri	RRIi	RRii	RrIi	Rrii
rI	RrII	RrIi	rrII	rrIi
ri	RrIi	Rrii	rrIi	rrii

RrIi x RrIi

The parents would produce a offspring with Rh^+ and normal insulin

7. For this problem, use the information concerning the traits given in problem 6. The father is homozygous for Rh^{+} and has diabetes. The mother is Rh^{-} and is homozygous for normal insulin production. What phenotype would their offspring show?

They would have Rh+ with nomal insulin production

8. In certain breeds of dogs, two different sets of alleles determine the color pattern. Black color is dominant and red color is recessive; solid color is dominant and white spotting is recessive. A homozygous black-and-white spotted male is crossed with a red-and-white spotted female. What is the probability of their producing a solid black puppy?

No will be black they will all be spotted

9. In humans, a type of blindness is due to a dominant allele; normal vision is the result of a recessive allele. Migraine headaches are due to a dominant allele, and normal (no headaches) is recessive. A male who is heterozygous for blindness and does not suffer from headaches marries a woman who has normal vision and does not suffer from migraines. Could they produce a child with normal vision who does not suffer from headaches? If yes, can the probability of such a child be determined?

Bbmm x bbmm

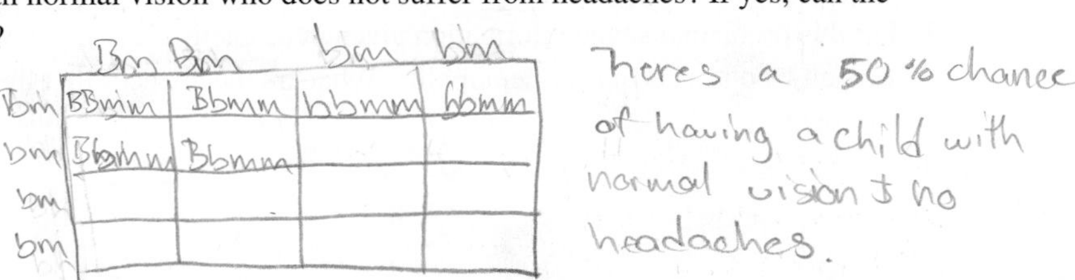

10. This problem is a little more challenging than the previous ones because it involves lack of dominance in both characteristics that are being followed. However, you use the same Punnett square method to determine the outcome of this cross as you would with any other double-factor cross.

In the radish plant, the long and round traits exhibit lack of dominance and the heterozygotes have an oval shape. The red and white color traits also exhibit lack of dominance and heterozygotes have a purple color. Two oval-shaped, purple plants are crossed. What phenotypic ratio would the offspring show?

Sex-Linked Problems
(Alleles Located on the X Chromosome)

For these problems, you need to remember that human males have one X chromosome and one Y chromosome; females have two X chromosomes. The Y chromosome does not carry the genes found on the X chromosome but carries genes that determine maleness.

11. In humans, the condition for normal blood clotting (*H*) dominates the condition for nonclotting (hemophilia) (*h*). Both alleles are linked to the X chromosome. A male hemophiliac marries a woman who is a carrier for this condition. (In this respect, a **carrier** is a woman who has an allele for normal blood clotting and an allele for hemophilia.) *If they have a son,* what are the chances he will be normal for blood clotting?

H = normal
h = hemopheliac

$X^hY \cdot X^hX^h$

	X^h	Y
X^H	X^HX^h	X^HY
X^h	X^hX^h	X^hY

There is a 50% chance of that happening

12. For this problem, use the information given in problem 11. A male who has normal blood clotting marries a woman who is a carrier for hemophilia. What are the chances that they will have a son who is normal for blood clotting?

$X^HY \times X^HX^h$

	X^H	Y
X^H	X^HX^H	X^HY
X^h	X^HX^h	X^hY

75% chance

13. Color blindness is a condition in which a person cannot distinguish specific colors from one another. For example, the person may not be able to distinguish red from green or blue from yellow. However, the person is able to distinguish some colors. Because color-blind people are not blind and they can see some colors, many people prefer to use the term *color-deficient.* In humans, the condition for normal vision dominates color blindness. Both alleles are linked to the X chromosome. A color-blind male marries a color-blind female. *If they have a daughter,* what are the chances she will have normal vision?

B = normal
b = Color blind

$X^bY \cdot X^bX^b$

	X^b	Y
X^b	X^bX^b	X^bY
X^b	X^bX^b	X^bY

Theres a 0% chance the daughter will have regular vision

14. For this problem, use the information given in problem 13. A male with normal vision marries a woman who is color-blind. She gives birth to a daughter who is also color-blind. The husband claims the child is not his. The wife claims the child is his. Can you support the argument of either parent? If yes, which one? Why?

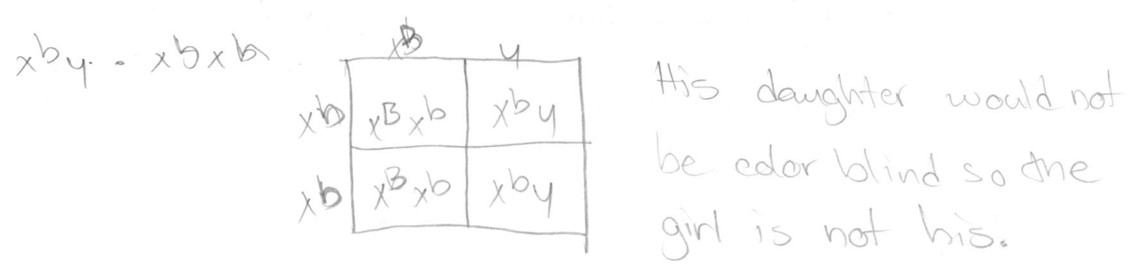

Multiple Allele Problems (Characteristics That Have More Than Two Possible Forms of the Same Gene)

15. In humans, there are three alleles for blood type: A, B, and O. The allele for blood type A and the allele for blood type B show incomplete dominance. A person with both alleles has blood type AB. Both A and B dominate type O. A person with alleles for blood types A and O marries someone with alleles for blood types B and O. List the types of blood their offspring could have and the probability for each blood type in the offspring.

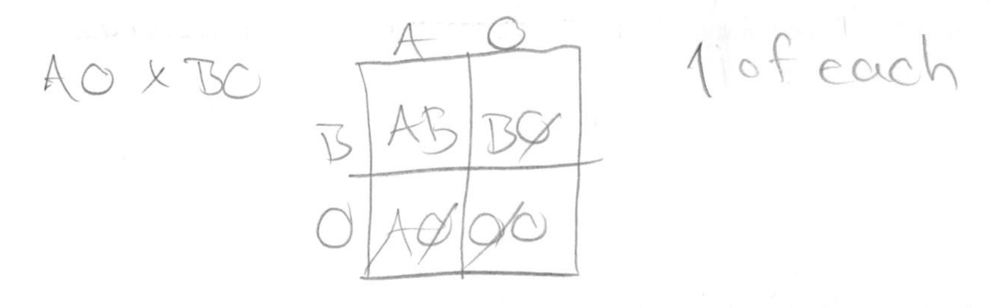

16. For this problem, use the information given in problem 15. A young woman with blood type O gave birth to a baby with blood type O. In a court case, she claims that a certain young man is the father of her child. The man has type A blood. Could he be the father? Can it be proven on this evidence alone that he is the father?

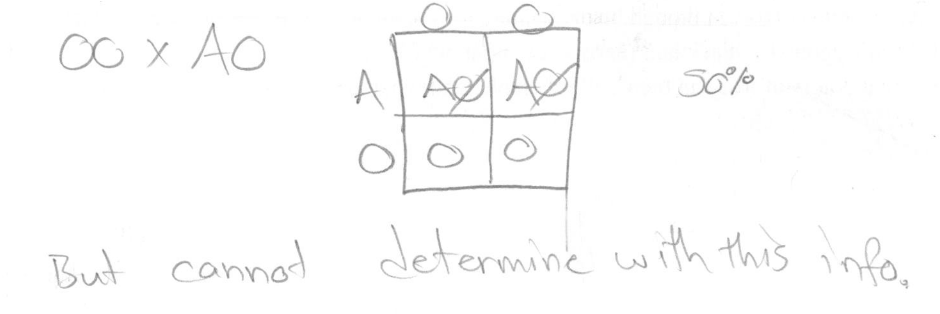

But cannot determine with this info.

17. In humans, kinky hair (H^{++}), curly hair (H^{+}), wavy hair (H), and straight hair (h) are dominant, in that order. Dark hair dominates red hair. A male with wavy red hair, whose mother has straight, dark hair, marries a dark-haired female with straight hair who has a father with curly red hair. What type of children can they produce, and what is the probability of producing these types of offspring?

Epistasis Problems

Epistasis occurs when one set of alleles can mask the presence of a different set of alleles. All the epistasis problems in this exercise involve two different traits; therefore, they are all similar to double-factor problems.

18. Normal pigmentation dominates no pigmentation (albino). For an organism to exhibit color, it must have an allele for normal pigment production as well as alleles for a specific color. In cattle, red color dominates black color. An albino bull that has a heterozygous genotype for red is crossed with a red cow. The cow is heterozygous for normal pigment production and for red coloring. What types of offspring will they produce, and what is the probability of producing these types of offspring?

19. In humans, normal pigmentation dominates no pigmentation (albino). Black hair dominates blonde hair. An albino person will have white hair, even though he or she may also have the alleles for black or blonde hair. An albino male who is homozygous for black hair marries a woman who is heterozygous for normal pigmentation and has blonde hair. What colors of hair can their children have, and what is the probability for each hair color?

Polygenic Inheritance

Polygenic inheritance occurs when two or more genes combine their effects to determine the phenotype.

20. In some types of wheat, color is caused by two sets of alleles. To produce a red color, both dominant alleles, *R* and *B,* are needed. White results from having both recessive alleles in the homozygous state, *rrbb.* Any other combination produces brown wheat grains. A strain with a genotype of *Rrbb* is crossed with a strain of wheat with a genotype of *rrBb.* What is the color of each of the parent strains?

    Rrbb color _______________; *rrBb* color _______________

    What colors of wheat result from this cross, and what is the probability for each color?

Determination of Genotypes (Genetic Detective Work)

Not all genetics problems deal with determining the genotype and phenotype of the offspring. A common problem is to determine the genotype of all individuals involved when only the phenotypes are known.

In humans, free earlobes are dominant and attached earlobes are recessive. Two free-earlobed people marry and produce one free-earlobed child and two attached-earlobed children. What are the genotypes of the parents and each of the children?

Because both parents have free earlobes, they must have at least one allele for free earlobes, so their genotypes are *E* __ (the __ means the allele could be either the dominant allele or the recessive allele). The genotype for the attached-earlobed children must be *ee.* Because each parent contributed one allele to each attached-earlobe offspring, you know that each free-earlobed parent has an allele for attached earlobes. Therefore, the __ must be *e* and the genotype for each parent is *Ee.* All you know about the free-earlobed offspring is that the child has one allele for free earlobes. The genotype might be either *EE* or *Ee.*

In the remaining problems, try to determine the genotypes of the individuals. Take it slowly; put down an allele only when you are certain the individual has that allele. If you are not certain, show that you don't know by leaving a blank (__).

21. Normal pigmentation (*A*) dominates no pigmentation (albino = *aa*). Dark hair coloring (*D*) dominates light hair coloring (*d*). Two people with normal pigmentation produce one child with dark hair, two children with light hair, and two albino children. What are the possible genotypes for the parents?

22. A red bull, when crossed with white cows, always produces roan-colored offspring. Explain how the colors for red, white, and roan are inherited.

23. In rabbits, short hair is due to a dominant allele, *S,* and long hair to its recessive allele, *s.* Black hair is due to a dominant allele, *B,* and white hair to its recessive allele, *b.* When two rabbits are crossed, they produce 2,518 short-haired, black offspring and 817 long-haired, black offspring. What are the probable genotypes of the parents?

24. In humans, the condition for normal blood clotting dominates hemophilia. Both alleles are sex-linked to the X chromosome. Two parents produce daughters who are all carriers and sons who are all normal. What are the probable genotypes of the parents?

25. In humans, deafness is due to a homozygous condition of either or both recessive alleles *d* and *e.* Both dominant alleles *D* and *E* are needed for normal hearing. Two deaf people marry and produce offspring who all have normal hearing. What are the probable genotypes of the children and parents?

Answers

1. a. Six fingers; six fingers
   b. *Ff*; *Ff*
   c. father *F* or f; mother *F* or f
   d. 3/4
   e. 1/4
2. 1/2; 1/2
3. 1/4 red; 1/2 pink; 1/4 white
4. 1/2 pink; 1/2 red
5. *BbTt;* all black trotters
6. 9/16 Rh^+, normal insulin; 3/16 Rh^+, diabetic; 3/16 Rh^-, normal insulin; 1/16 Rh^-, diabetic
7. All Rh^+, normal insulin
8. None
9. Yes; 1/2 normal and no headache
10. 1/16 long, red; 2/16 long, purple; 1/16 long, white; 2/16 oval, red; 4/16 oval, purple; 2/16 oval, white; 1/16 round, red; 2/16 round, purple; 1/16 round, white
11. 1/2
12. 1/4
13. None
14. Yes. Father. A female requires two color-blind alleles to have the condition and the father does not have any color-blind alleles.
15. 1/4 A, 1/4 B, 1/4 O, 1/4 AB
16. Yes. No. Blood typing can only disprove paternity.
17. 1/4 straight, red-haired; 1/4 wavy, red-haired; 1/4 straight, dark-haired; 1/4 wavy, dark-haired
18. 8/16 albino, 6/16 red, 2/16 black or 1/2 albino, 3/8 red, 1/8 black
19. 1/2 albino (white hair), 1/2 black hair
20. Brown; brown. 1/4 red, 1/2 brown, 1/4 white
21. *Aa_d* × *AaDd*
22. Incomplete dominance
23. *SsBB* × *Ss* _ _
24. Normal mother; hemophiliac father
25. Parents' genotype *ddEE DDee;* children's genotype *DdEd*

LABORATORY

Genetic Ratios and Chi-Square Analysis

15

+ Safety Box

- No unusual hazards are associated with this laboratory experience. Please follow standard laboratory safety procedures.

Objectives

Be able to do the following:

1. Work genetics problems to determine common phenotypic ratios.
2. Use a chi-square test to analyze a set of data to determine if an expected genetic ratio is represented by the data.
3. Use phenotypic ratios of offspring to determine the genotypes of the parents.

Introduction

Geneticists are often required to make judgments about the genetic makeup of parents and perhaps grandparents by examining the offspring from their matings. This is done by observing the specific traits of the offspring and determining whether the data fit a known genetic ratio.

When you deal with data from actual crosses, you don't usually expect the numbers to fit a known ratio exactly. Usually, it is necessary to do a simple statistical test to determine whether the data "fit" the expected ratio. We use the chi-square test to make this determination. See appendix A for a discussion of the chi-square test of statistical significance.

Preview

During this lab exercise, you will

1. determine how common phenotypic ratios are derived.
2. examine families of offspring exhibiting particular contrasting, segregating traits.
3. count the number of members in the family exhibiting each trait.
4. select a genetic ratio you hypothesize best fits your data.
5. use a chi-square test to analyze your data statistically to see if the data fit the ratio you selected.
6. determine the genotype of the two parents.

Procedure

You are required to examine families of offspring, determine the phenotypic ratio for a given trait, and work genetics problems backwards to find out the genotype of the parents. Therefore, first review typical genetics problems to see how the common phenotypic ratios are derived from crosses between parents of known genotype. (If you have not done genetics problems previously, see exercise 14 for instructions.)

Review of Genetics Problems

Problem 1. Single-Factor Cross

Make a cross between two parents who are heterozygous for one pair of alleles (A = dominant allele; a = recessive allele).

Parent's genotype *Aa* × *Aa*

Gametes ______ ______ and ______ ______

	Gametes from parent 1	
Gametes from parent 2		

Offspring genotypes __________ __________ __________

What is the phenotypic ratio among the offspring?

__________ dominant; __________ recessive

If you were to examine a family of offspring that has a phenotypic ratio of three dominant to one recessive, what could you conclude about the genotypes of the parents?

Problem 2. Single-Factor Cross

Make a cross between a heterozygous parent and one that is homozygous recessive.

Parent's genotype *Aa* × *aa*

Gametes ______ ______ and ______

	Gametes from parent 1	
Gametes from parent 2		

Offspring genotypes __________ __________

What is the phenotypic ratio produced by this cross?

__________ dominant; __________ recessive

Problem 3. Double-Factor Cross

Make a cross between parents who are both heterozygous for both of two unlinked alleles.

Parent's genotype *AaBb* × *AaBb*

Gametes ______ ______ ______ ______ and ______ ______ ______ ______

	Gametes from parent 1			
Gametes from parent 2				

Phenotype	Number/Ratio
A?B?	________
A?bb	________
aaB?	________
aabb	________

From your 16 offspring, what is the ratio of *A?* to *aa?*

__________ : __________

Ratio of *B?* to *bb?*

__________ : __________

Do you see that two 3:1 ratios multiplied together produce a 9:3:3:1 ratio (3:1) ∞ (3:1)?

Problem 4. Double-Factor Cross

Make a cross between two organisms with the genotypes given.

Parent's genotypes *Aabb* × *aaBb*

Gametes ______ ______ and ______ ______

Gametes from parent 2 \ Gametes from parent 1		

Offspring genotypes __________ __________ __________ __________

Phenotypic ratio __________ : __________ : __________ : __________

Using Corn Kernels as Examples of Families Resulting from a Specific Cross

If the pollination of corn is strictly controlled, the kernels of corn on an ear can represent the offspring of a cross between two parents. If you count the kernels of corn that demonstrate two segregating alleles (purple versus yellow), you can derive the parental genotypes.

Cross 1. Purple Versus Yellow Corn Kernels

Count the kernels of each color on the ear provided and record the data.

Purple __________; yellow __________; total __________

What common phenotypic ratio do these data come close to? __________

To find out if these results could have come from a 3:1 ratio, you need to subject your data to the chi-square test. Interrupt your counting and complete the introductory exercise on the use of the chi-square test in appendix A. Then, do a chi-square analysis of your corn data.

Chi-Square Test on Corn Color Data

a. Observed counts

Purple __________; yellow __________; total __________

b. Expected count based on 3:1 ratio

Purple (3/4 of total); yellow (1/4 of total)

O = observed; E = expected

Classes	Observed Number	Expected Number	O − E	$(O - E)^2$	$\frac{(O - E)^2}{E}$
Purple					
Yellow					

Chi-square value = __________

Degrees of freedom = __________

Probability = __________

c. Take the chi-square value to table A.2 in appendix A. Use the 1 degree of freedom row and determine the probability value. High probability values indicate that your expected ratio and the data you obtained are in close agreement. Probability values of 0.05 or less indicate that your data do not fit the ratio you think describes the data you obtained. The deviation from expected was not due to chance alone.

d. Does the probability value indicate that you have a 3:1 ratio? __________

e. If so, give the genotype of both parents. __________ × __________

Cross 2. Starchy Versus Sweet Corn Kernels

Using the same procedure, count the number of starchy (full) kernels and the number of sweet (shriveled) kernels. Predict a ratio and subject it to the chi-square test.

Starchy __________; sweet __________; total __________

Classes	Observed Number	Expected Number	O – E	$(O - E)^2$	$\frac{(O - E)^2}{E}$
Starchy					
Sweet					

Chi-square value = __________

Degrees of freedom = __________

Probability = __________

Do you have a 3:1 ratio? ________

If so, give the genotype of both parents. ________ × ________

Cross 3. Green Versus Albino Corn Plants

Examine the flat of corn seedlings. What is the segregating characteristic? ____________

Again, using the same procedure, count the green versus white seedlings, predict the ratio, and subject it to the chi-square test.

Green __________; white __________; total __________

Classes	Observed Number	Expected Number	O – E	$(O - E)^2$	$\frac{(O - E)^2}{E}$
Green					
White					

Chi-square value = __________

Degrees of freedom = __________

Probability = __________

Do you have a 3:1 ratio? _____

If so, give the genotype of both parents. _____ × _____

Cross 4. Segregation of Purple or Yellow and Starchy or Sweet Corn Kernels

In double-factor crosses, you follow two segregating traits at the same time. In this ear, there are four classes of seeds. Count and record them.

Purple, starchy __________

Purple, sweet __________

Yellow, starchy __________

Yellow, sweet __________

Total __________

Predict a phenotypic ratio and complete the table.

Classes	Observed Number	Expected Number	O − E	$(O - E)^2$	$\frac{(O - E)^2}{E}$
Purple, starchy					
Purple, sweet					
Yellow, starchy					
Yellow, sweet					

Chi-square value = ________________

Degrees of freedom = ________________

Probability = ________________

If you accept your predicted ratio, determine the genotype of the parents.

Parents' genotype __________ × __________

Cross 5. Purple Versus Yellow Corn Kernels

On this ear, count purple versus yellow.

Purple __________; yellow __________; total __________

Predict a ratio and complete the table.

Classes	Observed Number	Expected Number	O − E	$(O - E)^2$	$\frac{(O - E)^2}{E}$
Purple					

Chi-square value = ________________

Degrees of freedom = ________________

Probability = ________________

Parents' genotype __________ × __________

Cross 6. Segregation of Purple or Yellow and Starchy or Sweet Corn Kernels

Count the four classes of kernels on this ear (purple, starchy; purple, sweet; yellow, starchy; yellow, sweet), predict a ratio, do a chi-square test, and determine parental genotypes.

Hint: If the ratio doesn't come out like anything you have seen before, do a ratio for each characteristic—for example, starchy, sweet as one ratio and purple, yellow as another.

LABORATORY

16 Human Variation

+ Safety Box

- No specific hazards are associated with this exercise. Please follow standard laboratory safety procedures.

Objectives

Be able to do the following:

1. Define these terms in writing.

dominant	multiple alleles	epistasis
recessive	incomplete dominance	sex-limited
homozygous	allele	polygenic characteristics
heterozygous	locus	independent assortment

2. Apply the Mendelian principles of segregation and independent assortment to the inheritance of human facial characteristics.
3. Recognize the role of probability in the inheritance of common human traits.
4. Compute the genotypic and phenotypic ratios from the data generated.
5. Use chi-square analysis to determine if deviation from expected Mendelian ratios could be due to chance alone.

Introduction

Have you ever wondered why people vary so much in appearance, even when they are closely related? These differences exist because all individuals inherit unique combinations of genes from their parents. This variety is possible because there are often many different **alleles** (alternative forms of a gene) for a specific characteristic within a population and they are mixed in new combinations during sexual reproduction. Each child receives one-half of its genetic information from each parent. More accurately, each parent contributes one allele for each genetic **locus** (the specific location of a gene on a chromosome). However, because of meiosis, even children of the same parents will receive a different set of genetic information from each parent. (Identical twins are an exception, because they are the result of a single fertilization event.)

Several different patterns of inheritance have been documented. Because each individual gets one allele for each characteristic from each parent, the two alleles must be either the same or different. When both alleles for a characteristic are identical, the individual is said to be **homozygous.** If the two alleles for a characteristic are different, the individual is said to be **heterozygous.** Many alleles are *dominant,* which means they mask, or hide, the expression of *recessive* alleles. Put another way, **dominant** alleles are expressed, even when present in just a single dose. **Recessive** alleles can be expressed only when dominant alleles are absent. In some cases, two alleles may not show dominance or recessiveness and both express themselves, a condition known as **incomplete dominance** (lack of dominance). Often, such heterozygous cases result in phenotypic characteristics that are intermediate between the two homozygous genotypes. Many characteristics have only two alleles available in the population. However, many cases of **multiple alleles** exist, in which there are many alleles for a characteristic within a population, although an individual can only have a maximum of two of the alleles, one having been inherited from each parent. Other traits, such as beard growth, are **sex-limited,** which means their expression is limited to one of the sexes.

Other human traits are not determined by a single set of alleles and show a wide variety of phenotypic expressions among the individuals in a population. These traits are thought to be controlled by many sets of alleles (genes) that are located at different loci and these traits are called **polygenic characteristics.** Some traits are influenced by the action of genes at two or more loci. For example, the alleles that determine hair color or skin color are affected by another set of alleles that determines whether a person can produce pigment or is an albino. This phenomenon, genes at one locus influencing how the genes at another locus are expressed, is what geneticists call **epistasis.** This laboratory activity provides an opportunity to observe how alleles are passed from one generation to the next and how combinations of alleles determine the phenotype of offspring. This exercise will also illustrate how each of the topics discussed earlier (dominant-recessive, incomplete dominance, multiple alleles, sex-limited, polygenic, and epistasis) influences the pattern of inheritance.

In this exercise, you will choose a partner of the opposite sex. The two of you will be the parents of an offspring. (If there are unequal numbers of males and females in the class, your instructor may designate you to be a particular sex for the purposes of this laboratory.) "Parents" will be assumed to be heterozygous for characteristics and will flip coins to simulate segregation of alleles during the formation of gametes. Alleles that are on different chromosomes are segregated independently of one another. This **independent assortment** of chromosomes during the formation of gametes will be simulated by additional flips of coins. We will use facial characteristics in this exercise and you will draw a picture of what the child would look like in his or her teens.

The exact manner in which human facial characteristics are determined is often difficult to determine. Some of the characteristics are inherited in exactly the manner suggested and are indicated with a footnote. Others are assumed to be inherited in the manner suggested and are indicated with an asterisk (*). Others are useful for the laboratory exercise and are assumed to be inherited, but the manner of inheritance is not understood; these characteristics will not have a footnote or an asterisk.

Preview

During this lab exercise, you will

1. flip coins to determine which alleles are passed to your offspring and determine the genotype of your offspring.
2. record your data on the data sheet.
3. make a drawing of your offspring's phenotype based on the genotype obtained.
4. share your data with the class and compute the genotypic and phenotypic ratios for certain characteristics using the data from the entire class.
5. examine different modes of inheritance, including simple dominance and recessiveness, incomplete dominance, multiple alleles, and sex-limited, polygenic, and epistatic traits.
6. compute a gene frequency from a set of data.
7. use chi-square analysis to determine if deviation from the expected Mendelian ratios could be due to chance alone.

Procedure

Pair up with a classmate, who will play the role of your spouse. Begin the simulation with the assumption that each of you has one dominant and one recessive allele for each of the facial features illustrated on the following pages. In other words, each of you is *heterozygous* for each trait. To determine which allele you will pass on to your child, both you and your spouse flip a coin. "Heads" determines that a dominant allele is present in the gamete and is passed on to the offspring. "Tails" determines that a recessive allele is present in the gamete and is passed on to the offspring. Thus, if your partner flips heads and you flip tails, the child's genotype for that trait is heterozygous (*Aa*). If you both flip heads, the child's genotype is *AA,* and, if you both flip tails, the child's genotype is *aa.*

1. First, we should determine the sex of the child. Females have two X chromosomes and males have an X and a Y chromosome. Which parent determines the sex of the child? All egg cells have X chromosomes, but half of the sperm have Y chromosomes and half have X chromosomes. If a sperm cell bearing a Y chromosome fertilizes an egg, a male (XY) child will result; if a sperm bearing an X chromosome fertilizes an egg, a female (XX) child will result. Therefore, in this simulation, only the father needs to flip the coin to determine the sex of the child. If heads is flipped, your child is a boy (Y-bearing sperm) and, if tails is flipped, your child is a girl (X-bearing sperm).
2. Give your child a name and record this information on the data sheet on page 177.
3. Both you and your partner should flip a coin for each facial feature. Record (a) the alleles contributed by each parent, (b) the genotype of the offspring, and (c) the phenotype of the offspring on the data sheet on page 177.

Facial Traits

1. Face shape: *RR* or *Rr* = round face; *rr* = square face

Round face (*RR*, *Rr*)

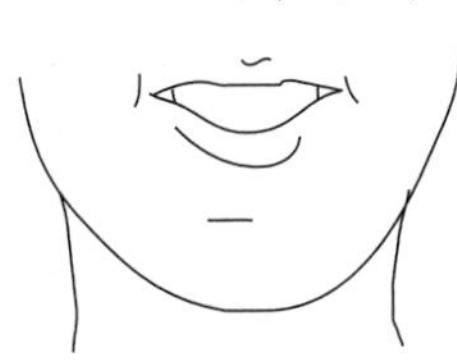

Square face (*rr*)

2. Chin size: *PP* or *Pp* = very prominent chin; *pp* = less prominent chin

Very prominent chin (*PP*, *Pp*)

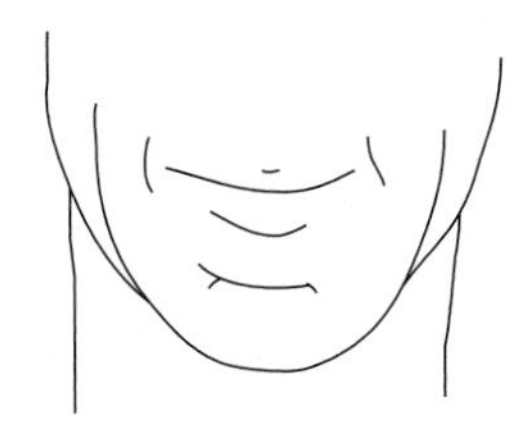

Less prominent chin (*pp*)

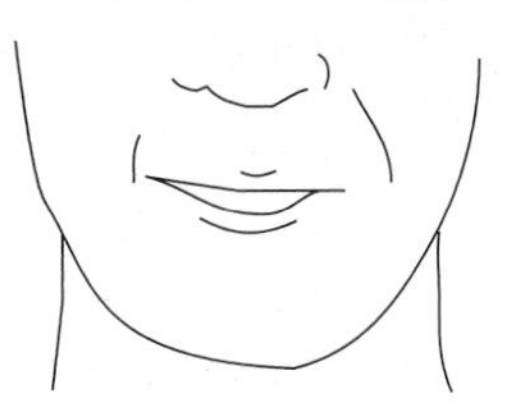

3. Chin shape: Only flip coins for this trait if chin size genotype is *PP* or *Pp*. The genotype *pp* prevents the expression of chin shape. *CC* or *Cc* = round chin; *cc* = square chin.

Round chin (*CC*, *Cc*)

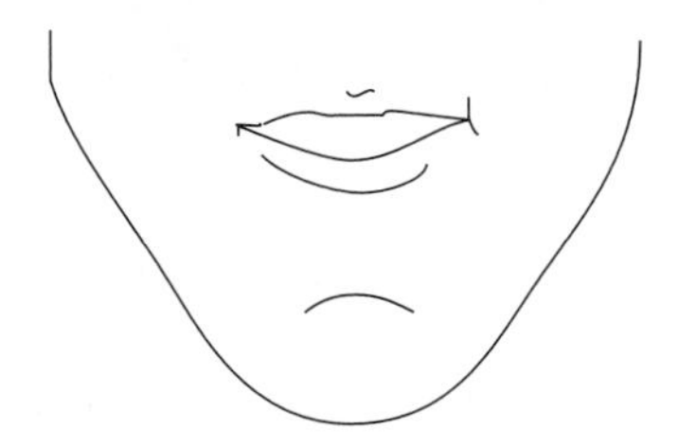

Square chin (*cc*)

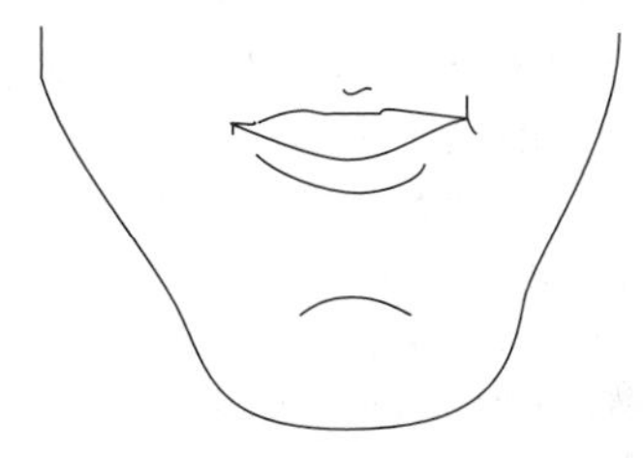

4. Cleft chin: 11900[1] *AA*, or *Aa* = cleft chin; *aa* = no cleft chin

Present (*AA*, *Aa*)

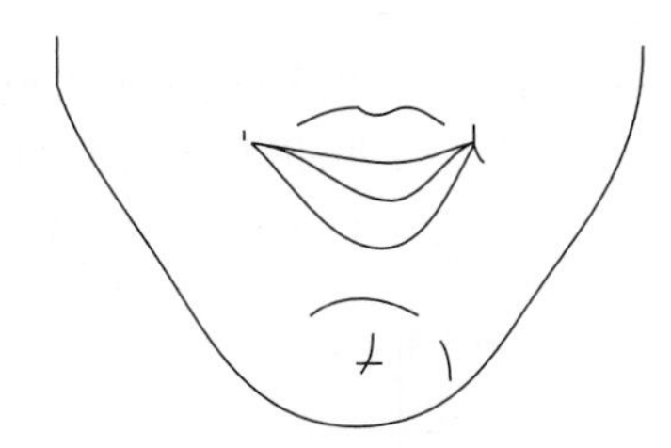

Absent (*aa*)

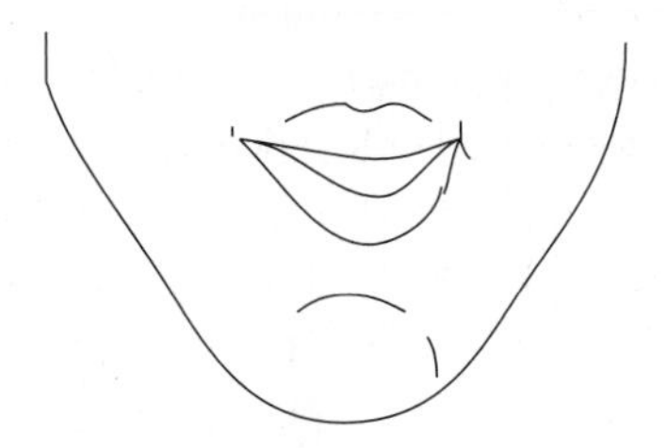

Skin Color

5. Skin color*: Although it is an over-simplification, assume that skin color is determined by three sets of alleles, located at three different loci, and that each allele contributes to the amount of pigment produced. Both you and your partner should flip coins to determine the genotype of the first pair of alleles (*AA, Aa, aa*). Then, flip again to determine the genotype of the second pair of alleles (*BB, Bb, bb*). Flip for the last time to determine the third pair of alleles (*CC, Cc, cc*). Determine the phenotype of your offspring based on the following polygenic model:

[1]Reference number to Victor A. McKusick, M.D., *Mendelian Inheritance in Man,* 8th ed. (Baltimore and London: Johns Hopkins University, 1988).

Six dominant alleles—very dark black
Five dominant alleles—very dark brown
Four dominant alleles—dark brown
Three dominant alleles—medium brown
Two dominant alleles—light brown
One dominant allele—light tan
Zero dominant alleles—fair skin

Example: If you flipped heads for the first two sets of alleles and tails for the third set of alleles, and your partner flipped tails for the first set of alleles and heads for the second and third sets of alleles, your offspring's genotype would be *AaBBCc* and your offspring's phenotype would be dark brown.

Hair Traits—Next Four Flips

6. White forelock*: A tuft of hair over the forehead is white.
   FF or *Ff* = white forelock; *ff* = no white forelock
7. Widow's peak*: The hairline comes to a point in the center of the forehead.
   WW or *Ww* = widow's peak present; *ww* = widow's peak absent

Present (*WW*, *Ww*)

Absent (*ww*)

8. Hair type and hair color*: Hair type and hair color are determined by multiple alleles. To determine which alleles you have and are able to contribute to your offspring, your instructor will have designated each individual in class as belonging to one of five groups. From the Special Characteristics Chart, determine your genotype and which alleles you may pass to your offspring. The alleles for hair shape are S^K = kinky, S^C = curly, S^W = wavy, and S^{ST} = straight. They exhibit dominance and recessiveness in the order given. Kinky is dominant to all other alleles; straight is recessive to all other alleles. Curly is recessive to kinky but dominant to wavy. Wavy is recessive to kinky and curly and dominant to straight.

   The alleles for hair color are C^{BK} = black, C^{BR} = brown, C^{BD} = blond, and C^R = red. They exhibit dominance and recessiveness in the order given. Determine the hair shape and hair color of your offspring by flipping coins.

Special Characteristics Chart			
Group Assigned by Instructor	Hair Type Genotype	Hair Color Genotype	Albinism Genotype
Group 1	S^KS^C	$C^{BK}C^{BR}$	*aa*
Group 2	S^KS^{ST}	$C^{BK}C^R$	*AA*
Group 3	S^CS^W	$C^{BR}C^{BD}$	*AA*
Group 4	S^CS^{ST}	$C^{BR}C^R$	*Aa*
Group 5	S^WS^{ST}	$C^{BD}C^R$	*Aa*

Once you have determined your genotype for hair type and color, complete the following chart.

Trait	My Genetic Information		My Partner's Genetic Information	
	Genotype	Phenotype	Genotype	Phenotype
8a. Hair type				
8b. Hair color				

Now, flip coins to determine which hair type and hair color alleles each of you will pass on to your offspring.

9. Albinism: Albinos are not able to produce pigment. Therefore, none of the skin color, hair color, or eye color genes can express themselves if an individual is an albino. The recessive gene for albinism is rare in the population. To determine your genotype for the ability to produce pigment, refer to the Special Characteristics Chart.

   Once you have determined your genotype and phenotype, complete the following chart.

Trait	My Genetic Information		My Partner's Genetic Information	
	Genotype	Phenotype	Genotype	Phenotype
Pigment production				

Determine if your child is an albino by flipping coins based on the genotype you were assigned. If you are homozygous, you do not need to flip a coin, because you can pass on only one type of allele. If your child is an albino, all color characteristics will be altered: Eyes will be pink because there is no pigment in the iris, the skin will be very pale white/pink because there is no pigment, and all hair (on head, eyebrows, eyelashes, etc.) will be white regardless of the other color alleles he or she may have inherited. Also, white forelock will not show because all hair will be white and freckles will not show because no pigment is produced in the skin. Albinism is inherited in the following manner:

AA or *Aa* = normal pigment; *aa* = no pigment (albino)

Eyebrow Traits—Next Two Flips

10. Eyebrow thickness: *BB* = bushy; *Bb* = intermediate; *bb* = very thin

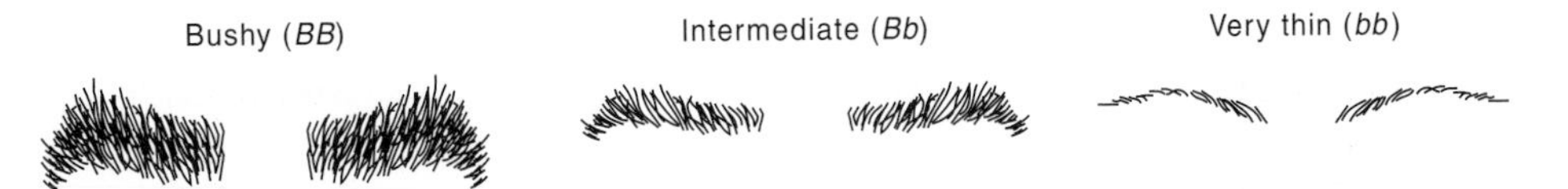

11. Eyebrow placement: *NN* or *Nn* = not connected; *nn* = connected

Not connected (*NN*, *Nn*)

Connected (*nn*)

Eye Traits

12. Eye color: Dark brown eyes are produced by the presence of a layer of brown pigment, which covers the entire surface of the iris. Lighter eyes (hazel) are caused by patches of brown pigment on the iris. Blue eyes are caused by the absence of pigment on the iris. In this situation, the dominant allele *S* is for solid pigment and the recessive allele *s* is for patchy pigment. The dominant allele *B* is for the presence of brown pigment and the recessive allele *b* is for the absence of brown pigment.

    To determine eye color, both you and your partner will need to flip twice. Determine the genotype of the first pair of alleles (*SS, Ss, ss*) and then the second pair of alleles (*BB, Bb, bb*). Determine the phenotype of your offspring according to the following guidelines:

SSBB, SSBb, SsBB, or *SsBb*	brown
ssBb or *ssBB*	hazel
SSbb, Ssbb, or *ssbb*	blue

13. Eye distance: *EE* = close together; *Ee* = average distance; *ee* = far apart

Close together (*EE*) | Average distance (*Ee*) | Far apart (*ee*)

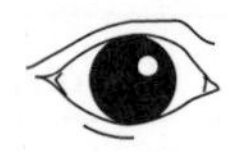
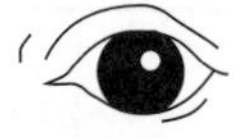

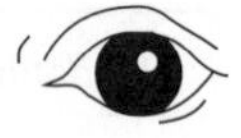
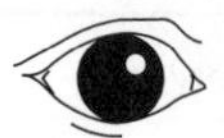
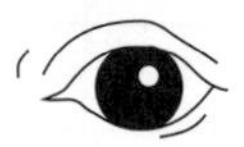

14. Eye shape: *AA* or *Aa* = almond-shaped; *aa* = round

Almond (*AA, Aa*)

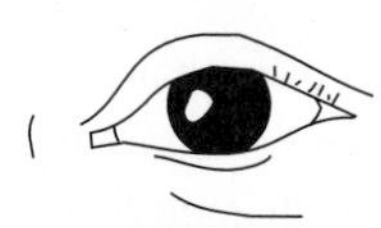

Round (*aa*)

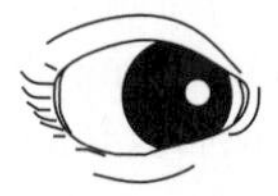

15. Eyelashes: *LL* or *Ll* = long; *ll* = short

Long (*LL, Ll*)

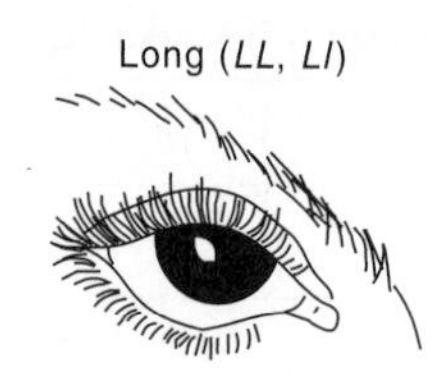

Short (*ll*)

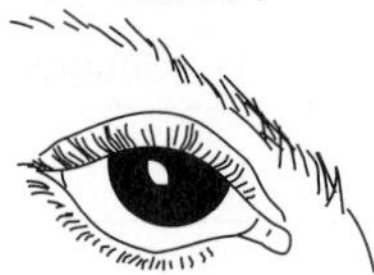

Mouth and Lip Traits

Determine the phenotype with respect to all three characteristics before drawing the mouth.

16. Mouth size: *MM* = long; *Mm* = average; *mm* = short

Long (*MM*)

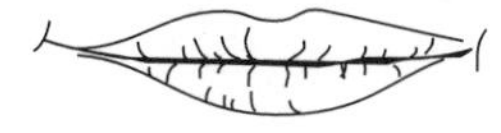

Average (*Mm*)

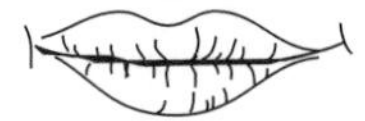

Short (*mm*)

17. Lip thickness: *TT* or *Tt* = thick; *tt* = thin

Thick (*TT, Tt*)

Thin (*tt*)

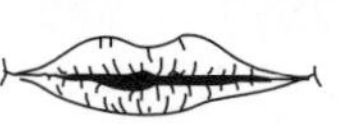

18. Dimples on side of mouth: *DD* or *Dd* = present; *dd* = absent

Present (*DD, Dd*)

Absent (*dd*)

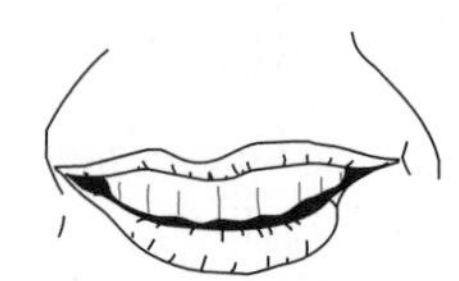

Nose Traits

19. Nose thickness: *BB* or *Bb* = broad; *bb* = narrow

Broad (*BB, Bb*)

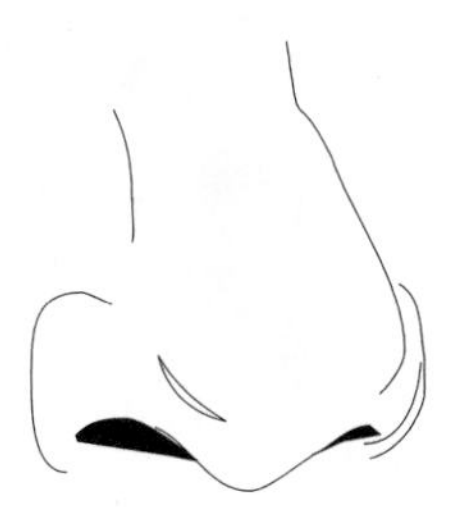

Narrow (*bb*)

20. Nose shape: *RR* or *Rr* = tip rounded; *rr* = tip pointed

Rounded (*RR, Rr*)

Pointed (*rr*)

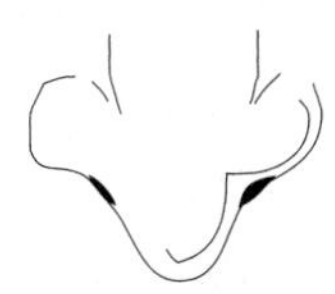

Ear Traits

21. Earlobe attachment: *FF* or *Ff* = free; *ff* = attached

Free (*FF, Ff*)

Attached (*ff*)

22. Darwin's earpoint: *DD* or *Dd* = present; *dd* = absent

Present (*DD, Dd*)

Absent (*dd*)

23. Ear pits: 12870[2] *PP* or *Pp* = present; *pp* = absent

Present (*PP*, *Pp*)

Absent (*pp*)

24. Hairy ears: Hairy ears is a recessive allele located on the Y chromosome, so it is contributed only by the male, and only males can display the characteristic. If your child is female, it cannot show the characteristic, so you do not need to do anything. If your child is a male, the father must flip a coin to decide if the child will have hairy ears. Tails denotes hairy ears.

Absent

Present

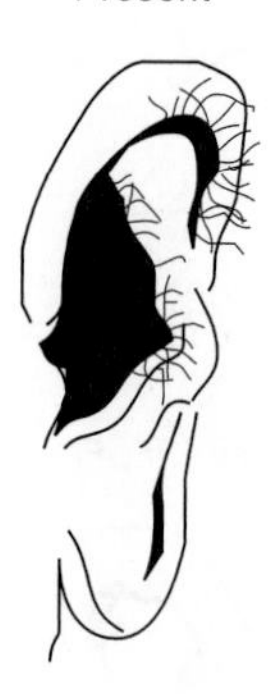

Freckles

25. Freckles on cheeks: *CC* or *Cc* = present; *cc* = absent

Present (*CC*, *Cc*)

Absent (*cc*)

26. Freckles on forehead: *FF* or *Ff* = present; *ff* = absent

Present (*FF*, *Ff*)

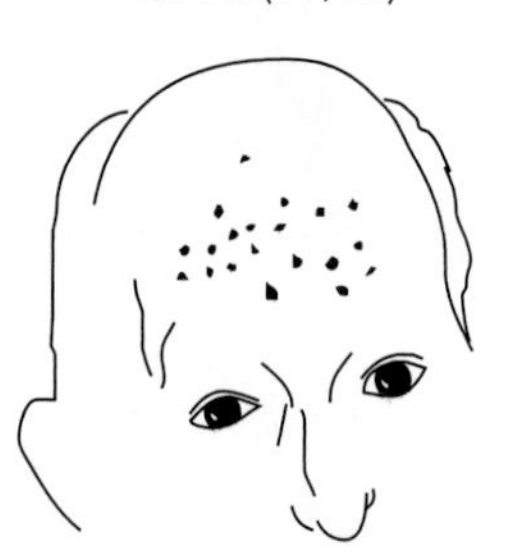

Absent (*ff*)

[2]Reference number to McKusick, 1988.

Data Sheet

Parents' names ______________________ and ______________________

Child's name ______________________

Child's sex ______________________

Traits Shown by Your Child

Trait Number	Trait	Allele from Mother	Allele from Father	Child's Genotype	Child's Phenotype
			Example		
9	***Albinism***	***A***	***a***	***Aa***	***The child would have normal pigmentation***
1	Face shape				
2	Chin size				
3	Chin shape				
4	Cleft chin				
5	Skin color				
6	White forelock				
7	Widow's peak				
8a	Hair type				
8b	Hair color				
9	Albinism				
10	Eyebrow thickness				
11	Eyebrow placement				
12	Eye color				
13	Eye distance				
14	Eye shape				
15	Eyelashes				
16	Mouth size				
17	Lip thickness				
18	Dimples				
19	Nose thickness				
20	Nose shape				
21	Earlobe attachment				
22	Darwin's earpoint				
23	Ear pits				
24	Hairy ears				
25	Freckles on cheeks				
26	Freckles on forehead				

16 Human Variation

Name ______________________________ Lab section ____________

Your instructor may collect these end-of-exercise questions. If so, please fill in your name and lab section.

End-of-Exercise Questions

1. Compare your child with the child of another couple. List five traits that both of the children have. Geneticists use the word *concordant* to refer to individuals that show the same phenotypes.

2. Using the same children as in question 1 for comparison, list five traits that are different. Geneticists use the word *disconcordant* to refer to individuals that show different phenotypes.

3. For most of the characteristics in this exercise, both parents are heterozygous. What is the probability that both parents will contribute a recessive allele for any given trait?

4. Refer to your child's data sheet and complete table 16.1. Place an X in the dominant column in table 16.1 if your child received at least one copy of the dominant allele. Place an X in the recessive column if your child received two copies of the recessive allele. Total the number of dominant and recessive phenotypes for these traits. What is the ratio of dominant to recessive traits? What ratio would you expect? Why?

Table 16.1 **Ratio of Dominant to Recessive Phenotypes**

Trait		Dominant	Recessive
1	Face shape		
2	Chin size		
4	Cleft chin		
6	White forelock		
7	Widow's peak		
11	Eyebrow placement		
14	Eye shape		
15	Eyelashes		
17	Lip thickness		
18	Dimples		
19	Nose thickness		
20	Nose shape		
21	Earlobe attachment		
22	Darwin's earpoint		
23	Ear pits		
25	Freckles on cheeks		
26	Freckles on forehead		
	Total		

5. There are three characteristics that demonstrate incomplete dominance in this exercise. What are they?

6. If the genes for earlobe shape and dimples were located close to one another on the same chromosome, how would their location influence how these two genes were passed on to the next generation?

7. List two examples of epistasis described in this exercise.

8. What type of inheritance pattern best describes how skin color is determined? Ignore the cases of albinism.

9. What impact do cases of multiple alleles have on the number of kinds of phenotypes displayed in the population?

10. Analyze the number of dominant versus recessive phenotypes recorded in table 16.1. Do your results agree with a Mendelian mode of inheritance? Use a chi-square test to determine if the deviation from the expected results could be attributed to chance. Make up your own chi-square table. Refer to appendix A if you need help.

LABORATORY

Sensory Abilities 17

+ Safety Box

- No specific hazards are associated with this exercise. When using solutions, however, take care that they do not splash into your eyes. If this occurs, flush your eyes with cold water. When using forceps and pencils to map sensory nerve endings, be careful not to puncture your skin.
- If you are allergic to monosodium glutamate (MSG) do not participate in that part of the exercise.

Objectives

Be able to do the following:

1. Determine if the sensations of sweet, sour, salt, bitter, and umami are located at specific places on the tongue.
2. State the importance of solubility to the function of the sense of taste.
3. Determine how accurate a subject is at localizing the sense of touch.
4. Determine the difference in density of touch receptors of different parts of the skin.
5. Determine the distribution of hot and cold receptor organs in the skin.
6. Recognize that some temperature-sensitive receptors in the skin respond to changes in temperature, not to absolute temperature.
7. Explain why colors are not readily recognized in poorly lighted areas.
8. Explain why it is difficult to identify the color of objects in your peripheral vision.
9. Demonstrate the presence of the blind spot.
10. Explain why persons can locate the direction from which a sound originates.
11. Describe the difference between airborne sound and bone-conducted sound.

Introduction

This laboratory exercise gives you an opportunity to study how we sense changes in our surroundings. Your ability to sense changes in your surroundings involves (1) the specific ability of sense organs to respond to stimuli (detection), (2) the transportation of information from the sense organ to the brain by way of the nervous system (transmission), and (3) the decoding and interpretation of the information by the brain (perception). In order for us to sense something, all three of these links must be functioning properly. For example, a deaf person might be unable to detect sound because (1) there is something wrong with the ear itself, (2) the nerves that carry information from the ear to the brain are damaged, or (3) the portion of the brain that interprets information about sound is not functioning properly. Although this laboratory activity focuses on the function of sense organs, it is important to keep in mind that the peripheral and central nervous systems are also important in determining your sensory ability. All sense organs contain specialized cells that are altered in some way by changes in their environment (stimuli). The sensory cells depolarize and, because they are connected to nerve cells, they cause the nerve cells to which they are attached to depolorize as well, and information is sent to the brain for interpretation over nerve pathways.

Preview

During this lab exercise, you will

1. make a map of the location of different kinds of taste buds on your tongue.
2. determine several characteristics of the sense of touch.
3. locate different kinds of temperature sensors in the skin.
4. study several aspects of visual acuity.
5. study several aspects of hearing.

Procedure

Taste

Taste involves several kinds of sensory cells located on the tongue and pharynx. Each kind of sensory cell responds to specific kinds of chemicals, so there is not just one sense of taste; there are several. We recognize at least five kinds of taste senses: sweet, sour, salt, bitter, and umami (meaty).

Mapping the Sense of Taste on the Tongue

1. Work with a lab partner.
2. Obtain a cotton swab and dip it into one of the solutions. The solutions are labeled sweet, sour, salt, bitter, and umami (meaty).
3. Have your lab partner touch the swab to the tongue at the following five locations: (*a*) the tip, (*b*) right side, (*c*) left side, (*d*) center, and (*e*) back.

Place an *X* on the following drawings of the tongue to indicate where you detected each chemical.

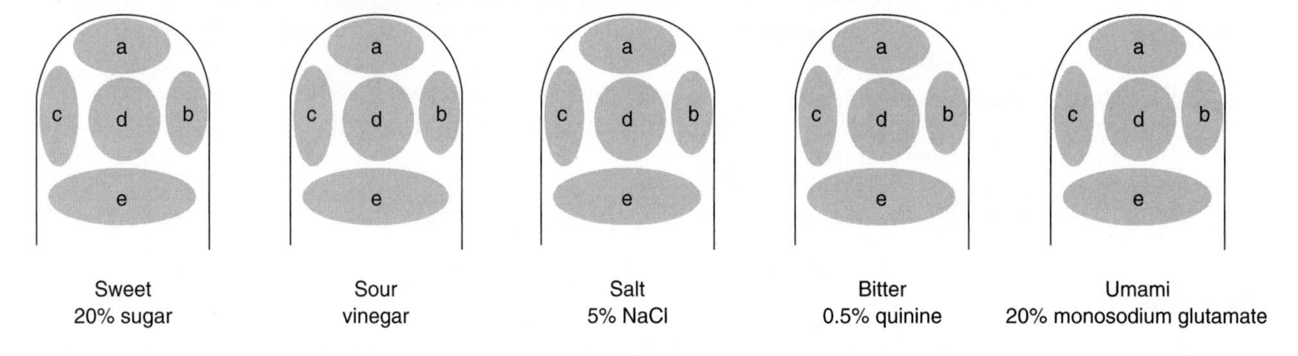

Sweet
20% sugar

Sour
vinegar

Salt
5% NaCl

Bitter
0.5% quinine

Umami
20% monosodium glutamate

4. Test the other four solutions in the same manner, but *be sure to rinse your mouth with water after each solution.*
5. When you have tested each of five chemicals, switch positions with your partner.

Results

Can you detect each chemical at all places on the tongue?

Compare your results with those of your partner and other people in class. Do they detect the same chemicals in the same place?

What does this tell you about the sense of taste?

The Role of Solubility in Detecting Taste

1. Dry off the tip of your tongue with a clean paper towel. Place a few grains of table salt (NaCl) on the tip of your tongue. Record the time interval from the time you place salt on the tip of your tongue until you first taste the salt.

 ____

2. Dissolve a few grains of salt in a small amount of water. Place this on the tip of your tongue. Record the time interval from the time you place the salt solution on the tip of your tongue until you first taste the salt. __________

 Were the two time intervals different? What does this tell you about the ability to taste salty materials?

Touch

The sense of touch is made up of a number of types of receptor organs. Pressure, pain, heat, and cold are all aspects of the sense of touch. We will experiment with some of them here.

Localization of Touch

You need a partner for this exercise.

1. The subject should keep his or her eyes closed throughout the exercise.
2. Touch the skin of the back of the hand of the subject lightly with the pointed end of a soft lead pencil. Be sure to leave a mark.
3. Then, ask the subject (with eyes still closed) to use a blunt probe to locate the place on the skin where the stimulus was received.
4. Use a ruler to measure as closely as possible the error in locating where the stimulus was applied. Measure the error in millimeters. Repeat five times at different locations on the back of the hand.
5. Change roles with your partner and repeat the experiment.

Results and Conclusions

In the space provided, write a short paragraph that states your findings and conclusions.

Density of Sense Organs

You need to work in pairs.

1. Have the subject keep his or her eyes closed.
2. Use a pair of forceps or calipers to touch the subject's skin gently, so that the two points of the instrument touch with the same light pressure and at the same time. Test the palm of the hand and two other regions of the body. Other regions that may be tested are the back of the hand, the tip of the index finger, the forearm, the tip of the nose, the forehead, and the back of the neck. Not all of these need to be tried, but a selection should be made.
3. Ask the subject to state whether one or two points of the instrument are felt. Repeat this procedure five times for each area of the body chosen. (To keep the subject from guessing, the experimenter should occasionally touch the skin with only one point. However, do not record the result of the response in your data.)
4. Record your data in the following manner: Record a minus sign (–) whenever two points were felt as one and a plus sign (+) whenever the two points were actually felt as two.
5. Begin with the points 20 millimeters apart and systematically decrease the distance between the points from 20 mm to 15 mm to 10 mm to 5 mm. Find the smallest distance at which the subject can still distinguish two points for each portion of the body tested.
6. Change roles. Record the data made on yourself as the subject.
7. From the data, estimate the comparative densities of touch receptors of the different parts of the body.

Area I: ______________

Distance between points of forceps in mm	Trial Number 1	2	3	4	5
20 mm					
15 mm					
10 mm					
5 mm					

Area II: ______________

Distance between points of forceps in mm	Trial Number 1	2	3	4	5
20 mm					
15 mm					
10 mm					
5 mm					

Area III: ______________

Distance between points of forceps in mm	Trial Number 1	2	3	4	5
20 mm					
15 mm					
10 mm					
5 mm					

Results and Conclusions

1. What is the smallest distance the subject can still recognize two points for each of the three areas tested? Are they the same? Explain.

2. Place a sketch of your "two-point device" on the drawing to indicate why two points are sometimes felt as one.

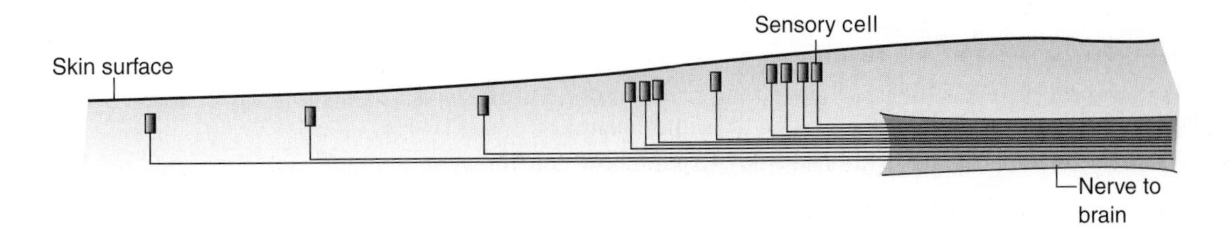

3. Which of the regions of the skin that you tested is represented by the left side of the drawing and which is represented by the right side of the drawing? Explain your answer.

Temperature Sense—Detecting Hot and Cold

Work with a partner.

1. With a pen, draw a square with 20-mm sides on the back of the subject's hand; then subdivide this square into 16 smaller squares by dividing each of the sides into 5-mm segments.
2. Have the subject keep eyes closed and place his or her hand flat on the table.
3. Obtain a nail that has been in ice-cold water.
   Dry it off with a paper towel.
   Lightly touch each of the squares of the grid on the hand at random.
   The subject should respond by saying "cold" if such a sensation is actually felt; otherwise, the subject remains silent. *It is important for the subject to ignore the sense of touch and concentrate on the sensation of cold.*
4. For every positive response, the experimenter marks a plus sign (+) on the following grid at a point corresponding to the point tested on the skin.
5. Be sure that the nail is very cold when you make each test.
6. Repeat this exercise with a hot nail and record your results on the second grid.

Cold

Hot

7. Switch roles with your partner and repeat the exercise.
8. Answer the following questions.

   Do you detect hot in every square? _______________

   Do you detect cold in every square? _______________
   Are hot and cold receptors always located in the same squares?
   Do the same receptors respond to hot and cold? Explain how you know.

Temperature Sense—Detecting Changes in Temperature

1. Dip one finger into a beaker of hot water and at the same time put a finger from the other hand into cold water.
2. After 30 seconds, transfer both fingers into a third beaker of warm water.

Results and Conclusions

Describe the sensations of both fingers in the beaker of warm water and explain why there is a difference in sensation.

Vision

The eye is a complex structure, focusing light on the cells of the retina, which respond to changes in light. There are two kinds of light receptors, rods and cones. Rods are very sensitive to light and only respond to differences in light intensity. The cones are less sensitive to light. There are at least three kinds of cones, each of which responds to specific colors of light. The rods and cones are located in different places in the retina of the eye. In this part of the lab activity, you will make a number of observations about the eyes and their response to various stimuli.

Light Intensity and Color Vision

Work with a partner.

1a. Take three pieces of different but similarly colored paper that are about 100×100 cm into a nearly dark room. Show only one square at a time and ask your partner to identify the color of the paper. Determine the distance at which your partner can tell the color of the squares of paper. It is not necessary to measure the distance exactly. Simply count the number of paces between you and your partner.
1b. Change roles and have your partner show you the squares of paper.
2. Return to a well-lighted area and determine the distance at which your partner can still identify the colors. Explain your results by discussing the function of the rods and cones in the retina.

Determining the Location of Rods and Cones

Rods and cones are not located in the same place on the retina of the eye. When you look at things from directly in front of the eye, the cornea and lens of the eye focus the light on a region known as the fovea centralis. When you look at things with your peripheral vision, the light is focused on regions of the eye other than the fovea centralis.

Work with a partner.

1. Choose three similarly colored squares of paper about 100 × 100 cm.
2. Have your partner stare at a distant object directly in front of him or her.
3. Start behind your partner (out of the field of vision) and slowly move each piece of paper forward at eye level about 30 cm to the side of the head.
4. Ask your partner to tell you when the piece of paper is first seen and when the color of the paper can be detected. Use the information about the location of rods and cones and the results you just obtained to answer the following questions.
   Which sense organs (rods or cones) are most common in regions *outside* the fovea centralis? ____________________
   Which sense organs (rods or cones) are most common *within* the region of the fovea centralis? __________________
   Explain how this experiment allows you to answer these questions.

Detecting the Blind Spot

Use the + and dot below in the following manner. Close your left eye. Place the page close to your face. Stare at the + with your right eye. Slowly move the page away from you. What happens to the dot?

+ ●

In order to detect the presence of an object, light must fall on the retina of the eye and stimulate either rods or cones. There are no rods or cones at the point where the optic nerve goes out of the back of the eye. Use this information to explain what you observed when looking at the + and ●.

Hearing

The sense of hearing involves the detection of sound vibrations. Airborne sounds cause the eardrum to vibrate. The eardrum is attached to a series of three small bones: the malleus, incus, and stapes. The stapes is attached to a membrane over a small opening in the cochlea. The cochlea is fluid-filled. Thus, the vibrations of the air are transferred to the fluid of the cochlea. When the fluid in the cochlea vibrates, cells in the cochlea are stimulated. When these cells depolarize, they send a signal by way of the auditory nerve to the brain. In this part of the lab activity, we will explore some aspects of hearing.

Work with a partner.

1. Strike a low-frequency tuning fork (100 cps) and hold it near one ear of your partner. Determine how far from the ear the subject can hear the tuning fork. Repeat with the other ear. Are both ears the same?

2. Strike the tuning fork and touch the *base* of the vibrating tuning fork to the skull just in front of the ear. Does the volume change?

   How is this sensation of hearing different from when the tuning fork is held near the ear?

3a. Have the subject sit with closed eyes.
Strike the tuning fork.
Have the subject point to the position of the tuning fork.
Repeat three times from different positions.
Can the subject correctly identify the position of the tuning fork?

3b. Now have the subject keep eyes closed and plug one ear with a finger.
Have the subject point to the tuning fork as it is struck at different positions.
Is the subject able to locate the position of the tuning fork accurately?
Why is there a difference between the two trials?

17 Sensory Abilities

Name ____ Lab section __________

Your instructor may collect these end-of-exercise questions. If so, please fill in your name and lab section.

End-of-Exercise Questions

1. Describe the regions of your tongue that are most sensitive to sweet, sour, salt, bitter, and umami.

2. How is solubility important to the sense of taste?

3. Determine the average distance, in mm, between points on the palm of the hand at which persons in the class correctly identified that they were being touched by two points.

4. Using the data you collected for different parts of the skin, rank them according to which had the greatest density of touch receptors and which had the lowest density.

5. Write a paragraph describing what you learned about the receptors that respond to temperature. How many kinds of receptors are there? Explain how you know there are different kinds of receptors.

6. Some people can see well in bright light but are not able to see in dim light. This condition is called night blindness. What kinds of sensory cells do not function to capacity in individuals who have night blindness?

LABORATORY

Population Demographics 18

+ Safety Box

- There are no safety hazards associated with this exercise.

Objectives

Be able to do the following:

1. Define these terms in writing.

   crude birthrate   crude deathrate   total fertility rate

   demography

2. Calculate annual rate of population change and doubling time.
3. Construct graphs to illustrate the relationships among population variables.

Introduction

The impact humans have on the Earth is often characterized by the following equation:

$$I = P \times C \times T$$

I = Impact on natural Earth systems per unit time

P = Human population

C = Consumption per person per unit time

T = Technology factor (high for environmentally destructive technology, low for environmentally friendly technology)

It makes sense that larger numbers of people will have a greater effect on Earth systems than fewer people. The equation $I = P \times C \times T$ captures this relation in a simple expression. Any change in the population *(P)* will have a major effect on the impact *(I)* unless consumption *(C)* or technology *(T)* change significantly.

The study of population demographics concentrates on the descriptive characteristics of human populations, including issues such as changes in the size and structure of populations and rates of population change. The worldwide human population is currently experiencing rapid growth. As human population growth continues, the equation $I = P \times C \times T$ indicates that the resulting impact on Earth's natural systems will also increase (unless consumption per person is reduced or environmentally friendly technology replaces current technology).

Demography is the study of the characteristics of human populations, such as size, growth, density, distribution, and vital statistics. Demographers use the following equations to calculate the key factors that describe changes in population characteristics:

$$\text{crude birthrate (births/1,000)} = \left(\frac{\text{live births per year}}{\text{mid-year population}}\right) \times 1{,}000$$

$$\text{crude deathrate (deaths/1,000)} = \left(\frac{\text{deaths per year}}{\text{mid-year population}}\right) \times 1{,}000$$

$$\text{annual rate of population change \%} = \left(\frac{\text{birth} - \text{deathrate}}{10}\right)$$

$$\text{population doubling time (years)} = \left(\frac{70}{\text{annual rate of population change [\%]}}\right)$$

$$\text{Total fertility rate} = \text{number of children born per woman}$$

The population growth rate has been correlated to a number of factors that affect the birthrate, the deathrate, or both. The following factors appear to influence the birthrate:

1. Level of education and wealth
2. Importance of children for family labor purposes
3. Urbanization—higher birthrates in rural areas
4. Cost of raising children
5. Education and employment opportunities
6. Average age at marriage
7. Availability of birth control
8. Cultural norms

The following factors influence deathrate:

1. Nutrition
2. Sanitation
3. Advances in available health care
4. Ability to afford medical care

18 Population Demographics

Name ______________________________ Lab section ____________

Your instructor may collect these end-of-exercise questions. If so, please fill in your name and lab section.

Procedure

Calculate annual rate of population increase and doubling time. Use the information in the following table and the formulas on page 194 to calculate the annual rate of population increase and the doubling time for the populations of the countries listed.

Country	Population Density (People per Square Kilometer)	Total Fertility Rate	Crude Birthrate (Births per 1,000)	Crude Deathrate (Deaths per 1,000)	Annual Rate of Population Change (%)	Doubling Time (Years)
1. Afghanistan	119	6.8	47	21		
2. Bangladesh	1,035	3.0	27	8		
3. Dominican Republic	192	2.9	24	5		
4. France	112	2.0	13	9		
5. Hungary	108	1.3	10	13		
6. Mexico	54	2.4	21	5		
7. Netherlands	394	1.7	11	8		
8. Russia	8	1.3	10	15		
9. United States	31	2.1	14	8		
10. Zimbabwe	34	3.8	31	21		

Source: Data from 2007 World Population Data Sheet, Population Reference Bureau, Inc., Washington, DC.

Plot the relationship between annual rate of population growth and population density on the following graph.

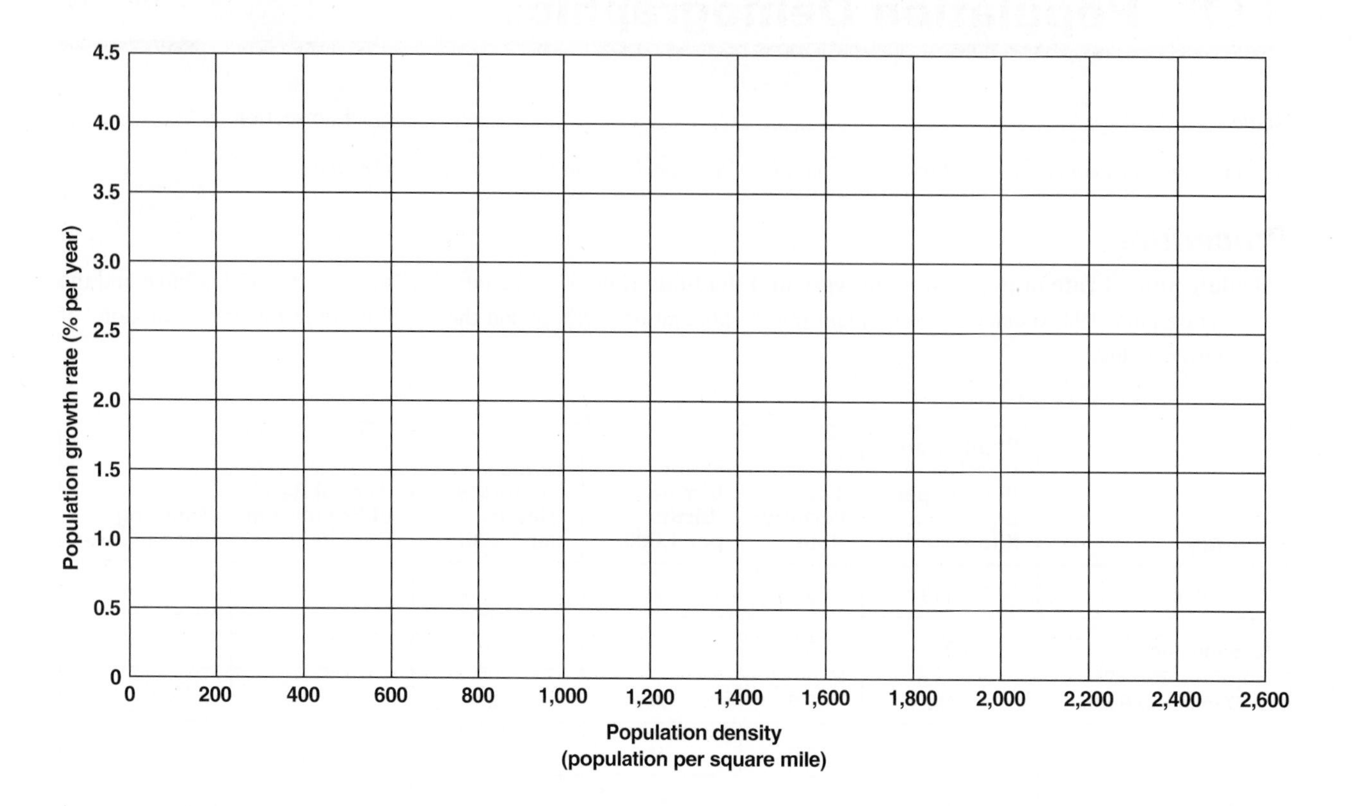

1. Does the annual rate of population change appear to be related to population density? (*Hint:* The two factors, population density and annual rate of population change, are very strongly correlated if you can draw one straight line and connect all the points. If you can draw a line that most of the points cluster near, the two factors are correlated less strongly. If there is no straight line that the points cluster near, the two factors are not correlated.)

Draw a graph of the relationship between the doubling time and the crude birthrate on the graph provided. On the same graph, plot the relationship between crude deathrate and doubling time. Plot the two sets of data with different colors.

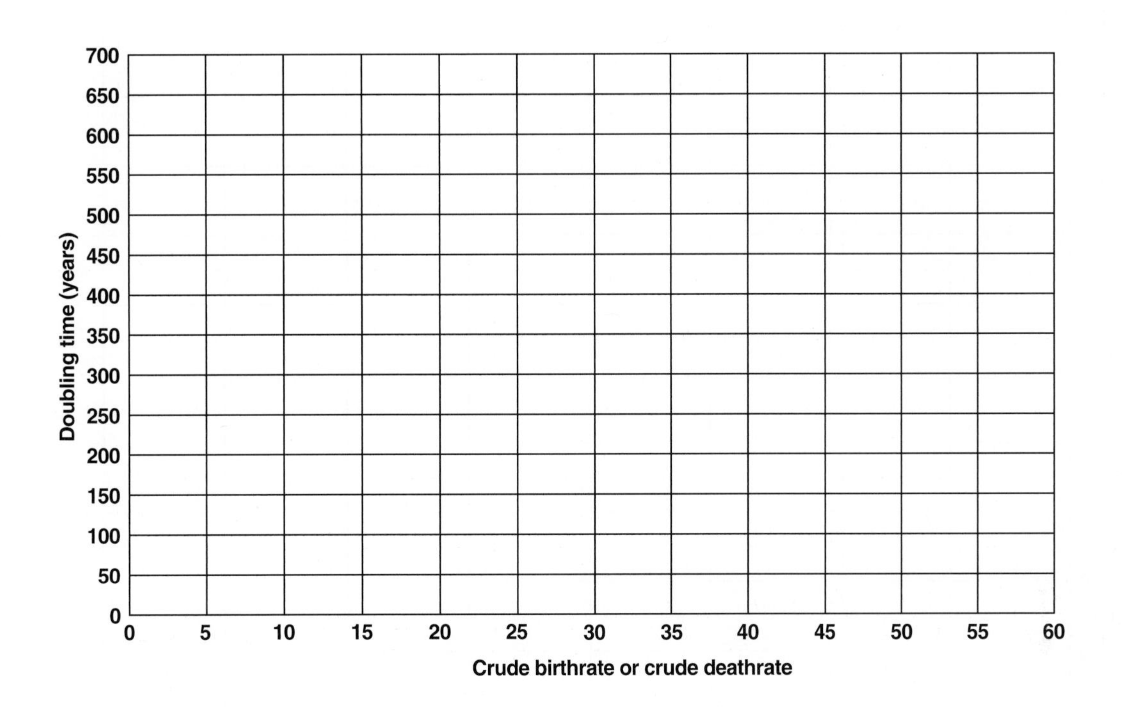

2. Which of the two (birthrate or deathrate) is most closely related to doubling time (i.e., which of the two sets of data is closer to a straight line)?

Plot the relationship between total fertility rate and annual rate of population growth on the following graph.

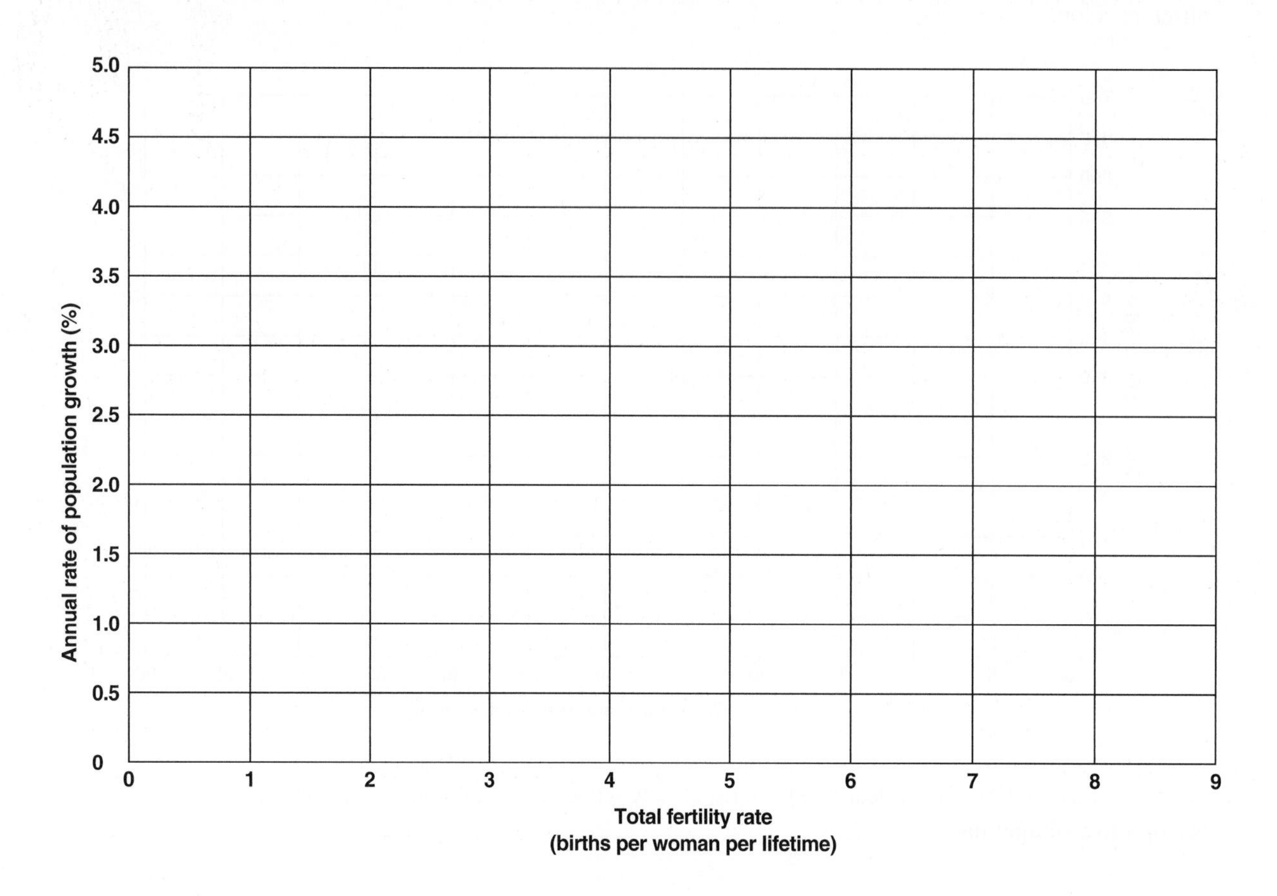

3. Is there a strong correlation between the number of births per woman and the annual rate of increase?

LABORATORY

Population Genetics Simulation 19

+ Safety Box

- The beads used in this exercise become hazardous if scattered on the floor. Students could fall and injure themselves seriously. Be careful to keep the beads in their proper containers at all times.

Objectives

Be able to do the following:

1. Define these terms in writing.

   natural selection
   recessive lethal allele
   allele frequency
   mutation
   simulation
   gene pool
   population genetics
   homozygous recessive

2. Relate changes in allele frequency to natural selection.
3. Explain why it is difficult to eliminate recessive alleles from a population.

Introduction

Natural selection deals with the idea that those individuals within a population with genes that make them better adapted to their surroundings are more likely to survive, reproduce, and pass their genes on to the next generation. In biological terms, *nature selects* those individuals *most fit* to survive. Over time, the process of natural selection leads to an increase in the frequency of "good," or beneficial, genes in a population and a decrease in the frequency of "bad," or harmful, genes. New alleles for a gene come into being as a result of **mutation,** which is a change in the DNA of an individual. Mutations may have no effect, be beneficial, or cause harm.

There are many degrees of injury caused by harmful alleles. Some, such as nearsightedness, may cause minor inconveniences but could shorten life because of an inability to see well (auto accidents). Others, such as the allele for sickle-cell anemia, may cause severe symptoms and shorten life for many people. Some harmful alleles, such as the allele that causes Tay-Sachs disease, are lethal in the homozygous recessive condition. **Homozygous recessive** individuals contain 2 recessive alleles for a particular characteristic. All homozygous recessive individuals with 2 Tay-Sachs alleles die. Severely harmful alleles are always rare in a population; they have a low allele frequency. The **allele frequency** is a mathematical statement about how common a specific allele is in the population. The study of the change in allele frequency in populations is known as **population genetics.**

In this exercise, we will use an artificial population to follow the frequency of a particular harmful allele over several generations. We will assume that there are two alleles present in the population: a dominant normal allele and a **recessive lethal allele.** Those individuals who inherit 2 recessive alleles must get 1 from each of its parents. When organisms receive 2 recessive lethal alleles, they die without reproducing and do not pass on their alleles to the next generation. Therefore, you might expect the number of recessive lethal alleles in the population to decrease over several generations.

The allele frequency of the recessive lethal allele should decrease as the recessive lethal allele becomes less common and the dominant normal allele becomes more common. In this exercise, we will conduct a **simulation** (an artificial setup that approximates the real situation) to determine how natural selection will affect the allele frequency of a recessive lethal allele in a population.

Preview

During this exercise, you will

1. Use blue beads to represent the dominant normal alleles and red beads to represent the recessive lethal alleles in a population.
2. Simulate sexual reproduction and natural selection over 10 generations.

Procedure

Generation 1

Use blue beads to represent the normal dominant allele (one form of a gene) and red beads to represent the recessive lethal allele. Set up a gene pool of 100 genes. The **gene pool** is all of the genes in the population of a species. Use the box provided as a container for your beads. ***The percentage of recessive alleles will be assigned to you by your instructor and will differ from that of the other groups.***

Blue = A; red = a

1. After you have created your gene pool (counted out the beads), shake the beads in the container and tip it to allow the beads to form a single row along one side of the container. The mixing of the beads simulates the mixing of genes that occurs during sexual reproduction as a result of meiosis and random mating. The row of beads at the side of the container represents the individual organisms of the first generation of organisms.
   (*Note:* Instead of boxes, some instructors may use paper bags and have you randomly draw two beads at a time from the bag to simulate the allele combinations in the individuals.)
2. Starting at one end, count off the beads in pairs. We count the beads two-at-a-time because diploid individuals inherit 2 alleles, 1 from each parent. As you count, you will have three possible combinations: two blue beads (AA), a red and a blue bead (Aa), or two red beads (aa).

Genotype	Phenotype
AA	Organism lives
Aa	Organism lives
aa	Lethal; organism dies

   Remember that whenever a pair consists of two red beads it represents a lethal combination. This organism will die. Simulate the deaths of these homozygous recessive individuals by removing them from the population. (Take the two red beads out of the container and separate them from the remaining beads.) After you have examined all the pairs of beads,

   a. determine the number of red beads remaining.
   b. determine the total number of beads remaining.
   c. record your results in table 19.1 on page 203.
   d. calculate the percentage of red beads remaining in your container (number of red beads remaining divided by the total number of beads remaining).

$$\% \text{ red beads remaining} = \frac{\text{total red beads still in the box}}{\text{total of all beads still in the box}}$$

   e. record this new allele frequency in the last column of table 19.1.

Generations 2–10

1. To begin the next generation, we will assume that the survivors reproduce and as a result the population will grow. In many populations of organisms, the number of individuals being born and the number of those dying are about the same and the population is stable. We will assume that in our hypothetical "bead" population the reproductive capability of the organisms keeps the population of total alleles at 100 for each generation and that the surviving alleles (red and blue beads) reproduce equally well. You have already calculated the percentage of red beads remaining after the first generation (percentage means number out of 100). Therefore, your second generation starts with 100 beads but with the ***new percentage of red beads you have just calculated.*** To bring the total number of beads back up to 100, while maintaining the new percentage of red beads, do the following:
   a. Count the red beads remaining in the container.
   b. Add enough red beads to the container so that the ***total number of red beads*** is equal to the ***percentage of red beads you calculated in table 19.1.***
   c. After you have added the red beads, add enough blue beads to have a total population of red and blue beads equal 100. For example, if you calculated that you have 30% red beads remaining, add enough red beads to make 30 red beads and enough blue beads to make 100 total beads. If you are unsure, see the following example.
2. Starting with the newly reproduced population which is the new generation 2,
   a. Shake the beads and count off the beads, two by two.
   b. Remove the homozygous recessive lethal organisms (those with two red beads), as you did for the first generation.
   c. Calculate the percentage of red beads remaining among the survivors, as you did after generation 1.
   d. Record your data in table 19.1.
   e. Repeat this process until you have completed 10 generations.
3. Graph the results of your simulation on the grid provided on page 204, and compare your graph with the graphs of other groups.

Example

Percent of red beads at start	20% = 20 ÷ 100 (20 red + 80 blue = 100)
Red beads removed by random mating	4
Number of red beads left	20 – 4 = 16
Total number of beads left	100 – 4 = 96
Red beads ÷ total beads × 100 = % red	16 ÷ 96 = 0.166 × 100 = 16.6%, or 17* rounded off
	*Record value on table 19.1.

To prepare for the next round of mating,

1. add 1 red bead to bring the red total to 17.
2. add 3 blue beads to bring the overall total to 100.

19 Population Genetics Simulation

Name ______________________________ Lab section __________

Your instructor may collect these end-of-exercise questions. If so, please fill in your name and lab section.

Table 19.1 Allele Frequency Data

Generation	% Red Beads at Start of Run	Number of Red Beads at End of Run	Total Number of Beads at End of Run	New Allele Frequency % Red Beads at End of Run (Red ÷ Total × 100)
1				
2				
3				
4				
5				
6				
7				
8				
9				
10				

1. Plot the data on the frequency of the recessive lethal allele (red beads) on the following graph.

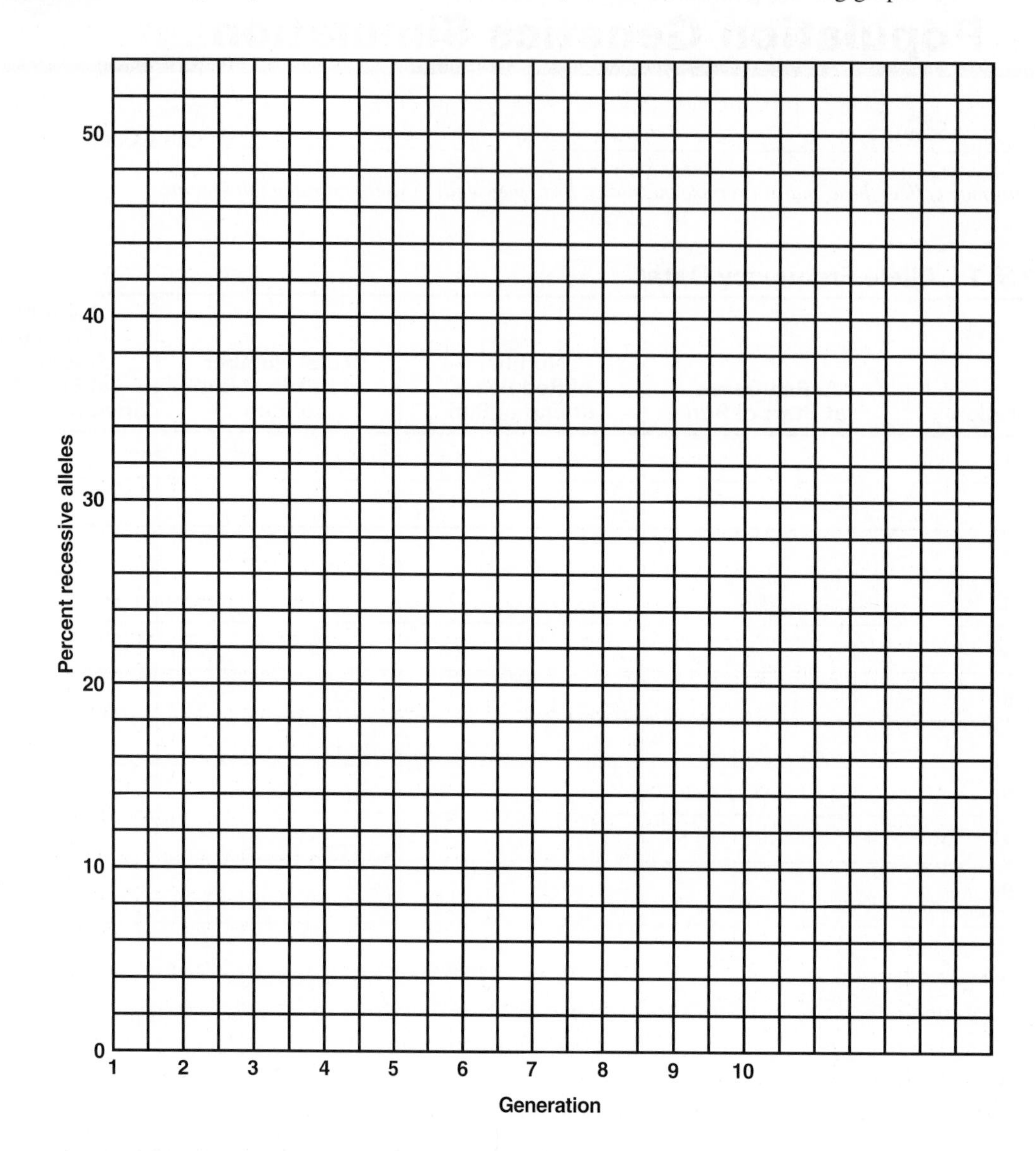

2. Compare your results with those of other students in the class. Did you eliminate the recessive lethal allele? Did anyone in class eliminate the recessive lethal allele?

End-of-Exercise Questions

1. We have been working with recessive lethal alleles in a situation in which none of the homozygous recessive individuals survives. What kind of allele frequency curve would you expect if only one-half the homozygous recessive individuals died without reproducing? Start at point *a* on the graph and sketch a line on the graph that would represent the data if half the homozygous recessive individuals survived and reproduced. The curve shown on the graph is one that would result if all homozygous recessive individuals died.

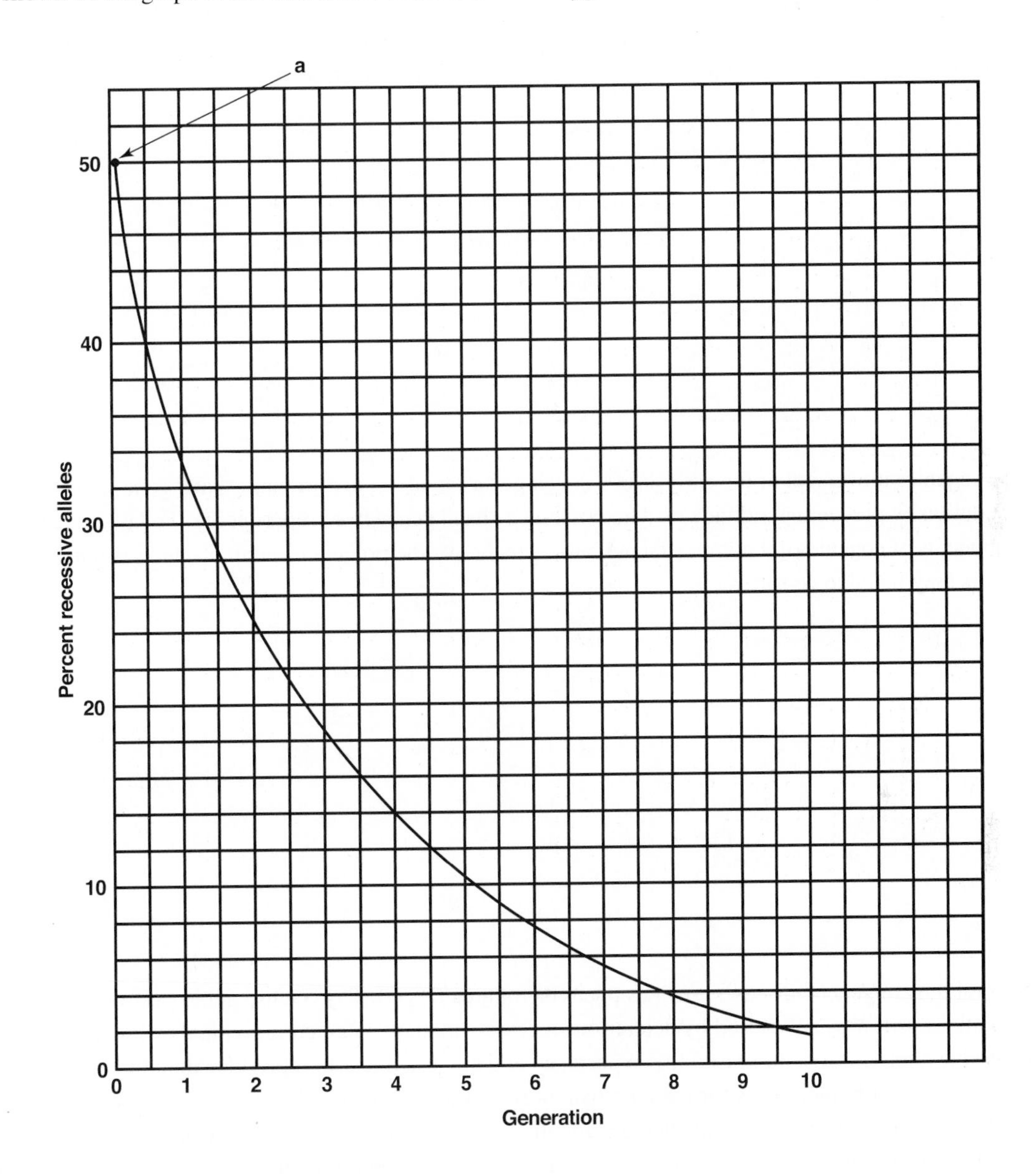

2. Suppose you are a cat breeder involved in breeding and showing exotic cats. You notice that one litter of kittens contains a curly-haired male. This is the first time you have noted this genetic characteristic, and no one in the Cat Breeders Association has ever seen it before. Because the other kittens in the litter are not curly-haired, you reason that this novel trait must be due to a mutant recessive allele that has been hidden in this breed until now. It is a desirable trait, because the coat is soft and the hair does not fall out as readily as normal hair. This cat could sell for about $600. The normal cost of this breed of cat is about $250. You would obviously like to obtain more of these curly-haired animals. What type of breeding program would give you large numbers of curly-haired cats in the shortest time? (*Note:* This time, you are trying to *increase* the number of recessive lethal alleles. Inbreeding is common among animal populations.)

3. The eugenics movement of the early 1900s developed as an effort to eliminate "bad" genes from the human population by preventing individuals who possessed "bad" genes from reproducing. Today, it is obvious that proponents of eugenics did not understand basic genetics or the process of natural selection. Discuss several factors they failed to consider.

4. Several human genetic diseases, such as phenylketonurea (PKU), Tay-Sachs, sickle-cell anemia, and muscular dystrophy, are caused by deleterious recessive alleles. Affected individuals who inherit 2 of the recessive alleles have a decreased life expectancy and are less likely to reproduce than those that are heterozygous or homozygous for the dominant normal allele. Describe the effect natural selection will have on the frequency of the alleles for such genetic disorders if the original frequency of the recessive allele is 1%. Will the frequency of the recessive allele (1) increase, (2) decrease, or (3) stay the same? Explain your answer. Think about a large population over 25 to 50 generations.

LABORATORY

20 Bacterial Selection

+ Safety Box

- Bacterial cultures should always be considered dangerous, because they contain large numbers of living cells, some of which may be mutants that are pathogenic (i.e., able to cause disease). Should you spill these organisms on your skin or have them transported to your eyes, nose, or mouth, they may establish an infection.
- If you spill or come in contact with the bacterial cultures, do not move. Immediately send someone to notify your instructor, so the proper disinfection procedures can be performed.
- Disinfect your work area before and after working with bacterial cultures.
- Always hold culture tubes by the tube and not the cap, because the cap may not be firmly attached.
- Be careful of the open Bunsen burner flame. Do not wear fuzzy sweaters. If you have long hair, take care to keep it out of the flame.
- Flame the open tops of all culture tubes to prevent other bacteria from falling into the tube.
- Mix the contents of bacterial culture tubes by rolling them between the palms of your hands. Do not shake them up and down.
- Sterilize inoculating loops and needles by placing them in a Bunsen burner flame. Be careful not to burn yourself on the hot wire.
- Dispose of your contaminated cotton swab in a container of appropriate disinfectant. Do not lay swabs on the table or other surfaces, because those surfaces would then have to be disinfected. If you contaminate your notes, books, or clothing, they will need to be disinfected.
- Keep all Petri plates containing bacteria taped shut. Do not open them unless you are instructed to do so.
- When heat-fixing a bacterial smear preparation, be careful not to keep the slide in the flame too long, because the glass will conduct the heat and burn your fingers. Keeping the slide in the flame too long will also result in its breaking.
- Stains used in this laboratory may stain your hands and clothing.

Objectives

Be able to do the following:

1. Define these terms in writing.

wild type	Petri plate	generation time
mutant	nutrient agar	Gram stain
pathogenic	selecting agent	sterile

2. Recognize that antibiotics can serve as selecting agents and prevent the reproduction of some individual bacteria while not affecting the reproduction of others.
3. Stain a sample of a bacterial culture using the Gram stain technique.
4. Recognize that commonly studied spherical bacteria are about 1 μm in diameter, one-seventh the size of a red blood cell.

Introduction

Bacteria are tiny prokaryotic organisms that are usually 1 micrometer or less in diameter. They lack a nucleus and other complex cellular organelles and have a great many biochemical differences from eukaryotic organisms, such as plants and animals. Bacteria are excellent laboratory organisms for showing evolution because they have such a short generation time. **Generation time** is the time required for an individual to grow to the point that it can reproduce. In bacterial populations, individuals under the best conditions can double in number in only 20 minutes. This means that, if natural selection and evolution are taking place in the population, we can detect changes in the frequency distribution of characteristics of the population in only 24 hours because several dozen generations have occurred in that time.

If the environment of a particular population of a bacterium changes, the new factor in the environment acts as a **selecting agent** and determines which individual bacterial cells will grow and reproduce. A few individuals in the population **(mutants)** may have the genes needed to survive in the new environment; others **(wild types)** do not and are killed or do not reproduce. Under the right conditions, it is possible to identify these genetically different bacteria.

One of the great advances of the twentieth century was the discovery and use of antibiotics to control the growth of bacterial populations. The use of antibiotics has greatly reduced the deathrate from infectious diseases. Because bacteria are prokaryotic organisms and plants and animals are eukaryotic organisms, there are significant differences in the way these two kinds of organisms function. Therefore, antibiotic chemicals can be chosen that attack prokaryotic bacteria but do not affect eukaryotic cells. Each kind of antibiotic has a specific effect on the function of certain bacterial cells. Some, such as penicillin, prevent bacteria from manufacturing protective cell walls. Thus, the bacteria are more susceptible to destruction. Other antibiotics, such as tetracycline, aureomycin, and streptomycin interfere with bacterial ribosomes and, therefore, their ability to produce proteins. Because they cannot produce enzymes efficiently, they eventually die from their inability to carry out essential biochemical reactions.

A person with a bacterial infection, such as strep throat (*Streptococcus pyogenes*) or tuberculosis (*Mycobacterium tuberculosis*), is actually infected with thousands, if not millions, of these bacteria. When we use antibiotics, we hope to kill all these cells. However, only the susceptible bacterial cells are killed, whereas the resistant cells of the same species of bacterium survive to reproduce and pass on their genes to the next generation. Therefore, antibiotics act as selective agents on populations of bacteria and ultimately resistant populations become so common that the antibiotics originally used to control them are ineffective. This exercise will explore the development of antibiotic resistance in common kinds of bacteria.

For this exercise to work successfully, it is necessary to work with only one species of bacterium. Therefore, all your equipment must be **sterile** (free of living organisms), so that no other bacteria from the air or table accidentally become mixed with the population you use.

> ***Caution:*** *Because mutant bacteria exist in all populations, any bacterial culture is potentially* ***pathogenic****—or able to cause disease. This type of mutant does not occur very often, but once is too often. Your instructor will demonstrate safe methods to use when handling the bacteria. Practice the exercise without the bacteria before you start, using a tube with plain water. If you are unsure of a step in the procedure, ask for help.*

Bacterial cells are extremely small (about 1–5 micrometers in diameter); therefore, it is necessary to use a microscope that can magnify approximately 1,000 times. The lens on your microscope that magnifies × 1,000 is called an oil-immersion lens because it must be submerged in a small drop of special oil on the slide for you to obtain a clear view of the bacterial cells. Only rarely are bacteria viewed alive. Most of the time, the cells are stained to make them more visible under the microscope.

One common staining technique is known as the **Gram stain.** The technique is used to stain the cell walls of bacterial cells so they may be more easily seen and identified. Bacteria are categorized into two groups based on the Gram stain procedure, Gram-positive (+) and Gram-negative (–). The cell walls of Gram-positive (+) bacteria have a different chemical composition and retain the violet-colored dye (crystal violet) used in the Gram stain despite an attempt to take out this color (decolorize) with an alcohol rinse. Gram-negative (–) bacteria lose their violet color when rinsed. Therefore, it is possible to use this staining technique to help identify the specific kind of bacterium. Your instructor will review with you the procedures necessary to stain and view cells with the oil-immersion lens.

Preview

This exercise has a dual purpose. First, you will grow bacteria and use antibiotics as selecting agents to identify the presence of some bacteria in the cultures that are genetically different (mutants) and capable of surviving in an environment containing antibiotics. Second, you will work with bacteria, learn how to handle them, stain them, and identify some of their characteristics.

During this lab exercise, you will use the aseptic technique demonstrated by your instructor to

1. pour warm nutrient agar into a labeled Petri plate and let it cool (this step might be done for you).
2. spread bacteria over the surface of the agar with a sterile swab to make a "lawn" of bacteria.
3. place two antibiotic disks on the hardened surface of the agar.
4. close and tape the Petri plate closed and incubate the bacteria at 37°C for 24 hours.
5. remove plates from the incubator and examine them for growth and clear areas around the antibiotic disks (this might be done on your own time or during the next lab session).
6. make a smear slide preparation for the examination of the bacterial cells.
7. Gram stain the slide.
8. observe the bacterial cells with the oil-immersion lens.

Procedure I—Demonstrating Antibiotic Resistance

Petri Plate Preparation (Figure 20.1)

Obtain two **Petri plates** and label the top of each dish with your name, lab time, and date. Keep the dish closed. The **nutrient agar** is melted in a test tube. It must be kept warm (46°C) or it will begin to solidify.

1. Remove the cap of the tube of nutrient agar and pass the top of the test tube through the flame of the Bunsen burner to heat the open end.
2. Lift the cover (the larger section) of the Petri plate without completely uncovering the bottom.
3. Pour the contents of the tube into the bottom (smaller section) and gently swirl the plate without lifting it from the surface of the table to make sure the agar completely covers the bottom of the plate.
4. Let this plate set on the table for about 15 minutes or until the agar is solid.

Figure 20.1 Preparation of agar plates.

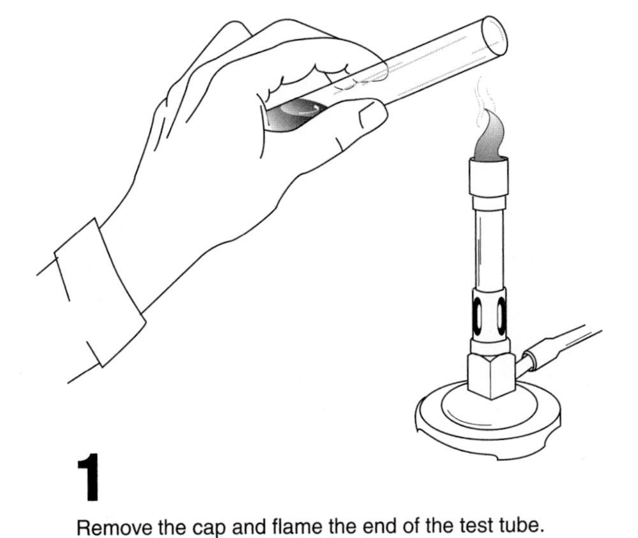

1

Remove the cap and flame the end of the test tube.

2

Pour the medium into the bottom of the Petri plate and let it cool until hard.

Inoculation Procedure (Figure 20.2)

Before doing this procedure, rehearse it using the technique shown in figure 20.2. You will be working with two species of bacteria. The following procedure should be performed twice—once for each species of bacteria. From the stock table get a tube containing a culture of bacteria, a test tube rack, a packet of sterile swabs (cotton-tip applicator), and a beaker of disinfectant for disposal of the swabs.

1. Open the swab pack and remove a swab, but ***do not put the swab on the table.*** Keep the swab in your hand at all times. If it touches *anything,* it is no longer sterile. With the other hand, pick up the tube of bacteria (see figure 20.2).
2. Roll the tube between the palms of your hands or gently shake from side to side to put the organisms in suspension. ***You should never shake the tubes in a way that will cause the bacterial culture to slosh near the top of the tube.***
3. Remove the cap with your little finger. ***Do not set the cap down.*** Flame the top of the test tube.
4. Dip the swab into the bacterial culture, and allow it to drip excess culture back into the tube. Use caution not to touch the side of the tube while removing the swab. Hold the swab in your hand at all times.
5. Flame the opening of the tube again.
6. Replace the cap. Replace the tube in the test tube rack.
7. Open the Petri plate, holding the top over the agar to protect it from contamination. Spread the bacteria on the swab evenly over the *entire* surface of the agar, so that all portions of agar are inoculated with bacteria. Close the top of the dish.
8. Place the swab in the disinfectant container provided.
9. Place a paper disk containing penicillin on the surface of the agar about 1/2 inch from one edge of the dish. Just lay the disk on the surface of the agar. Place a second disk containing Aureomycin at the opposite side of the dish about 1/2 inch from the edge. Write the name of the species of bacterium on the outside of the Petri plate. Place the plate on the tray labeled “incubate,” and your instructor will grow the bacteria at 37°C for 24 hours.

Repeat steps 1–9 with the second species of bacteria.

Figure 20.2 Inoculation procedure.

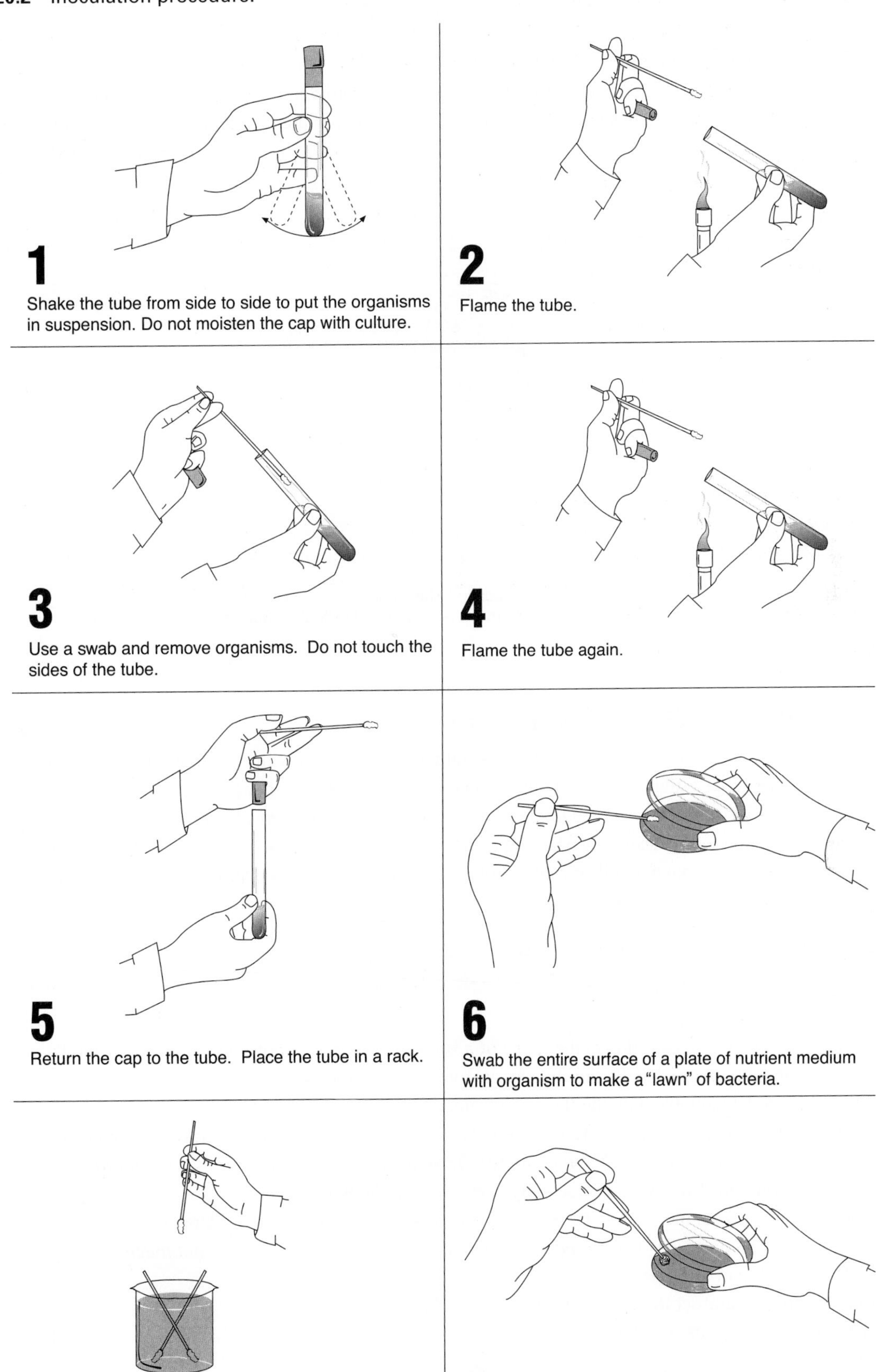

Examination of Incubated Plates (Figure 20.3)

After 24 hours, examine your Petri plates. Wherever bacteria have grown, the nutrient agar will be cloudy. Any areas where bacteria were prevented from growing will be clear. Measure the zone of bacterial death (clear area around the antibiotic disk) for each of the two antibiotic disks on both plates. Record your data on the data sheet on page 215.

If instructed to examine the bacteria on your own, proceed. If you are to examine the plates at the next lab session, your instructor will remove the plates and refrigerate them until then. The lower temperatures will greatly slow the growth of the bacteria and allow you to see the result of the exercise.

Figure 20.3 Examination of incubated plates.

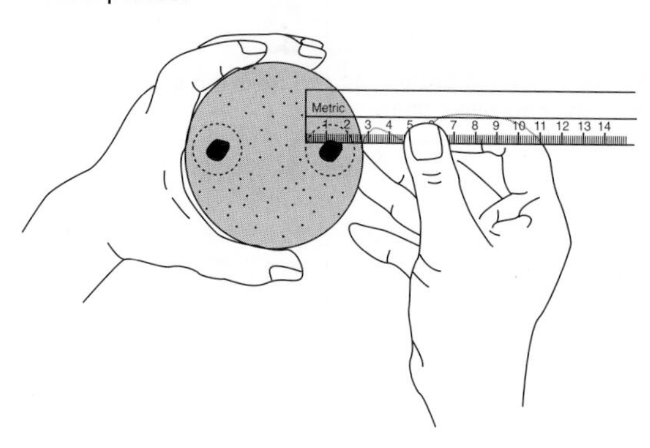

After 24–48 hours incubation, the clear zone of bacterial death is measured from the edge of the disk to the edge of the lawn.

Procedure II—The Gram Stain Technique

Because bacteria are so small, it is necessary to use special techniques to view them through the microscope. First, the bacteria must be attached to a slide by a technique known as a smear. The technique ensures that the bacteria will be killed and fastened to the surface of the slide, so that they will not be washed off during staining. The bacteria are stained to make them more easily seen. The staining technique we will use is the Gram stain. Once the slide is stained, the bacteria can be viewed through an oil-immersion objective lens of a microscope. Follow the directions to prepare a stained slide of the bacteria you are to view.

Preparation of a Smear Slide (Figure 20.4)

The following procedure should be performed twice—once for each of the two species of bacteria you are using in this lab.

1. Gently heat one side of a clean slide in the flame of a Bunsen burner to remove any grease present. To avoid recontamination with grease, handle the slide by grasping it at the edges.
2. Place the slide on the lab table with the flamed side up.
3. Gently roll the tube containing the bacterial culture between your hands or gently shake the tube from side to side to put the organisms in suspension. ***Do not moisten the stopper at the top of the tube with culture.***
4. Heat the inoculation loop red hot to destroy any organisms on the wire.
5. Remove the stopper from the stock culture by grasping it with your small finger. Flame the top of the tube.
6. *After the loop has cooled for at least 5 seconds,* remove a loopful of organisms. *Do not touch the sides of the tube.*
7. Again, flame the top of the tube.
8. Replace the stopper, and set the culture in a test tube rack.
9. Touch the loop to the slide. Use the inoculating loop to spread the bacterial suspension over an area about the size of a dime.
10. Again, flame the inoculating loop before setting it down on the lab table. Make it a rule to always flame the inoculating loop immediately before and after use.
11. After the slide has dried, fix the culture by quickly passing the slide through the flame five or six times. This prevents the film from being washed off during the staining process and kills the bacteria. The smear is now ready for staining after the slide has cooled.

Figure 20.4 Smear slide preparation routine.

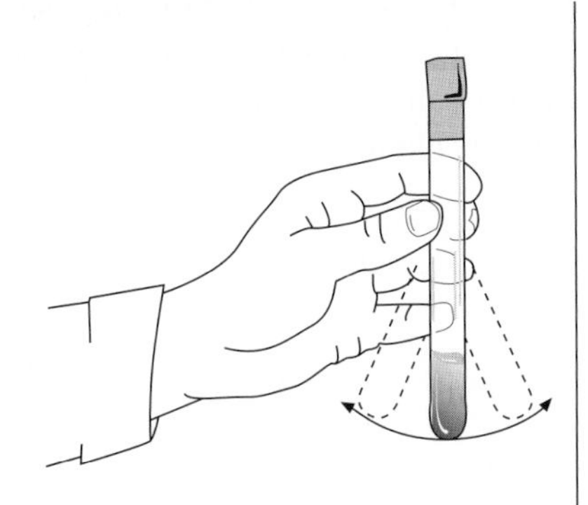

1

Shake the tube from side to side to put the organisms in suspension. Do not moisten the cap with culture.

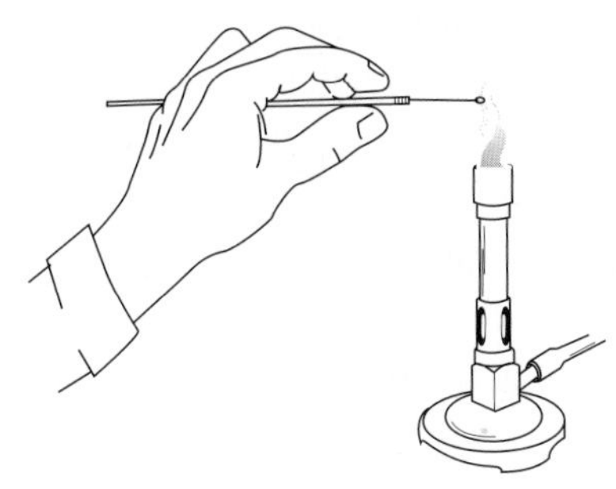

2

Flame the entire loop and wire until it becomes red hot. Flame the handle slightly also.

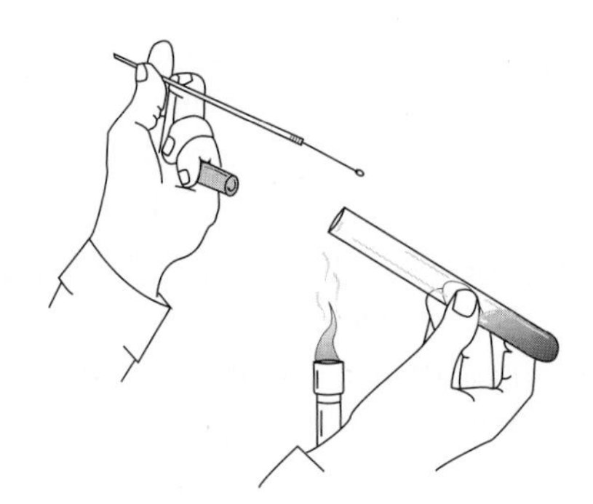

3

Remove the cap and flame the neck of the tube. Don't contaminate the cap by placing it on the table.

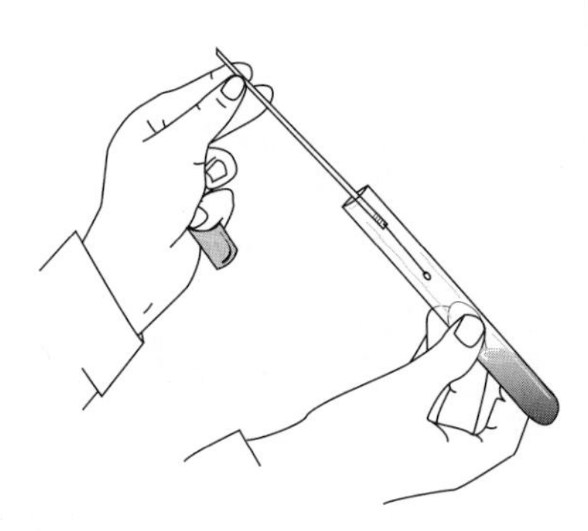

4

Remove a loopful of organisms after the loop has cooled for at least 5 seconds. Do not touch the sides of the tube.

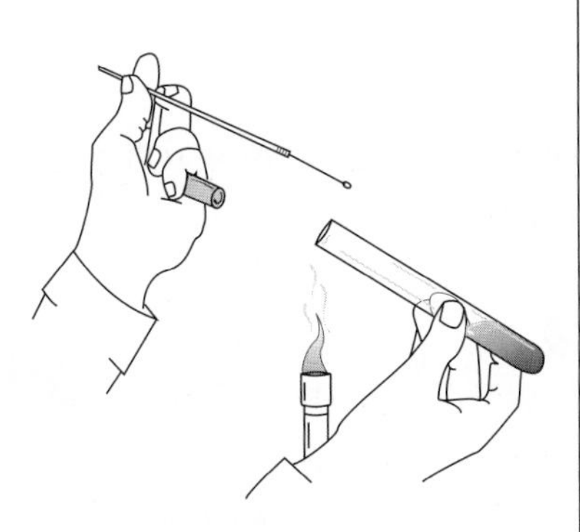

5

Flame the tube again.

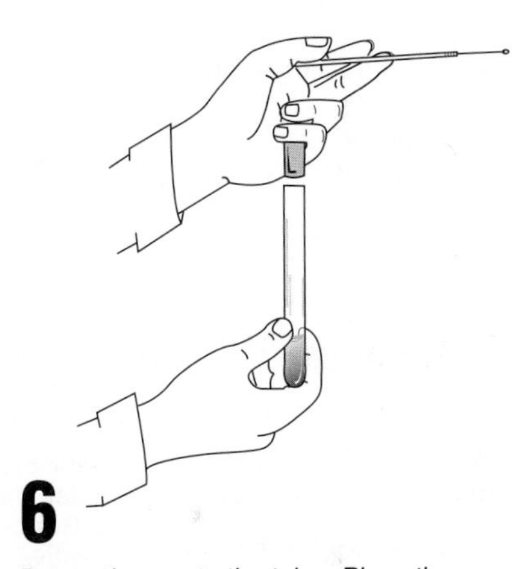

6

Return the cap to the tube. Place the tube in a rack.

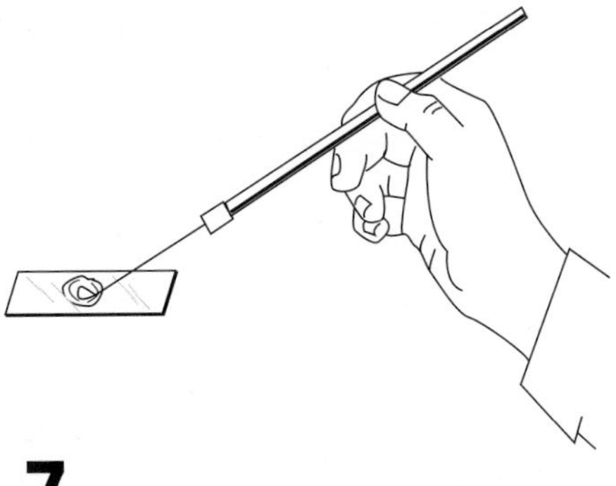

7

Place the loopful of organisms on the center of the clean slide.

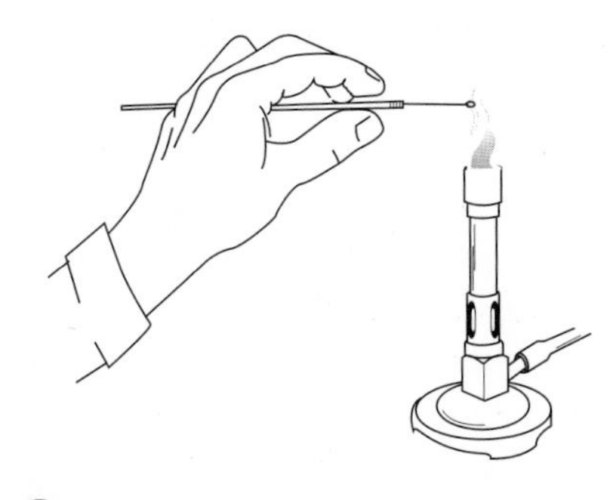

8

Flame the loop before removing another loopful from the culture or before placing it down.

Gram Stain Procedure (Figure 20.5)

The **Gram stain** is probably the most important differential stain technique used by bacteriologists. The procedure separates bacteria into two groups, depending upon whether the original stain is retained or lost when the stained smear is treated with an iodine solution and then washed in alcohol. Organisms that retain the stain are termed *Gram-positive;* those that lose the original stain are called *Gram-negative.* The following procedure should be performed twice—once for each of the two species of bacteria.

1. Cover the smear with crystal violet stain and allow it to remain on the smear for 30 seconds.
2. Pour off the excess stain and wash the slide in tap water.
3. Apply Gram's iodine solution and allow it to remain for 60 seconds.
4. Wash the slide in tap water.
5. Wash with ethyl alcohol for 20 seconds.
6. Wash the slide in tap water.
7. Counterstain with red safranine for 30 seconds.
8. Again, wash the slide in tap water and let it air dry.
9. The stained smear is now ready to view, with a microscope. When you have the bacteria in sharp focus, ask your instructor to check it.

Figure 20.5 Gram stain.

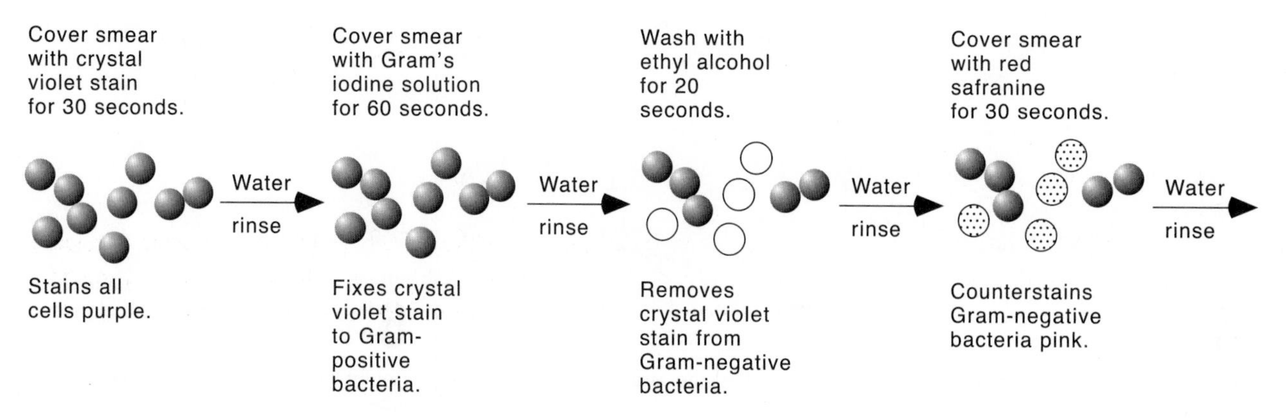

20 Bacterial Selection

Name ____________________ Lab section ____________

Your instructor may collect these end-of-exercise questions. If so, please fill in your name and lab section.

Data Table for Effectiveness of Antibiotics

Size of Zone of Bacterial Death in Millimeters		
Species of Bacteria	**Penicillin**	**Aureomycin**
1.		
2.		

End-of-Exercise Questions

1. Which of the two antibiotics is the best inhibitor of growth of bacterial species 1? Of bacterial species 2? How do you determine this?

2. All areas of your Petri plate were inoculated with bacteria, but they were prevented from growing in the clear area around the antibiotic disk. You probably have some patches of bacteria growing in the clear area around the antibiotic disks in some cases. How are these bacteria different from other members of the same species of bacteria?

3. If you do not have any patches of bacteria growing in the clear areas around the antibiotic disk and your neighbor does, could the bacteria on his or her plate grow in your clear areas? Why or why not?

4. If the bacteria growing in the clear zone were used in a repeat of this whole experiment, what do you think the result would be?

5. What is the selective agent in this experiment? Explain how this experiment demonstrates that natural selection can select for bacteria with specific characteristics and result in a change in the frequency of a specific gene in the population of a bacterial species.

6. It is clear that, when certain antibiotics are introduced into the food of cattle, chickens, and other livestock, the animals grow faster. Because the cost of the antibiotics is low, the use of antibiotics is an easy way to achieve faster weight gain in animals. Explain how this practice could affect the probability of developing populations of antibiotic-resistant bacteria that cause human disease.

7. Tuberculosis is caused by the bacterium *Mycobacterium tuberculosis*. The incidence of the disease is increasing throughout the world, primarily because many strains of the bacterium are resistant to the commonly used antibiotics and some strains are resistant to more than one antibiotic. The World Health Organization recommends that patients be treated in the following manner.
   a. The patients should be directly observed to take their antibiotics for a period of 6–8 months.
   b. More than one antibiotic must be used.

   Explain why each of these two rules is important in controlling drug-resistant strains of tuberculosis.

8. Colds and flu are caused by viruses. In general, viruses are not affected by antibiotics. At one time, physicians routinely prescribed antibiotics for patients with these diseases in order to control possible bacterial infections that might follow the viral infections. For example, bacterial pneumonia is often a secondary infection that follows viral diseases, such as the flu. Most physicians are discontinuing the practice of prescribing antibiotics for flu patients. How is this change related to the problem of antibiotic drug resistance in bacteria?

9. Which of the two organisms is Gram-positive? Which is Gram-negative?

LABORATORY

The Effect of Abiotic Factors on Habitat Preference

21

+ Safety Box

- Handle hypodermic syringes with care. Do not squirt their contents anywhere. They are not to leave the lab.
- Heat lamp bulbs are designed to get very hot. Be careful when moving them.

Objectives

Be able to do the following:

1. Define these terms in writing.

   abiotic factors

   habitat

   control

   variable

   gradient

2. Demonstrate how to conduct an experiment designed to determine the kind of habitat an organism prefers.
3. Collect and interpret data to determine which environmental variables are significant.
4. Describe the function of a control in an experimental procedure.

Introduction

The individuals within a population are able to detect and respond to certain features of their environment. Many characteristics of a **habitat** (the space an organism inhabits) are **variable** from time to time or at different locations within the habitat. **Abiotic factors** are physical factors, such as *temperature,* quantity of *light, gravity,* and *pH.* These often vary in aquatic habitats. When a specific environmental factor varies continuously over a distance, a **gradient** exists. Light intensities can range from absolute darkness to extreme brightness. A shady spot may be a few degrees cooler than a position in direct sunlight only a few meters away. The pH of a lake or stream may also vary from place to place. When a gradient exists, it is possible for an animal to detect when the stimulus is getting stronger and either move toward or away from the stimulus. For example, if you hear a sound and want to go toward the sound, you can walk in a particular direction. If the sound is getting louder, you continue walking in that direction. However, if the sound is getting fainter, you change your direction until it gets louder. In this manner, you can follow a sound gradient to its source. It seems logical to expect that certain abiotic conditions would be more suitable for an organism to thrive and that organisms would migrate to places where the abiotic conditions were most favorable.

If we want to determine the significance of a specific variable, we need to isolate it from other variables. Then, we can present a population of organisms with a gradient for only that one environmental factor and allow the organisms to choose where along the gradient they prefer to be. If the organisms collect at certain positions along the gradient, we can see that the particular variable is significant to the organism.

In this exercise, you will apply the scientific method. You will test the hypothesis that animals will respond to environmental gradients by congregating at specific positions along the gradient. You will set up an experimental situation and carefully and accurately collect data. You will then analyze the data you collect to determine which environmental variables are significant to the organism and where along the gradient they prefer to be.

Preview

During this exercise, you will work in an assigned group. Each group works with one variable, such as light, pH, gravity, or temperature, and determines how the organism (brine shrimp) responds. Each group

1. places brine shrimp in the test apparatus (plastic tubing or trough).
2. adjusts the apparatus to establish the specific environmental gradient assigned to it.
3. allows the brine shrimp sufficient time to move to their preferred position along the gradient.
4. collects data concerning population density at five positions along the gradient.
5. reports its data to the class.
6. records data collected by other groups.
7. interprets all the data reported.

Experimental Design and Data Collection

Although it may appear simple to count the organisms present at each point along the gradient, several problems may cause inaccurate counts. Also, the number of organisms will be very large. Therefore, it may be desirable to count a random sample of the organisms from each of the five positions along the gradient. Therefore, your instructor will discuss possible ways to collect the samples and count the number of organisms in each sample. The following items need to be considered.

1. How should the organisms be removed from the apparatus, so that one sample is totally isolated from others?
2. Should you count every individual, or should you sample your populations from each section?
3. How do you make sure that you are counting only living organisms?

Procedure—Setting Up the Experiment

1. The class will be divided into five groups: group 1 (control), group 2 (pH), group 3 (temperature), group 4 (light), and group 5 (gravity).
2. Each group needs to obtain its test apparatus, which will be either a specific piece of tubing or a plastic trough.
3. Fill the test apparatus with brine shrimp from a well-mixed culture.
4. Adjust your specific variable as described in the following section and allow the container to remain undisturbed for 30 minutes.

Specific Instructions for Each Group

Group 1 (Control)

The **control** group should have no gradients from one end to the other. Other groups will compare their data to yours. You want to make sure that all regions of the test apparatus have exactly the same conditions. The apparatus should be horizontal, have no access to light, no difference in temperature along its length, and no difference in pH from one end to the other. Leave the container undisturbed for 30 minutes; then collect your data by the method agreed to at the beginning of the lab.

Group 2 (pH)

You are working with one variable, pH. You want to establish a pH gradient along the length of your test apparatus. Your apparatus should be horizontal, not have access to light, and be the same temperature from one end to the other. To establish the pH gradient, use a hypodermic syringe to slowly inject 0.5 mL of 1% HCl into one end of the container. Next, use a different syringe to inject 1 mL of 1% KOH into the other end. *Be careful when removing the needle from the tube. A small amount of liquid may spray out.* Leave the container undisturbed for 30 minutes; then collect your data by the method agreed to at the beginning of the lab. You will also need to determine the pH of each of the five samples you collect.

Group 3 (Temperature)

You are working with one variable, temperature. You want to establish a temperature gradient along the length of your test apparatus. Your apparatus should be horizontal, not have access to light, and be the same pH from one end to the other. To establish the temperature gradient, cover the left end of the container with a plastic bag of crushed ice and place an infrared heat lamp 30 cm above the other end. Leave the container undisturbed for 30 minutes; then collect your data by the method agreed to at the beginning of the lab. You will also need to record the temperature of each of the five samples you collect.

Group 4 (Light)

You are working with one variable, light. You want to establish a light gradient along the length of your test apparatus. Your apparatus should be horizontal and have the same temperature and pH from one end to the other. To establish the light gradient, place a source of light at one end of the apparatus. The apparatus will need to be shielded from the lights from the room, so that the only source of light available to the brine shrimp is coming from one end of the apparatus. The source of light should either be a fluorescent lamp or be placed far enough away from the apparatus that it does not heat up one end of the apparatus and accidently set up a temperature gradient. Leave the container undisturbed for 30 minutes; then collect your data by the method agreed to at the beginning of the lab. You will also need to record the distance each of the five samples is from the source of light.

Group 5 (Gravity)

You are working with one variable, gravity. You want to establish a gravity gradient along the length of your test apparatus. Your apparatus should not have access to light and should be the same pH and temperature from one end to the other. To establish the gravity gradient, position the apparatus so that one end is much higher than the other. Leave the apparatus undisturbed for 30 minutes; then collect your data by the method agreed to at the beginning of the lab. You will also need to record the elevation of each of the five samples you collect.

Data Gathering

After allowing 30 minutes for your brine shrimp to respond to the environmental gradient you established, do the following:

1. Divide the apparatus into five equal sections, so that organisms are unable to swim from one section to the next. Section your apparatus as follows.

Section I	Section II	Section III	Section IV	Section V

0 cm 20 cm 40 cm 60 cm 80 cm 100 cm

Tube Gradient Sections

2. Empty the contents of each section into a separate beaker.
3. Label your beakers so that you can identify which beaker came from each section of the apparatus.
4. Record the pH and temperature of each beaker.
5. Measure the volume of water in each of the beakers.
6. Count the number of individuals in each sample by the method agreed to at the beginning of the lab.
7. Divide the number of brine shrimp in your sample by the number of milliliters of water in that portion of the apparatus. This will give you the number of organisms per mL of water.

$$\text{brine shrimp per mL} = \frac{\text{number of brine shrimp counted in the section}}{\text{number of mL of water in the section}}$$

8. Report your data and record data from all other groups in table 21.1.
9. Interpret the data collected. Do brine shrimp respond to light, temperature, pH, or gravity? How do you know? Describe the preferred habitat of brine shrimp.

Table 21.1 Data Sheet

Team	Section of the Apparatus I	II	III	IV	V
1. Control	**Left End** pH ____ Temp. ____ mL H_2O ____ Organisms counted ____ org/mL ____	pH ____ Temp. ____ mL H_2O ____ Organisms counted ____ org/mL ____	pH ____ Temp. ____ mL H_2O ____ Organisms counted ____ org/mL ____	pH ____ Temp. ____ mL H_2O ____ Organisms counted ____ org/mL ____	**Right End** pH ____ Temp. ____ mL H_2O ____ Organisms counted ____ org/mL ____
2. pH	**Acid** pH ____ Temp. ____ mL H_2O ____ Organisms counted ____ org/mL ____	pH ____ Temp. ____ mL H_2O ____ Organisms counted ____ org/mL ____	pH ____ Temp. ____ mL H_2O ____ Organisms counted ____ org/mL ____	pH ____ Temp. ____ mL H_2O ____ Organisms counted ____ org/mL ____	**Base** pH ____ Temp. ____ mL H_2O ____ Organisms counted ____ org/mL ____
3. Temperature	**Cold** pH ____ Temp. ____ mL H_2O ____ Organisms counted ____ org/mL ____	pH ____ Temp. ____ mL H_2O ____ Organisms counted ____ org/mL ____	pH ____ Temp. ____ mL H_2O ____ Organisms counted ____ org/mL ____	pH ____ Temp. ____ mL H_2O ____ Organisms counted ____ org/mL ____	**Hot** pH ____ Temp. ____ mL H_2O ____ Organisms counted ____ org/mL ____
4. Light	**Dark** pH ____ Temp. ____ mL H_2O ____ Organisms counted ____ org/mL ____	pH ____ Temp. ____ mL H_2O ____ Organisms counted ____ org/mL ____	pH ____ Temp. ____ mL H_2O ____ Organisms counted ____ org/mL ____	pH ____ Temp. ____ mL H_2O ____ Organisms counted ____ org/mL ____	**Light** pH ____ Temp. ____ mL H_2O ____ Organisms counted ____ org/mL ____
5. Gravity	**Bottom** pH ____ Temp. ____ mL H_2O ____ Organisms counted ____ org/mL ____	pH ____ Temp. ____ mL H_2O ____ Organisms counted ____ org/mL ____	pH ____ Temp. ____ mL H_2O ____ Organisms counted ____ org/mL ____	pH ____ Temp. ____ mL H_2O ____ Organisms counted ____ org/mL ____	**Top** pH ____ Temp. ____ mL H_2O ____ Organisms counted ____ org/mL ____

Calculate the number of organisms per milliliter as follows:

$$\text{organisms/mL} = \frac{\text{organisms counted in the sample}}{\text{total milliliters in the section of the apparatus}}$$

21 The Effect of Abiotic Factors on Habitat Preference

Name ______________________________ Lab section ____________

Your instructor may collect these end-of-exercise questions. If so, please fill in your name and lab section.

End-of-Exercise Questions

1. What is the purpose of the control?

2. Why is it necessary to have large numbers of organisms in your sample?

3. Were the brine shrimp equally distributed in the five sections of the apparatus for the control group at the completion of the experiment? Should they have been the same? Explain why or why not.

4. Use the chi-square test to determine if the number of organisms in the five sections of the control are significantly different. Use the same test of statistical significance to evaluate the other four sets of data. (See appendix A for a discussion of the chi-square test.)

	O **Observed**	E **Expected**	**(O – E)**	**$(O-E)^2$**	**$\frac{(O-E)^2}{E}$**
Control					
pH					
Temperature					
Light					
Gravity					

Chi-square value = ____________________

Degrees of freedom = ____________________

Probability = ____________________

5. How would you modify your procedures if you were to repeat this exercise?

6. Using your data, describe how you think brine shrimp respond to the following:

pH ____________________

Temperature ____________________

Light ____________________

Gravity ____________________

LABORATORY

Successional Changes in Vegetation

22

+ Safety Box

- Because you will be outdoors, in the field, while you are collecting data, dress appropriately for the terrain and the weather.

Objectives

Be able to do the following:

1. Define these terms in writing.

   succession
   ecosystem
   pioneer plants
   niche
   climax community

2. List five differences between an early successional stage (old field) and a climax community (temperate deciduous forest).
3. Describe five changes that occur to the plant life as an area changes from an early successional stage to a climax community.
4. State several ways in which the plant life of an area influences the animal life of that area.

Introduction

The analysis of communities of organisms is a rather difficult task. One should be able to identify all the different kinds of plants, animals, and microbes, describe the relationships among different organisms, and measure the sizes of populations. We are not that experienced, so we will look only at plants, use artificial designations to identify types of plants, and not worry about the exact names. In this exercise, we need to estimate the size of some populations but not try to characterize all the kinds of organism interactions.

The various plant and animal populations interacting in an area receive their energy from the Sun and constitute an **ecosystem.** Each organism has a **niche**—a job to perform—in the ecosystem. **Succession** is a predictable pattern of change in the species that occurs within an ecosystem as it proceeds from a pioneer community to a climax community. **Pioneer plants** are the first to become established in a barren area. These plants shade the soil and lower the soil temperature, reduce the wind velocity at the soil surface, add organic material to the soil, and support various types of animals. Thus, the presence of the pioneer plants changes the environment of an area. A second group of plant types can then invade the area. These new varieties of plants will increase their populations and replace the pioneer species. The new varieties, in turn, also change the environment of the area. A third group of plant types will now colonize the area and grow and replace the second group of plants. These temporary communities are called successional communities. This constant change in the environment and the resulting changes in the plant and animal populations present in an area eventually result in a plant community that is able to maintain its species; new types will not invade the area. This last stage of succession is the climax stage. It may require several hundred years to undergo succession from pioneer plants to the **climax community.** Each region of the world had a climax community typical for the locality. Your instructor will discuss the specific kinds of communities characteristic of your region of the world and the typical successional stages that lead to the climax community.

Preview

Your instructor will provide you with additional information regarding the various stages of succession found in your area.

During this lab exercise, you will examine at least two stages of succession and characterize each by

1. determining the types of small plants present and their approximate numbers.
2. determining the types of larger plants present and their approximate numbers.
3. determining the physical environmental conditions of an area.
4. determining the height of the dominant vegetation.
5. interpreting the data gathered in the field.

Procedure

Your instructor will have stretched out 100 m of string in a line so that approximately 50 m is in one stage of succession and 50 m is in a different stage of succession.

This string has a knot every 10 m. These knots denote places where you will collect information (sampling stations). There are 11 sampling stations.

The class is divided into groups. Each group has a specific job to do. The groups collect data and record them on the data sheet provided (see table 22.1 on page 227).

Group 1 (Small Herbaceous Plants)

1. You need the following equipment:
   a. Wire hoop
   b. Meterstick
   c. Data sheet
   d. Pencil
2. At each of the knots on the string, place the hoop on the ground and count the number of plants in each of the following categories:
   a. Narrow-leaved plants (grasses and similar plants)
   b. Herbs under 50 cm tall (broad-leafed, not woody)
   c. Herbs over 50 cm tall (broad-leafed, not woody)
3. If there are a lot of plants, try to make an accurate estimate.
4. Record this information on your data sheet.

Group 2 (Large, Woody Plants)

1. You need the following equipment:
   a. String, 2 m long with clamp on the end
   b. Pencil
   c. Data sheet
   d. Meterstick
2. At each knot on the string, attach your 2-m length of string. Travel in a circle around the knot, and count all the woody plants.
   a. Shrubs—if the plant is divided into two or more major stems below 1 m, record it as a shrub.
   b. Trees—if a plant does not divide into two or more branches or does so above 1 m, record it as a tree.
   c. If only a portion of a tree or shrub falls inside the circle, record the portion that does (i.e., 1/2, 1/8).
3. Record all data on the data sheet provided.

Group 3 (Estimate Average Height of Dominant Vegetation)

1. You need a meterstick.
2. At each sampling station, determine the height of the tallest plant within 2 m of the sampling station. If there are tall trees, you may need to determine the height by the method shown in figure 22.1.
3. Record data on the data sheet provided.

Figure 22.1 Tree height.

To estimate the height of a tree, measure the distance from your eye to your hand held at eye level. Form a right triangle by holding the meterstick vertically with your hand at eye level. The distance from eye to hand and the distance the meterstick protrudes above your hand must be equal. Position yourself so that you can sight over the top of the meterstick to the top of the tree. The height of the tree is equal to the distance from where you are standing to the base of the tree plus the distance from the ground to your eye.

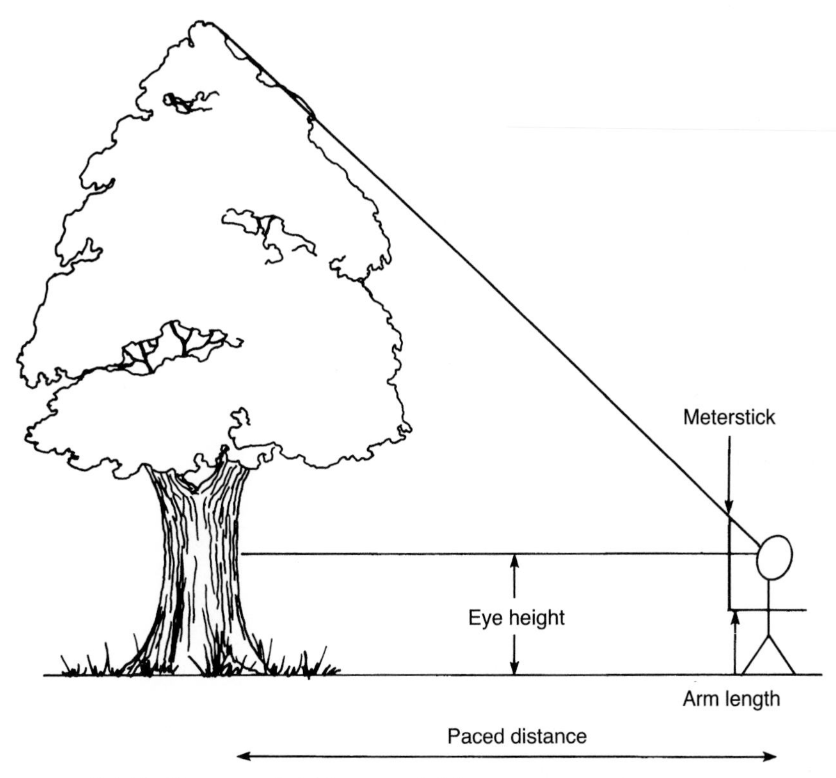

Group 4 (Physical Conditions)

1. You need the following equipment:
   a. Two soil thermometers
   b. Two regular thermometers
   c. Wind speed indicator
   d. Light meter
   e. Relative humidity apparatus
2. At each of the sampling stations, record the following information:
   a. Soil temperature (you must leave the thermometer in the ground for 5 minutes)
   b. Air temperature (hold the thermometer in such a way that it does not get wet, the Sun doesn't shine directly on it, and the wind doesn't blow directly on it)
   c. Wind speed (hold wind speed indicator 1 m above the ground; if the wind is variable, measure peak gusts)
   d. Light reading (hold light meter 1 m above the ground and always have it oriented in the same direction)
   e. Relative humidity (your instructor will explain how to use the apparatus)
3. Record all data on the data sheet.

22 Successional Changes in Vegetation

Name ______________________________ Lab section ____________

Your instructor may collect these end-of-exercise questions. If so, please fill in your name and lab section.

Table 22.1 **Data Sheet**

	Station 1	Station 2	Station 3	Station 4	Station 5	Station 6	Station 7	Station 8	Station 9	Station 10	Station 11
Narrow-leaf plants (e.g., grass)											
Herbs under 50 cm tall											
Herbs over 50 cm tall											
Shrubs (branches at 1 m or less)											
Trees (branches above 1 m)											
Average height of dominant vegetation											
Air temperature											
Soil temperature											
Wind speed											
Light meter reading											
Relative humidity											

End-of-Exercise Questions

1. Use the information gathered to describe differences in the physical environment of the two communities you visited.

2. Was there a difference in stages of succession in the various stations where you collected data? At which station was succession at its earliest? At which station was it closest to a climax community?

3. What human influences did you notice that have interfered with the normal process of succession? What is their effect?

4. Which station do you think showed the greatest complexity and variety of organisms? Why do you think this is true?

LABORATORY

Behavioral Differences in Small Mammals

23

+ Safety Box

- The animals may bite if they are not properly handled. Choose a person confident in handling animals to place them onto the arena. You may want to wear gloves.

Objectives

Be able to do the following:

1. Construct a hypothesis relating to the behavior of small mammals toward objects in their environment.
2. Test the hypothesis by using an arena that has walls as physical objects.

Introduction

Most animals make use of objects in their environment. Birds use twigs and grass to build nests, small mammals build burrows in grass and earth, birds of prey use trees and electrical lines as observation places, and many insects hide under leaves and bark. The use of physical objects for cover by small mammals is common but must be coupled with the typical life history of the animal. *Diurnal* (active during the day) animals have a greater chance of being observed by predators than do *nocturnal* (active at night) animals. Small forest or prairie mammals can conceal themselves more easily as they go about their normal activities than can desert animals that live in areas with little grass or other vegetation to conceal them. Does this mean that small desert mammals behave differently than other small mammals?

The common laboratory mouse has been used for years as a research animal. It has been bred to do well in laboratory conditions, to show little genetic variability, and to be easily handled by people. Does this mean that laboratory mice have behavior patterns that differ from those of other mice of the same species? Would laboratory mice respond to physical objects in their environment differently than wild or free-living mice?

Preview

During this lab exercise, you will

1. form a hypothesis relating to the behavior of small mammals toward their environment.
2. use an arena with walls to test the behavior of one or more species of small mammals.
3. use the chi-square test to determine if the results of the test are significant.

Procedure

Your instructor will probably divide you into groups of about three persons. One person needs to be a timekeeper, one person needs to be a recordkeeper, and another person needs to be an animal handler.

Your instructor will give you two kinds of small mammals. They may be different strains of the same species or they may be of different species. Your instructor will also provide some background information on the animals: what kinds of ecological niches they occupy in nature, whether they are wild or laboratory animals, and so forth.

Notes

Species 1:	Species 2:

From the information you are given, construct a hypothesis concerning how these two kinds of small mammals differ in the amount of time they spend adjacent to objects. Write your hypothesis in the space provided.

Hypothesis:

Experimental Design

To test your hypothesis, we use an arena that consists of a flat sheet of material with 36 squares drawn on it. The 4 squares in the middle are occupied by boxes or are walled off from the two rows of squares surrounding them. A diagram of this arena is shown in figure 23.1. There are no walls at the periphery of the arena. Therefore, the arena should be elevated, so that the small mammals do not wander off the arena. If one of them does jump off, gently capture the animal and return it to the same square from which it left the arena.

Because you want to study the behavior of the animals toward vertical surfaces in their environment, you must prevent other kinds of stimuli from influencing their behavior. Dim the lights in the laboratory and refrain from loud noises or sudden movements during the experiment.

Place an animal on the arena in one of the peripheral squares. Every 15 seconds for 10 minutes, record whether the animal is in a peripheral or an interior square on the arena. The timekeeper should quietly say "now" every 15 seconds, so that the recordkeeper knows when to record the location of the animal. Record your data in table 23.1.

Figure 23.1 Diagram of arena to use to test your hypothesis.

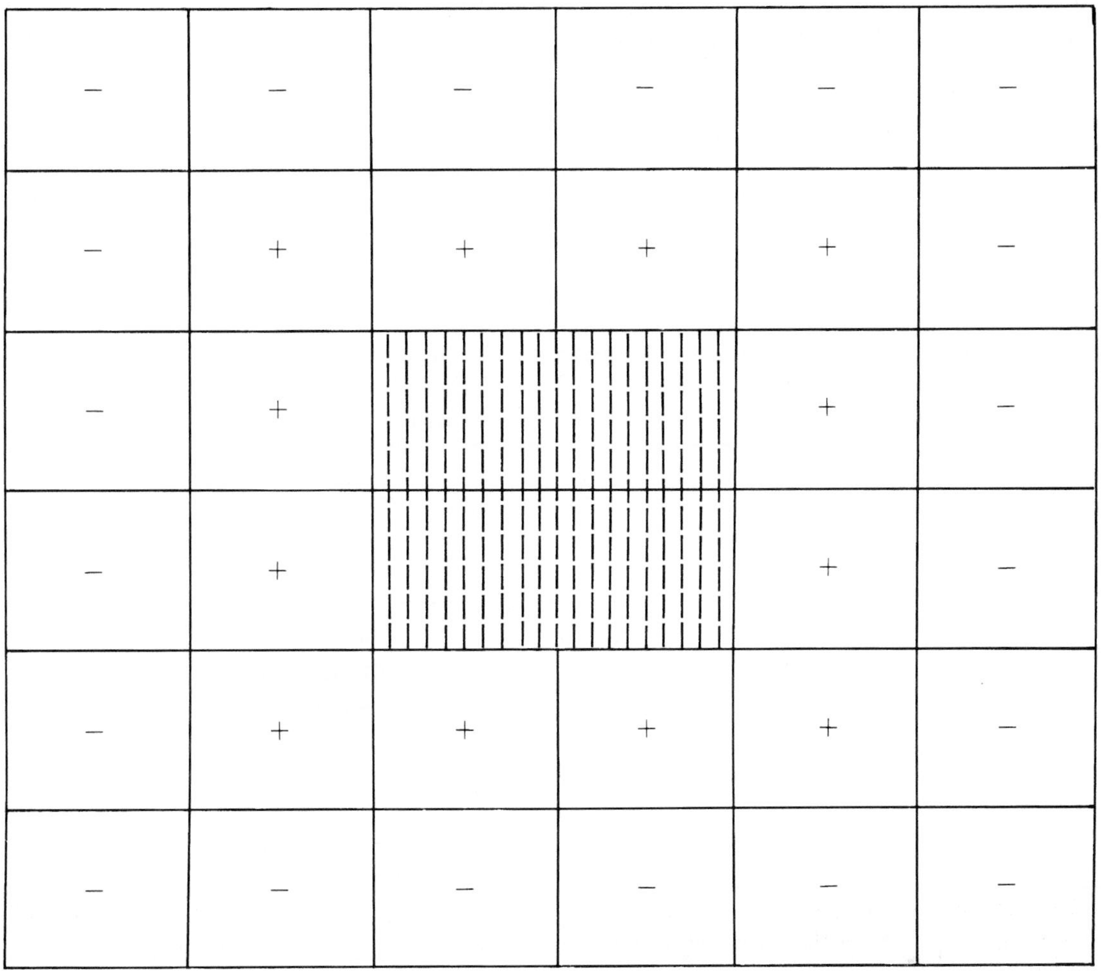

Table 23.1 **Location of Small Mammal**

Species of Small Mammal	Number of Times in Peripheral Squares	Number of Times in Interior Squares

Analysis of Results

You want to pool all the data collected by the members of your class before beginning the statistical analysis of your results. To determine if the hypothesis is supportable or invalid, we can apply the chi-square test of statistical significance (see appendix A). Suppose our hypothesis was that species 1 would move randomly around the arena. After collecting and pooling data from the observations of all groups in the class, your next task is to determine if a statistical test of your hypothesis supports or disproves your hypothesis. If you expect species 1 to move about at random, you can use a statistical test to analyze your data as follows (see table 23.2):

1. The number of peripheral squares (20) divided by the total number of squares (32) indicates that, if the animals moved about the arena randomly, we would expect them to spend 62.5% of their time in the peripheral squares and 37.5% of their time in the interior squares.
2. Therefore, use the following formulas to determine the expected values based on the total number of observations made by the class. Record these numbers in the "expected" column of table 23.2.

$$\text{total observations} \times 0.625 = \text{expected in outside squares}$$

$$\text{total observations} \times 0.375 = \text{expected in interior squares}$$

3. Enter the number of times you observed the animal in the interior and peripheral squares in the "observed" column of table 23.2.
4. Use appendix A to determine the chi-square values.

Comparing the Behavior of the Two Species

You can also use the chi-square test to determine if the two kinds of animals behaved differently than one another. To do this choose one of the kinds of animals observed and make it the "expected" category. If the same number of observations are recorded for each organism, you can use the data from the second species as the "observed" and determine if the two species behaved similarly or not. See table 23.3.

Table 23.2 Chi-Square Analysis for Hypothesis 1—Species 1

Classes	O Observed	E Expected	(O – E)	$(O - E)^2$	$\frac{(O - E)^2}{E}$
Total					

Chi-square value = ________________

Degrees of freedom = ________________

Probability = ________________

Chi-Square Analysis for Hypothesis 1—Species 2

Classes	O Observed	E Expected	(O – E)	$(O - E)^2$	$\frac{(O - E)^2}{E}$
Total					

Chi-square value = ________________

Degrees of freedom = ________________

Probability = ________________

Table 23.3 Chi-Square Analysis for Similar Behavior

Classes	O Observed	E Expected	(O – E)	$(O - E)^2$	$\frac{(O - E)^2}{E}$
Total					

Chi-square value = ________________

Degrees of freedom = ________________

Probability = ________________

23 Behavioral Differences in Small Mammals

Name ____ Lab section ____________

Your instructor may collect these end-of-exercise questions. If so, please fill in your name and lab section.

End-of-Exercise Questions

1. Did either of your animals prefer the interior squares? To answer this question, state the hypothesis you used and refer to the results of your chi-square analysis in your response.

2. Did the two kinds of animals differ significantly in their movement around the experimental arena? State a hypothesis that would allow you to answer this question.

3. Did you need to revise any of your hypotheses? How did you know?

4. Are the results of the experiment consistent with the ecological niche of each organism? Explain.

LABORATORY

24 Plant Life Cycles

+ Safety Box

- No unusual hazards are associated with this laboratory experience. Please follow standard laboratory safety procedures.

Objectives

Be able to do following:

1. Define these terms in writing.

   diploid
   haploid
   sporophyte generation
   gametophyte generation
   gamete
   alternation of generations
   zygote
   heterogamous

2. Describe the *Spirogyra* life cycle.
3. Describe the *Oedogonium* life cycle.
4. Describe the *Polytrichum* life cycle.
5. Describe the fern life cycle.
6. Describe the *Pinus* life cycle.
7. Compare the differences and similarities in the reproduction methods used by the organisms in these exercises.
8. Locate the following on slides of *Spirogyra:*

   chloroplast
   pyrenoid
   nucleus
   conjugation tubes
   zygote
   gamete
   zygospore

9. Locate the following on slides of *Oedogonium:*

   chloroplast
   pyrenoid
   nucleus
   oogonium
   egg
   antheridium
   sperm
   zygote

10. Locate the following on slides or intact moss plants:

    male gametophyte plant
    female gametophyte plant
    sporophyte plant
    antheridium
    sperm
    jacket cells
    archegonium
    egg
    capsule
    spores
    protonema

11. Locate the following on slides of fern gametophyte plants:

    archegonium
    antheridium

12. Locate the sori on the fronds of a fern sporophyte plant.
13. Identify the demonstration specimens illustrating the pine life cycle.

Introduction

All algae and plants that reproduce sexually follow a life cycle similar to the one shown in figure 24.1. Because the life cycles of nearly all algae and plants follow the same generalized life cycle, it is appropriate that we study this generalization. There are two distinctly different stages in the life cycle. One is haploid and the other diploid. **Diploid** *(2n)* cells contain two chromosomes of each kind, one from each parent. Therefore, they have two genes for each characteristic. **Haploid** *(n)* cells only have one chromosome of each kind and therefore only one gene for each characteristic. The haploid stage in the life cycle is called the **gametophyte generation.** The gametophyte generation can divide and grow by mitosis to produce a multicellular structure and produces haploid gametes by *mitosis*. Fertilization occurs when two haploid **gametes** join to form a single diploid cell known as a **zygote.** The zygote can divide by *mitosis* to produce a multicellular diploid **sporophyte generation.** The sporophyte generation also produces haploid spores by the process of *meiosis*. The haploid spores divide by *mitosis* to produce the gametophyte generation. Because in the typical life cycle these two generations alternate with one another, the life cycle shows **alternation of generations.**

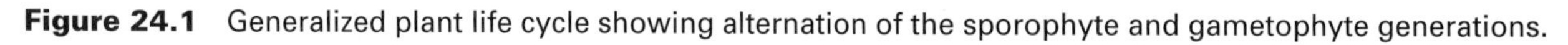

Figure 24.1 Generalized plant life cycle showing alternation of the sporophyte and gametophyte generations.

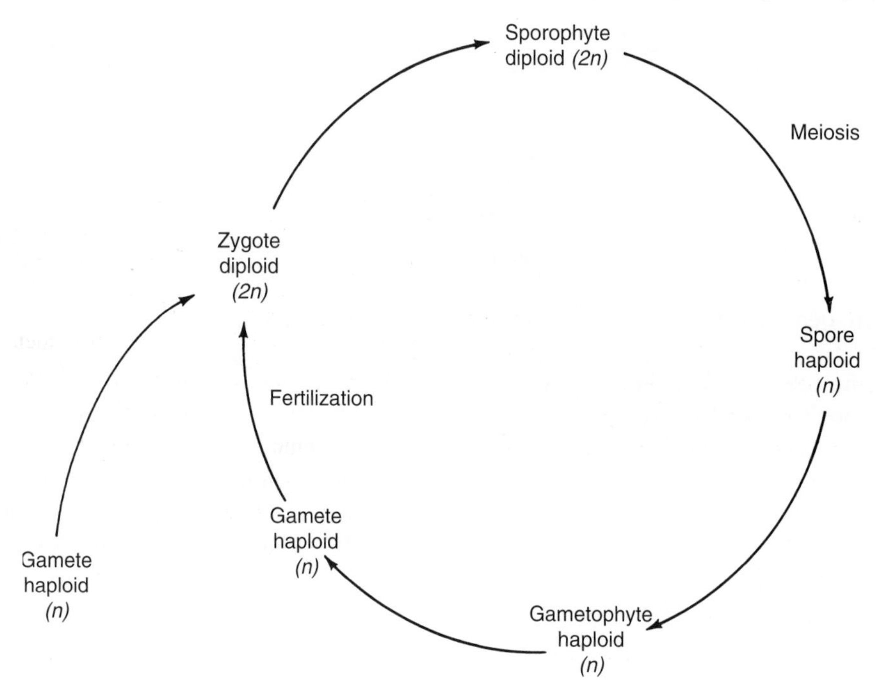

This exercise is constructed so that you can see the similarities and differences in the way an organism fits a generalized life cycle. When you are finished with the exercise, you should be able to answer such questions as "What do the gametophytes of *Spirogyra, Oedogonium, Polytrichum,* fern, and *Pinus* look like?"

Plants are thought to have evolved from the green algae, because the green algae and plants have the same kinds of chlorophylls and share other cellular characteristics. Among the plants, mosses are considered the most primitive, because they lack vascular tissue. The ferns are the next group of plants to develop, followed by the gymnosperms (pines and their relatives) and the angiosperms (flowering plants). As we look at the life cycles of algae, moss, ferns, and gymnosperms, notice the change in the importance of the two generations (gametophyte and sporophyte) in the life cycle.

Preview

To show the changes that take place in the life cycles of organisms, during this lab exercise, you will

1. examine the life cycle of *Spirogyra.*
2. examine the life cycle of *Oedogonium.*
3. examine the life cycle of *Polytrichum.*
4. examine the life cycle of the fern.
5. view the demonstration of the life cycle of *Pinus.*

Procedure

Spirogyra Life Cycle

Obtain a slide of *Spirogyra* and examine it under a microscope. This alga is made up of a number of cells that fit together end to end to form a long filament. Each cell within this filament is a vegetative cell that has the haploid number of chromosomes. Because gametes are also haploid cells, each cell within the filament is capable of acting as a gamete. You will see, inside the cells, a spiral chloroplast with a number of dark spots located along its length. These dark spots, called *pyrenoids,* are starch concentrations. If you look carefully, you can also see the nucleus located in the center of the cell.

What is the three-dimensional structure of a typical *Spirogyra* cell?

The filament you are looking at is the gametophyte generation of *Spirogyra.* Note that some of the cells have short tubes protruding at right angles to the long axis of the cell. These are called *conjugation tubes.* During sexual reproduction, two filaments come to lie side by side, and conjugation tubes are formed between cells of the two filaments. The protoplasm of one of the cells then flows through the conjugation tube into the other cell. This results in one cell being left empty and the other having the cytoplasm and nuclear material of two cells; consequently, this new cell is a diploid zygote. Fertilization of this type is called conjugation. Locate conjugation tubes, gametes, and zygotes on your slide. If you are not sure of the identity of any of the structures that you see, ask your instructor for assistance.

After the gametes have united, the zygote forms a structure resembling a football and develops a heavy wall around itself while it is still inside the cell wall of the previous gamete. When it takes this shape, the zygote is called a *zygospore.* This zygospore divides by meiosis, such that four haploid cells are produced from the single diploid cell. Three of these haploid cells disintegrate, leaving a single haploid cell to form a new filament. Therefore, as a result of meiosis, we return to the haploid gametophyte generation. Locate a zygospore on your slide. See figure 24.2 for the life cycle of *Spirogyra* and to review this basic information. Label each part, indicating whether it is haploid or diploid. Indicate the gametophyte and sporophyte generations.

Figure 24.2 *Spirogyra* life cycle.

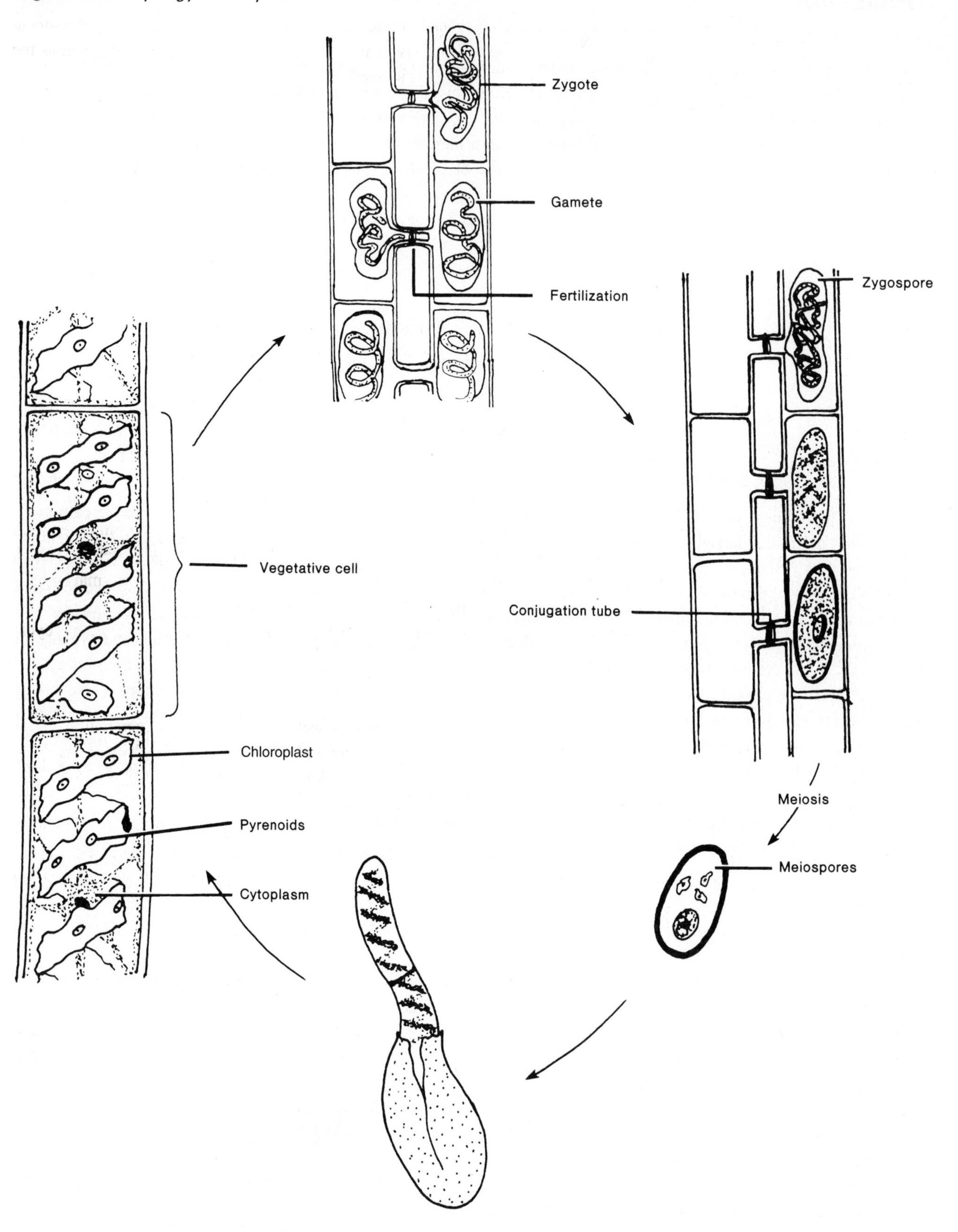

Oedogonium Life Cycle

Obtain a slide of *Oedogonium* and look at it under the compound microscope. Notice that again we have a filament made up of a number of cells as in *Spirogyra;* however, in this case, each cell is longer. These cells are known as the vegetative cells and are haploid. Note the *chloroplast, pyrenoids,* and *nucleus.*

What is the cell like in three dimensions?

By scanning the slide, you should locate some cells that appear to be swollen. These cells are called *oogonia.* Each oogonium contains an egg. Eggs are haploid and develop by modification of the protoplasm of a vegetative cell.

You should also be able to find some short cells about half as long as they are wide. These cells are *antheridia* and contain two male gametes, or sperm. The antheridia are developed from a vegetative cell also. The antheridium releases the sperm, which swim through the water to the egg and fertilize the egg inside the oogonium. Following fertilization, the zygote develops a heavy wall around the outside of it. This is the only diploid part of the life cycle of *Oedogonium.*

This zygote eventually divides by meiosis to give rise to four haploid cells. Each of these cells is capable of giving rise to a new filament by the process of mitosis.

As you review the life cycle of *Oedogonium,* note two major differences from what you saw in *Spirogyra.* First of all, *Oedogonium* is **heterogamous.** This means that you can recognize two different kinds of gametes. Second, note that the cells that produce the gametes are specialized. They are conspicuously different from the normal vegetative cells of the filament. As in *Spirogyra,* the only diploid part of the life cycle is the zygote.

Use figure 24.3, the *Oedogonium* life cycle, to label each structure and determine which parts of the life cycle are haploid and diploid.

Moss *(Polytrichum)* Life Cycle

Obtain a small bottle that contains male and female gametophyte plants as well as the sporophyte generation of the moss plant. You will see two leafy plants on the inside of the bottle. The one with the flattened top is the male gametophyte plant. Located in the top of this male gametophyte plant are sex organs known as antheridia. These antheridia are different from the antheridia you saw in *Oedogonium;* they are multicellular.

Obtain a slide of moss antheridia and focus on the slide. Use low power only. The antheridia appear as sausage-shaped structures. Enclosing the large number of sperm is a layer of cells called jacket cells. The antheridia are in different stages of development, so the size and color of the antheridia vary. Some are purple structures; others are bright pink and contain blue dots.

The female gametophyte plant in the small bottle is the one that does not have a flat top. This plant has archegonia present at its tip.

Examine a prepared slide of moss archegonia under the low power of a compound microscope. There are a number of archegonia on the slide. They are cut at a number of different angles; consequently, you may see only bits and pieces of any particular archegonium. The archegonium is flask-shaped; that is, it has a long neck with a swollen base. The long neck of the archegonium forms a tube running to the base of the archegonium, which contains the egg. The egg is produced by mitosis, because the gametophyte plant is already haploid.

When the male and female gametophyte plants are covered by water, either in the form of dew or rain, the sperm has the opportunity to break out of the antheridium and swim through a thin film of water to the archegonium. The sperm then swims down the neck of the archegonium and fertilizes the egg inside the archegonium. This fertilized egg is called a zygote. The zygote is the new diploid sporophyte generation and begins to develop immediately. The sporophyte never leaves the female gametophyte plant.

The third plant in the demonstration set is a female gametophyte plant with the long, thin sporophyte plant growing out of the top of it. Located at the tip of the sporophyte is a capsule that contains haploid spores. These spores were developed from the diploid tissues of the sporophyte by the process of meiosis. These spores give rise to the new gametophyte generation. The spores are shed from the spore capsule, and, if they happen to fall on a suitable substrate, they germinate. When they germinate, they give rise to a filament known as a *protonema.*

Figure 24.3 *Oedogonium* life cycle.

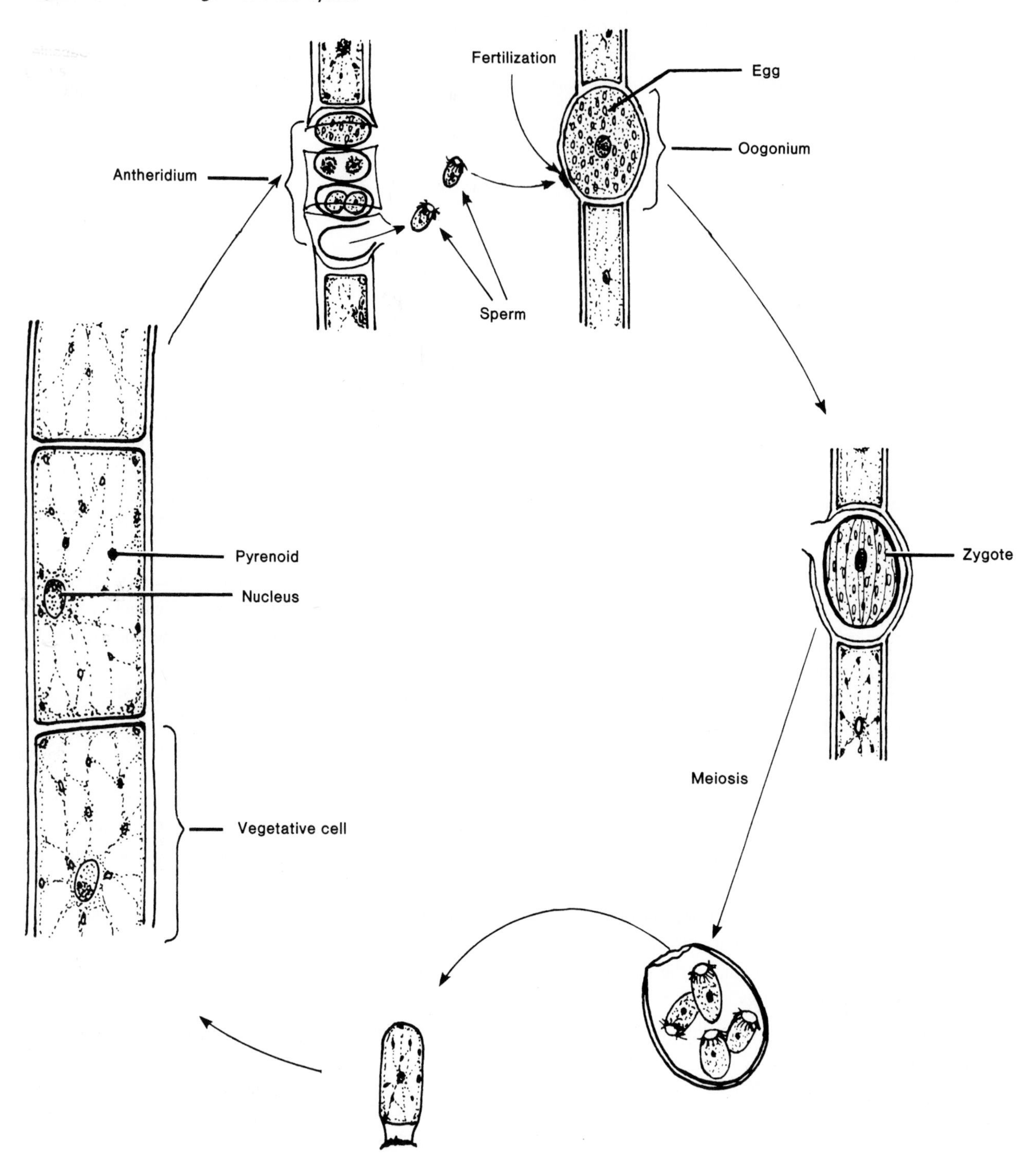

Examine a slide of moss protonema. Note its algaelike characteristics and the spores or germinated spores that you find on the slide. This protonema eventually develops into either a male or a female gametophyte plant. The protonema is part of the gametophyte generation and gives rise to the leafy gametophyte plant.

The moss life cycle is more complex than what you saw in the algae (figure 24.4). First of all, there is a multicellular sporophyte generation that was not present in the life cycle of the two algae that you looked at. Second, there is the development of multicellular sex organs that contain the eggs and sperm.

Some of the significant items at this point are (1) the moss gametophyte generation is the dominant generation and the sporophyte is dependent upon the gametophyte; (2) the moss is a heterogamous organism—in other words, sperm and eggs are present; and (3) the moss is an organism that has two different kinds of sex organs—an archegonium to house the egg and the antheridium to house the sperm.

Figure 24.4 Moss life cycle.

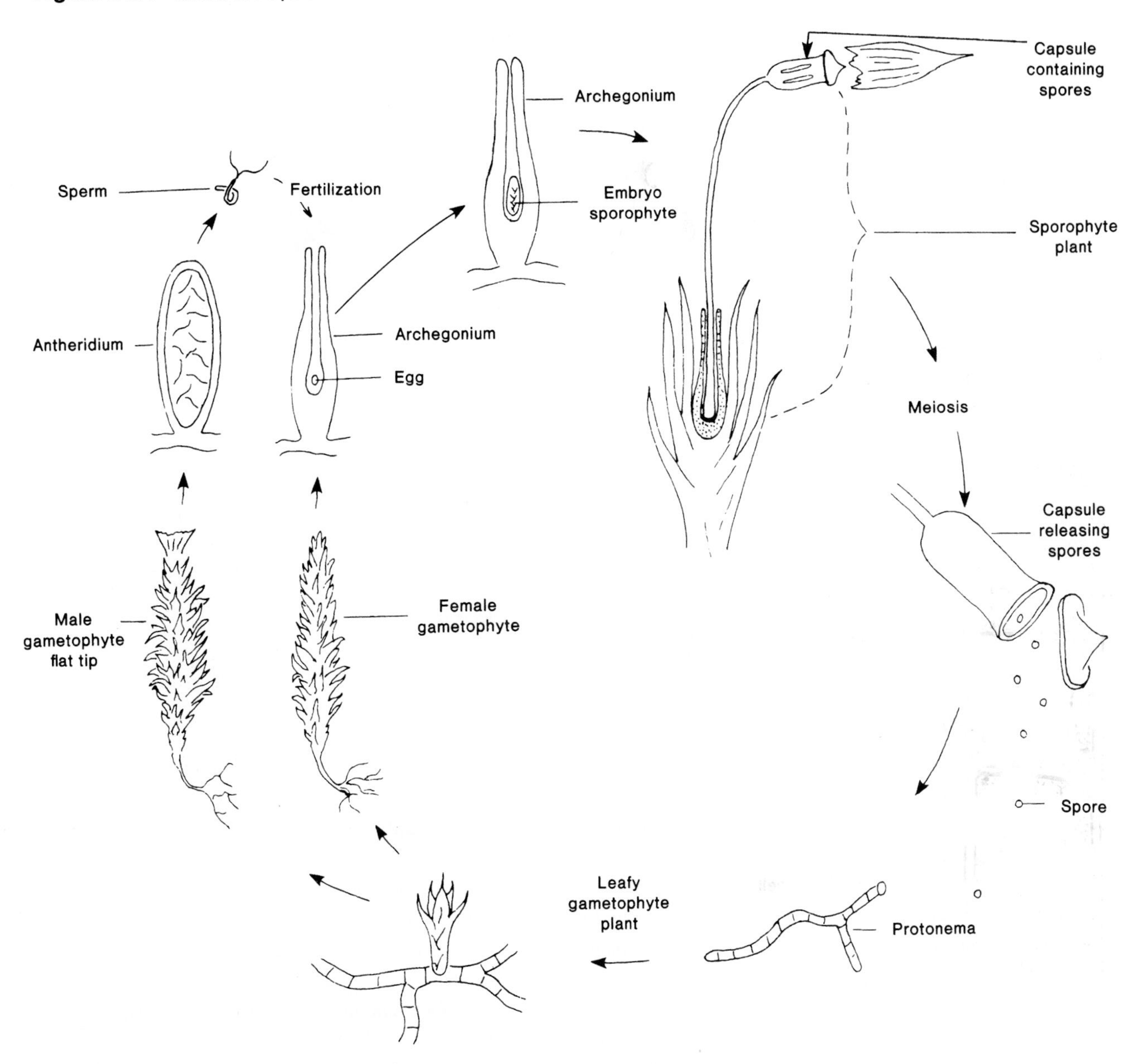

Fern Life Cycle

Examine the examples of fern plants in the lab (figure 24.5). The large plant with the fronds (leaves) is the sporophyte part of the life cycle. Ferns produce spores by meiosis on their leaves.

Examine the fronds of the ferns. The little brown spots on the underside of the leaves are called *sori,* and they contain structures that produce the microscopic spores. The haploid spores fall to the ground and grow into a heart-shaped gametophyte plant. The gametophyte plant is capable of carrying on photosynthesis and contains both *archegonia,* which produce eggs, and *antheridia,* which produce sperm.

Examine the slide of the fern gametophyte plant. The archegonia are located near the cleft of the heart-shaped gametophyte and the antheridia are located near the apex. As with the mosses, the sperm swim to the archegonium to fertilize the egg. Therefore, in order to reproduce, the fern gametophyte plant must be in a moist environment. Once the egg is fertilized, the zygote begins to divide and grow into a sporophyte plant with roots, stems, and leaves.

There are several significant changes between the life cycle of the moss and the life cycle of the fern. First of all, the sporophyte is now the most prominent part of the life cycle. It leads an independent existence and contains complex organs: roots, stems (the stems of ferns are usually underground), and leaves. Because the sporophyte has roots, stems, and leaves and has vascular tissue to carry water from the roots to the leaves, it is able to live in drier environments than mosses. Although the gametophyte must live in moist conditions it is relatively short-lived—fertilization occurs during a period of the year when conditions are moist. The ferns were the first plants to be successful in dry environments.

Figure 24.5 Fern life cycle.

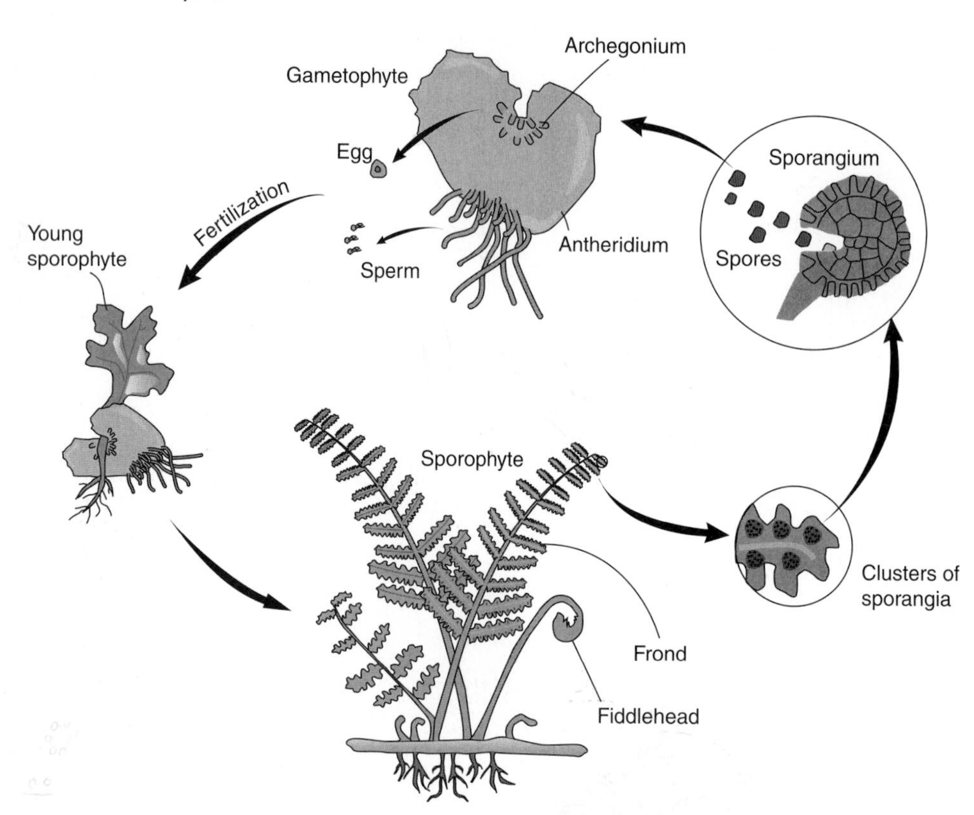

Gymnosperm (Pine) Life Cycle

The life cycle of the pine tree is quite different from that of the organisms you have looked at so far (figure 24.6). The pine tree is a sporophyte plant that produces two kinds of cones. The male cones are small structures, about a centimeter long, containing cells that undergo meiosis to produce haploid cells. These haploid cells develop into a male gametophyte plant known as a *pollen grain.*

The female cones are larger and normally stay on the tree for a year or more. Inside the female cone, some of the cells of this diploid structure go through meiosis and produce haploid spores. These haploid spores grow into female gametophyte plants, which produce eggs. The female gametophyte remains within the female cone.

In pines and similar plants, the pollen (male gametophyte plant) is transferred to the female cone by wind. Large amounts of pollen are produced because of the low probability of pollen landing in the correct place. The transfer of pollen from the male cone to the female cone is known as pollination.

Following pollination, the pollen grain begins to grow a long tube. Sperm nuclei are eventually released in the vicinity of the egg nucleus within the female gametophyte plant, which is located inside the female cone. When the two nuclei fuse, fertilization has occurred and the diploid zygote has been formed. The zygote grows into a tiny embryo sporophyte plant, which, along with some food material, is contained within a protective coat. The entire structure is called a seed. The seeds are produced within the female cone. Eventually, the seed is shed from the cone. Some seeds germinate and grow into large sporophyte plants.

1. Examine the demonstration material showing male cones. How big is each cone? ____________________
2. Examine some pollen grains (either fresh pollen in sugar water in a temporary wet mount or some prepared slides). How many cells are in each pollen grain? ____________________
3. Where is the nucleus of the long, narrow cell? ____________________
4. Now, look at several stages of female cones. How big is the mature cone? ____________________
5. See if you can locate some seeds on the "leaves" of the mature female cone. What is the purpose of the flap of tissue attached to the seed? ____________________
6. Look at some germinated pine seeds. How long will it take for the seedlings to complete their life cycles (produce seeds)? ____________________

Figure 24.6 Pine life cycle.

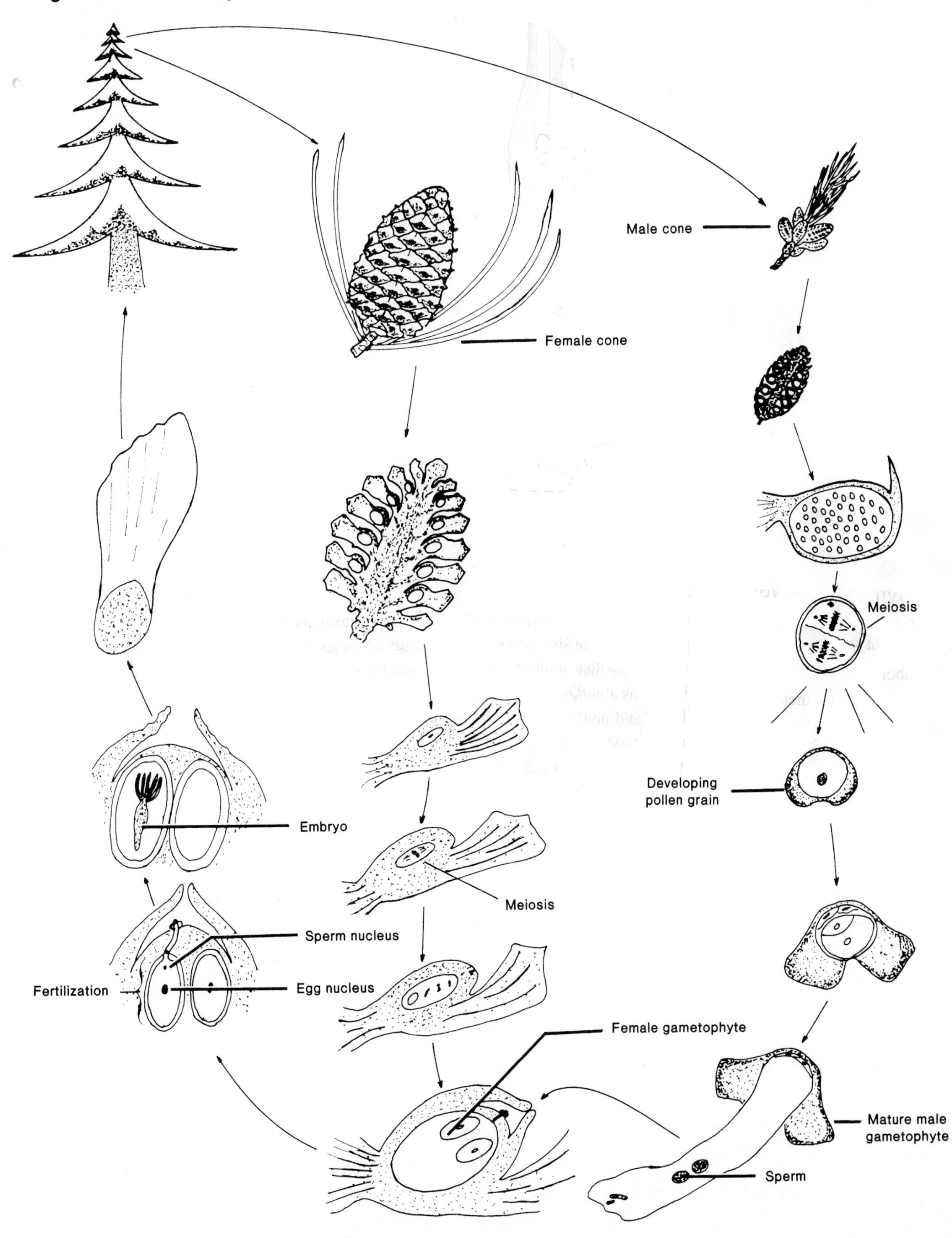

Antheridia

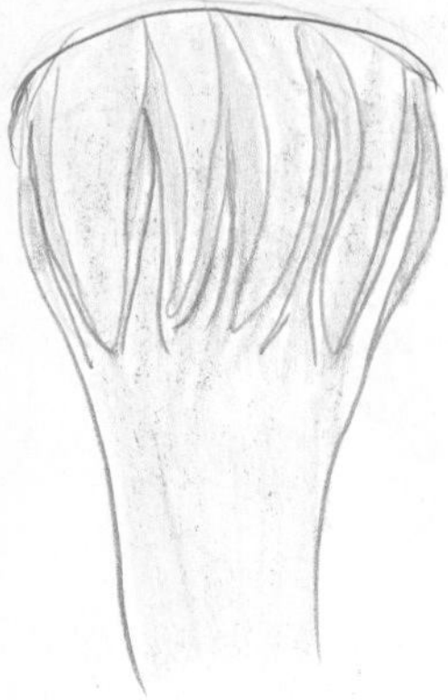

Archegonium

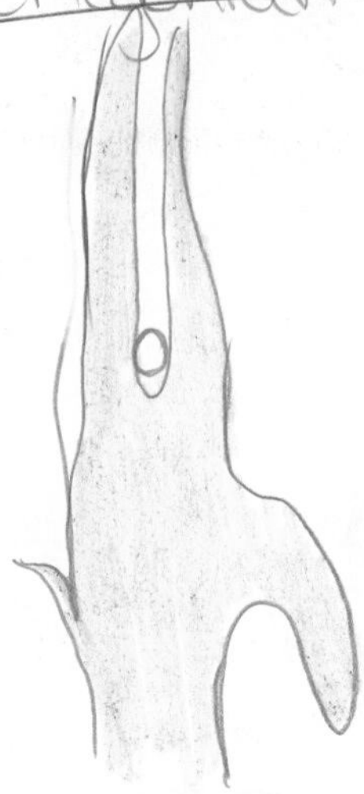

Pollen

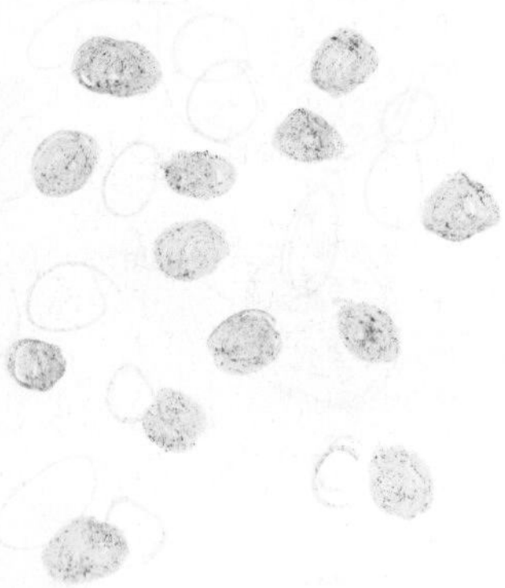

Spirogyra

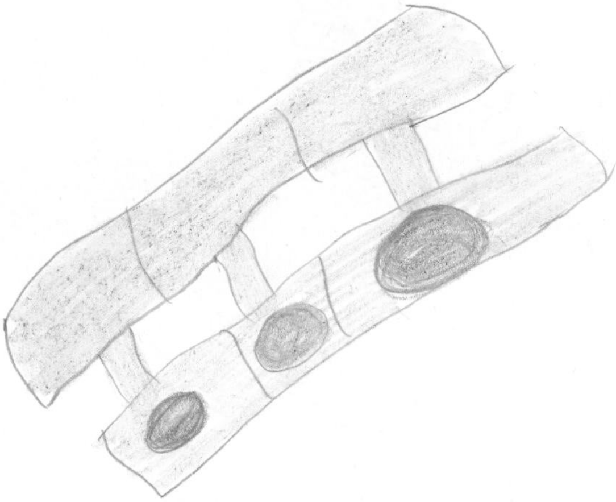

24 Plant Life Cycles

Name ______________________ Lab section __________

Your instructor may collect these end-of-exercise questions. If so, please fill in your name and lab section.

End-of-Exercise Questions

Compare the life cycles of the four organisms by completing table 24.1.

Table 24.1 Comparison of the Life Cycles of Algae and Plants

	Spirogyra	*Oedogonium*	Moss	Fern	Pine
Which generation is dominant, gametophyte or sporophyte?	G		G	S	S
Does the gametophyte carry on photosynthesis?	yes		yes	yes	N
Does the sporophyte carry on photosynthesis?	N		yes	yes	yes
Is the gametophyte multicellular?	NO		yes	yes	yes
Is the sporophyte multicellular?	NO		yes	yes	yes
Is water necessary for fertilization?	yes		yes	yes	yes
Are there two kinds of gametes?	NO		NO	yes	yes
In what structure are the spores produced?			capsule	spore cued	cones
Is vascular tissue present?	NO		NO	yes	yes
Are seeds present?	NO		NO	NO	yes
Is pollen present?	NO		NO	NO	yes

LABORATORY

Plant Structure and Function 25

+ Safety Box

- The stains you will use can permanently stain clothing.
- Be careful when using knives.

Objectives

Be able to do the following:

1. Define these terms in writing.

roots	seeds	veins	pistil
fruit	cortex	sepals	style
root hairs	petals	ovary	cotyledon
spongy layer	stamen	filament	xylem
anther	pollen	upper and lower epidermis	pith
stigma	vascular tissue	flowers	phloem
seed coat	leaves	guard cells	pollination
stems	stomates	palisade layer	fertilization

2. Be able to locate the following on intact plants or microscope slides and describe their function.

roots	stems	leaves	flowers
fruit	seeds	stomates	guard cells
root hairs	cortex	veins	palisade layer
spongy layer	petals	sepals	pistil
anther	stamen	ovary	style
stigma	pollen	filament	cotyledon
seed coat	vascular tissue	upper and lower epidermis	

Introduction

We deal with plants on a daily basis but often do not know much about their structure and the way they function. In this lab, we will examine the major structures found in plants: roots, stems, leaves, flowers, fruits, and seeds.

Preview

During this lab exercise, you will examine the major organs of plants: roots, stems, leaves, flowers, fruits, and seeds.

Procedure

Roots

Roots are the underground structures of plants, which

1. anchor the plant.
2. absorb water and nutrients.
3. store food.

We will briefly examine each of these functions.

Anchoring

You are aware that it is difficult to remove plants from the soil. The branching nature of many roots is responsible for this.

1. Gently remove the soil from around the roots of a potted bean plant. Note the branching nature of the root mass.
2. Cut the plant into the above-ground, green part and the below-ground, nongreen part. Gently remove the soil from the roots. Weigh the two parts separately and record the weights.

Weight of roots	_________ grams
Weight of stem and leaves	_________ grams

Which has the greatest weight? ______________________________

Absorption

1. Obtain a Petri dish containing germinating radish seedlings. Leave the lid on the dish and examine the roots under the dissecting microscope. *Do not remove the radish seedlings from the dish.* The tiny extensions of the central root are *root hairs.*
2. Choose a single root and count the root hairs.
3. Make a drawing of the root and root hairs in the following space.

4. How do these root hairs expand the plant's ability to absorb nutrients and water from the surrounding soil?

Food Storage

1. Obtain a parsnip and cut it crosswise, so that you have a very thin cross section. The central part of the root is the *vascular tissue,* which carries materials to and from the leaves. The cells surrounding the vascular tissue form a part of the root known as the *cortex.*
2. Most plants store food materials as the carbohydrate starch. When iodine reacts with starch, it turns the starch dark blue/black. Place a dilute solution of iodine or Lugol's solution *over the entire cut surface of the parsnip.* Allow a couple of minutes for the reaction to occur.
3. Examine the cross section on a dissecting microscope with light passing through the slice of parsnip from the bottom.
4. Where is starch most highly concentrated?

5. Use the same procedure to examine a radish and a turnip. Do they store starch in the same places? How can you tell?

Stems

Stems are the upright portions of the plant that connect the leaves and the roots. The primary functions of stems are to position the leaves and transport materials. However, some stems are also involved in storing materials. For example the potato is a modified underground stem. The "eyes" are buds from which leaves grow.

1. Examine the stem of the bean plant. Cut it at the point where the stem and roots join. Keep the leaves attached to the stem. Place the plant in a solution containing a red dye for 1 hour. Then, wash off the stem. Examine the leaf. Is there evidence that the red dye has migrated to the leaf? What kind of tissue would conduct the dye to the leaf?

2. Examine the prepared slides of the cross section of the stem of alfalfa *(Medicago)*. You should be able to see four different kinds of cells. (See figure 25.1.)

Figure 25.1 Herbaceous dicot stem, cross section.

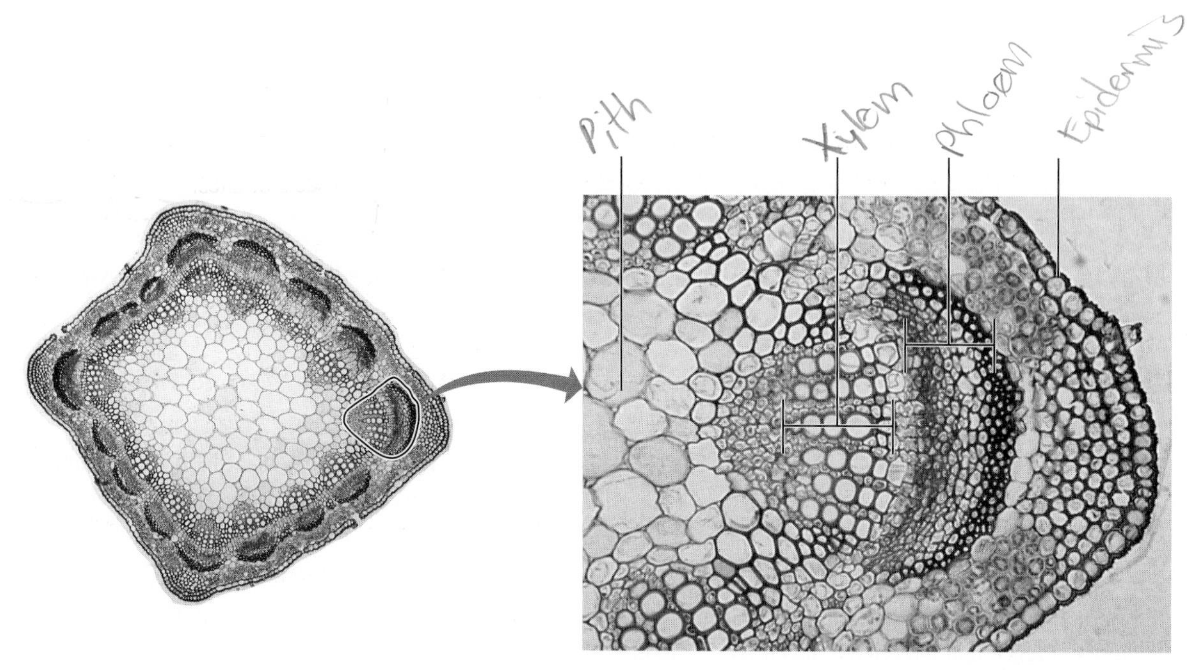

Courtesy of Jim Bidlack

Label the following:

1. *Pith,* a storage material in the center of the stem
2. *Xylem,* a vascular tissue that carries water and dissolved materials from the roots to the leaves
3. *Phloem,* a vascular tissue that carries sugars and other food materials from the leaves to storage places such as the roots
4. *Epidermis,* a protective outer layer

Leaves

The *leaves* are usually thin, flat structures, primarily designed as light-trapping organs. Most of the cells of leaves contain chloroplasts, which contain the green pigment chlorophyll. It is within the chloroplasts that the process of photosynthesis takes place.

1. Examine the leaves of your bean plant under a dissecting microscope. Describe how the top and bottom surfaces differ.

Top Surface	Bottom Surface

2. Notice the veins in the leaf. In addition to the function of distributing materials within the leaf, what other function is performed by the veins?

Duckweed *(Lemna)* is a small aquatic plant, with roots, stems, and leaves, that floats on the surface of water.

1. Obtain two duckweed plants and make a wet-mount slide so that one plant is right side up and the other is upside down. Use a compound microscope to examine the leaves. You should be able to see stomates. The banana-shaped cells found in pairs are *guard cells* and the space between them is known as the *stomate.* Gases (O_2, CO_2, and H_2O vapor) pass through these openings. The guard cells can change shape to open or close the stomate. Plants normally take up water through their roots and transport it to the leaves, where it exits as water vapor through the stomates.
2. Were there stomates on both the upper surface and lower surface? ___________
3. Now that you know what you are looking for, you will be able to see the stomates on the *underside* of the bean leaf. Attach a piece of clear tape to the bottom surface of a leaf. Peel the tape off the leaf in such a way that a layer of cells sticks to the tape. Place the tape with its attached leaf material on a slide and examine it under a compound microscope.
4. Examine the stomates of bean leaves that have been *subjected to drought.* Are the stomates open or closed?

   ____

Duckweed has stomates on the top of the leaf. Beans and most other plants have stomates on the bottom of the leaf. Why does duckweed have stomates on the top surface of the leaf?

5. Use a compound microscope to examine the slides of the cross section of a privet *(Ligustrum)* leaf. (See figure 25.2.)

Figure 25.2 Leaf cross section.

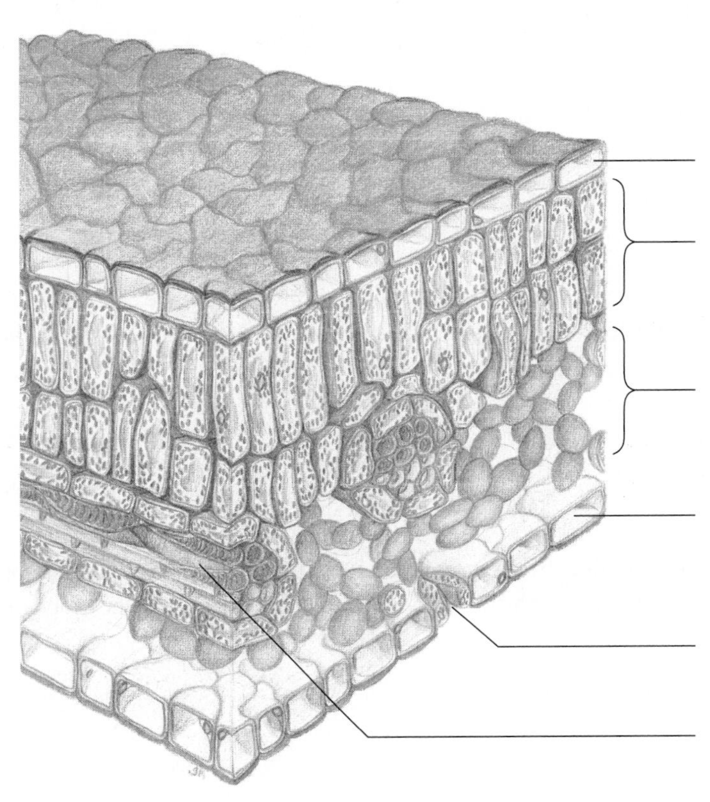

Identify each of the following layers or structures:

1. *Upper* and *lower epidermis,* which are protective, waterproof layers.
2. *Stomates,* which are located in the lower epidermis

   Do epidermal cells have chloroplasts? ___________

   Do guard cells have chloroplasts? ___________
3. *Palisade layer,* which is a tightly packed layer of cells near the upper epidermis containing the chloroplasts responsible for the photosynthesis that takes place in the leaf
   What value might the tightly packed nature of this layer of cells have?___________
4. *Spongy layer,* which is a layer of loosely arranged cells containing chloroplasts

   What value might the spaces between cells have? ___________
5. *Vascular tissue* of the veins that transport materials to and from the leaf and within the leaf

Flowers

Flowers are the sexual parts of flowering plants. They produce the functional equivalent of eggs and sperm. The eggs are inside a structure called the ovary and the sperm are within the pollen grains.

1. Examine the flowers provided and identify the following structures. You will need to be brutal. Tear the flowers apart to see the structures. The most obvious structures are the (a) showy *petals,* which often attract insects and birds. Usually, surrounding the petals on the outside, you will be able to identify (b) green, petal-like structures called *sepals.* In the middle of the petals will be a structure called a (c) *pistil,* which has a swollen base known as the (d) *ovary,* within which the eggs develop. Above the ovary is a portion of the pistil known as the (e) *style*. It is a long, thin structure with a sticky (f) *stigma* at the top end. The stigma traps pollen. It is the ovary that develops into the fruit in most flowering plants. Surrounding the pistil will be several (g) *stamens,* which consist of a long (h) *filament* and an (i) *anther* containing (j) *pollen.*
2. Make a diagram of your flower and label structures a–i.

3. Cut the ovary in cross section and examine it under a dissecting microscope. You should be able to see the tiny structures, called ovules, that contain the eggs. Draw the arrangement of ovules within the ovary in the following space.

4. Crush an anther on a slide in a drop of water and examine it under a compound microscope to see the pollen. The pollen consists of a few cells, one of which is a sperm cell. Draw or describe the shape of the pollen.

Pollination and Fertilization

Pollination involves the transfer of the sperm-containing pollen grains from the anther to the stigma. **Fertilization** occurs in the ovary after the pollen grain grows down the style to the ovary and releases the sperm to fertilize the egg, which is within the ovule.

Fruits

Fruits develop from the flower and contain *seeds,* which are essentially embryo plants with stored food and a protective covering around them. In some cases, the fruits are designed to protect the seeds, as in the hard covering of many nuts. In other cases, the fruit calls attention to its presence and makes it more likely that the seeds will be eaten and distributed by animals. In still other cases, the fruits provide modes of transport, as in the winged fruits of some trees and the sticky fruits of others.

1. Examine at least four of the fruits provided. What evidence do you see that these structures developed from flowers?

2. Cut the fruit in cross section. Describe the location and arrangement of seeds within the fruit.

   Describe how each of these fruits is involved in seed dispersal.

Seeds

Examine some of the seeds you found in your fruits or look at the seeds of bean plants that have been soaked in water overnight. Dissect them. Identify the tough, protective (1) *seed coat* on the outside and the two fleshy "halves" of the seed known as (2) *cotyledons,* which are the first two leaves that the seedling will produce. Between the two cotyledons, you may be able to see the beginning of an (3) embryonic stem and root. This is often referred to as the germ of the seed (wheat germ).

Some plants, such as grasses, have only one cotyledon.

Other Kinds of Plants

During this exercise, you have been looking at the structures of flowering plants. However, there are three other categories of plants.

1. Conifers share many characteristics with flowering plants but they do not have flowers. They have roots, stems, and leaves but produce the eggs and pollen within cones; the seeds are also formed inside the female cones.
2. Ferns have roots, stems, and leaves but do not produce seeds. They reproduce by producing spores on the surface of the leaf. The spores give rise to a tiny stage in the life cycle that eventually gives rise to the fern plant we know.
3. Mosses are the simplest plants and have tiny, leaflike structures on stemlike structures, but they lack roots and vascular tissue.

Examine the examples of these three kinds of plants.

Ficus

Guard Cell

Stomata

25 Plant Structure and Function

Name ______________________________ Lab section __________

Your instructor may collect these end-of-exercise questions. If so, please fill in your name and lab section.

End-of-Exercise Questions

1. Where do each of the following processes take place in a plant?

   a. Photosynthesis

   chloroplast
   Leaves

   b. Water transport

   xylem

   c. Absorption of water

   root tip

   d. Absorption of carbon dioxide

   Stomata or Leaves

   e. Food storage

   Roots

   f. Reproduction

   flowers

2. List four plant structures in which you would find vascular tissue.

   stems fruits
   petals leaves

LABORATORY

26 Natural Selection

+ Safety Box

- There are no specific hazards associated with this exercise.

Objectives

Be able to do the following:

1. Define these terms in writing.

   natural selection — phenotype

   selecting agents — evolution

2. Describe why the phenotype of an organism can be important in determining which genes are passed from one generation to the next.
3. Explain how differences in survival, reproductive success, and attractiveness to potential mates can influence the number of copies of one's genes passed from one generation to the next.

Introduction

The success of an individual within a species is determined by a variety of conditions and events in the life of the organism. The characteristics an individual displays determine whether it will survive and reproduce. The **phenotype** of an organism consists of the physical, behavioral, and physiological characteristics displayed by the organism. Most of these characteristics are determined by genes, which can be passed from one generation to the next. From an evolutionary perspective, the most successful individuals are those that reproduce and pass on the largest number of copies of their genes to the next generation. Those individuals that pass on many copies of their genes are selected for and those that pass on few copies of their genes are selected against. The individual environmental factors that affect an organism's ability to reproduce successfully are known as **selecting agents. Natural selection** involves all the processes that determine which individuals within a species have the opportunity to pass their genes to the next generation. The result of natural selection is *evolution.* **Evolution** is a change in the gene frequency of a population over time. Therefore, natural selection operates through selecting agents and results in evolution.

In this exercise, we will look at three mechanisms by which natural selection can influence which individuals pass their genes to the next generation: differential survival, differential reproductive rates, and differential mate selection.

Preview

During this lab exercise, you will

1. simulate conditions that will result in different rates of reproduction based on differences in survival.
2. simulate conditions that will result in different rates of reproduction based on differences in reproductive success.
3. simulate conditions in which female choice will determine which males will have the opportunity to reproduce.

Background Information

Several thousand years ago, the survival of any individual human depended on an ability to locate food and avoid predators. Food was often in short supply and individuals were forced to experiment. They would try new kinds of food if they were desperate. Many kinds of plants produce chemicals that taste bad and are toxic to the organisms that eat them. Humans were also preyed upon by other kinds of animals. Good eyesight was useful for avoiding predators and locating food. The ability to obtain adequate food was important for survival and reproduction, because poorly nourished individuals had a much lower chance of reproducing successfully.

In the following activities, we will simulate how selecting agents related to the ability to taste, see, and obtain food could influence the number of offspring a particular individual human could produce.

Procedure

Avoiding Toxic Plants

All individuals will be issued a piece of PTC paper to taste.

1. Place the PTC paper in your mouth. You are either a taster or a nontaster. It will be easy for you to determine this.
2. The significance of the ability to taste PTC is as follows:
   a. Tasters will be able to identify foods that contain certain common kinds of plant toxins. Therefore, they will have a greater number of children who survive, because they will not feed toxic plants to their children. During their lifetime, each taster will have 10 children who will live.
   b. Nontasters will not be able to identify the toxins. Therefore, they will feed toxic plants to their children, and many of their offspring will die. Thus, they will have only 5 offspring who live.
3. Record the number of offspring you will have in the space provided on the data sheet on page 261.

Spotting Danger and Locating Food

Because eyesight is important in spotting danger and locating food, the number of offspring produced will be related to how well people can see.

1. To simulate this situation, the instructor will briefly hold up a card with a word written on it. Write the word on the data sheet on page 261.
   a. Those who correctly identify the word have excellent eyesight and are likely to avoid dangerous situations and find food. Therefore, they will have 10 offspring.
   b. Those who get part of the word correct have poor eyesight and will have 5 offspring.
   c. Those who cannot identify the word have very poor eyesight and will have 0 offspring.
2. Record the number of offspring you will have on the data sheet on page 261.

Health and Nutrition

Nutritional status is important in determining how successful an individual will be at raising young. Individuals with poor nutritional status are less likely to conceive and, if young are born, are less likely to be able to feed the offspring well enough to have the offspring survive.

1. To simulate nutritional status, write down what you ate for breakfast this morning in the space provided on the data sheet on page 261.
   a. Individuals who had breakfasts that included at least three of the major kinds of food items—(1) cereals, (2) milk products, (3) fruit, (4) vegetables, and (5) meat—will have 10 children.
   b. Individuals who ate breakfast, but had only one or two of the major food groups, will have 5 children.
   c. Individuals who had no breakfast (coffee, soft drinks, etc. do not qualify as breakfast) will have no offspring.
2. Record the number of offspring you will have on the data sheet on page 261.
3. Determine your total score for all three sections of this part of the exercise and report it to the class in the manner determined by your instructor.

Background Information

The "fitness" of an individual organism is determined by the number of offspring the individual is able to produce. Each offspring produced by sexual reproduction is carrying half its genes from each parent. Therefore, each offspring produced by a parent allows the parent to pass its genes on to the next generation. Those individuals that pass more copies of their genes to the next generation are being selected for. The phenotype of an organism consists of the characteristics that can be observed. The phenotype is determined in part by the genes an organism has and in part by the environment of the organism.

Procedure

1. In this exercise, you will be given a playing card that will represent your phenotype. The playing card represents an important feature that will determine the likelihood you will reproduce.
2. Males and females in the class will have an opportunity to select mates and produce offspring. The object of the game is to produce the largest number of offspring in 10 generations.
3. No individual may have more than one mate. Unmated individuals will have no offspring. Any disputes about access to potential mates will be settled by the instructor based on the phenotypes (playing cards) of the disputing persons.
4. Move around the room and choose your mate based on your phenotype (playing card) and the phenotype of the potential mate. At this point, you don't know what the best combinations are, but some combinations of phenotypes will produce many offspring and others will produce none.
5. Once all individuals have had an opportunity to choose a mate, the instructor will tell each couple how many offspring they produced. The instructor will use a consistent method of determining the number of offspring based on a value given to particular cards and to the combination of cards held by the couple. Different combinations of cards will result in three, two, one, or zero offspring.
6. At the end of each round, record the number of offspring you produced on the data sheet on page 261.
7. At the end of each round, all individuals may stay with the same mate or choose new mates. Remember, your goal is to produce as many offspring as possible, so you might want to experiment by choosing different mates, particularly if you are not producing many offspring.
8. Continue this process through 10 rounds.
9. After the tenth round, the instructor will explain the rules, if they have not already been figured out.
10. Record your data on the data sheet on page 261.
11. Report to the class your total score for this part of the exercise in the manner determined by your instructor.

Background Information

In lek mating systems, males stake out territories, usually in the presence of other males. The females evaluate the characteristics (phenotypes) of the various males and choose which male they will mate with from among all the males present. Usually, only a small number of the males are chosen and the other males do not mate. Males can have several mates during any mating season. The males that are chosen are selected for and have a greater chance of passing their genes on to the next generation.

Procedure

1. Because in a lek mating system females decide which males show the most desirable characteristics, the females in the class will be given special rules by the instructor. These special rules will be used in making decisions about which males will be chosen for mating and how many offspring they will have.
   a. Each male will be issued a meterstick.
   b. Each person will be issued a playing card to be used in case disputes arise.
2. Males will distribute themselves around the room but must be at least 2 meters apart.
3. Females will choose mates by standing near them.
4. Any disputes about where a male may stand or which females have access to a specific male will be decided by the instructor based on the playing cards held by the individuals who are in dispute.
5. Each female will produce one offspring per year.
6. Each male can produce as many offspring as he has females, but the maximum number of females per male is five. Each individual will record the number of offspring he or she will have per year on the data sheet on page 261.
7. At the end of each year, the males may choose a different spot in the room and females must choose again.
8. Repeat for 10 rounds.
9. After the tenth round, the instructor will explain the rules by which the females choose males.
10. Record your data on the data sheet on page 261.
11. Report to the class your total score for this part of the exercise in the manner determined by your instructor.

26 Natural Selection

Name ______________________ Lab section __________

Your instructor may collect these end-of-exercise questions. If so, please fill in your name and lab section.

Data Sheet, Analysis, and End-of-Exercise Questions

Natural Selection Score Card

Differential Survival	
Offspring Produced	
Avoiding toxic plants ______	Write the word you saw here.
Spotting danger ______	
Health and nutrition ______	Write the breakfast you ate here.
Total ______	

Differential Reproductive Rates	**Differential Mate Selection**
Offspring Produced	*Offspring Produced*
Year 1 ______	Year 1 ______
Year 2 ______	Year 2 ______
Year 3 ______	Year 3 ______
Year 4 ______	Year 4 ______
Year 5 ______	Year 5 ______
Year 6 ______	Year 6 ______
Year 7 ______	Year 7 ______
Year 8 ______	Year 8 ______
Year 9 ______	Year 9 ______
Year 10 ______	Year 10 ______
Total ______	Total ______

Analysis of Results

Natural selection works on individuals in a population to cause changes in the gene frequency of populations. Therefore, we will pool all the data from all the individuals in the class, so that we can see which individuals had the greatest impact on the gene pool. Use the following charts to record the combined data from the class and answer the associated questions.

Differential Survival Analysis

The maximum number of offspring possible from this series of simulations is 30. Complete the following chart.

Number of Offspring	Number of People in Class with This Number of Offspring
30	
25	
20	
15	
10	
5	
0	

1. Identify the individuals in the class who were the most successful at producing offspring. What characteristics did they have that allowed them to be successful?

2. Identify the individuals in the class who were the least successful at producing offspring. What characteristics did they have that contributed to their lack of success?

3. Could individuals have done anything to improve their chances of reproducing?

Differential Reproductive Rates Analysis

The maximum number of offspring possible is 30. Complete the following chart.

Number of Offspring	Number of People in Class with This Number of Offspring
28–30	
25–27	
22–24	
19–21	
16–18	
13–15	
10–12	
7–9	
4–6	
1–3	
0	

1. Based on the cards they held, did all of the people in the class have the same opportunity to reproduce?

2. Remember that the cards represented aspects of an individual's phenotype. Could individuals have done anything to change their phenotype?

Differential Mate Selection Analysis

The maximum number of offspring possible is 50. Complete the following chart.

Number of Offspring	Number of People in Class with This Number of Offspring
46–50	
41–45	
36–40	
31–35	
26–30	
21–25	
16–20	
11–15	
6–10	
1–5	
0	

1. Were some males more successful than others?

2. How much of their success was determined by genes?

End-of-Exercise Questions

1. Describe two human characteristics, presumed to be determined by genes, that would lower a person's reproductive success.

2. Many eyesight characteristics are inherited (e.g., colorblindness, astigmatism, nearsightedness). Compared with 1,000 years ago, are these genes being selected against more or less strongly? Explain your answer.

3. In many studies, observers consider people with symmetrical facial features to be more beautiful than those who have some degree of symmetry. How might facial symmetry or lack of symmetry affect a person's reproductive success?

4. If an organism's reproductive fitness is determined by the genes it has inherited, can all individuals have an equal chance of reproducing? Explain your answer.

5. In a lek mating system, if the genes that determined the behavior of the females mutated so that the females behaved differently, would the same males be successful? Explain your answer.

LABORATORY

27 Species Diversity

+ Safety Box

- Your instructor will discuss the safety issues related to the areas you will visit.

Objectives

Be able to do the following:

1. Define these words in writing.

   species diversity

   sequential comparison index

2. Measure species diversity by using the sequential comparison index method.

Introduction

Natural ecosystems have a great deal of diversity. Many species interact with one another to create a stable functional unit. Disturbed ecosystems tend to have large fluctuations in population size of any one species and have fluctuations in the species of organisms present. **Species diversity** is a measure of the number of species of organisms present, in comparison with the number of each species present. A lawn with only one species of grass have a very large population but very low species diversity. In this exercise, we will use a simple method known as the **sequential comparison index** to estimate the species diversity of two different habitats. This method is particularly valuable because it is not necessary to specifically identify each kind of organism. All that is necessary is that you be able to tell if one organism is the same as or different from the one you looked at previously.

Preview

During this lab exercise, you will

1. collect organisms from two different habitats.
2. apply the sequential comparison index method to determine species diversity.
3. compare the species diversity of two different habitats.

Procedure

1. Choose two different but similar habitats to assess. One should be relatively undisturbed; the other should be disturbed. These can be suited to your local situation. The following are some suggestions:
   a. polluted stream or ditch and clean stream
   b. lawn and grassy field
   c. forest plantation and natural forest
   d. soil in a yard and soil in a forest

2. Determine the kinds of organisms you are going to sample in each habitat and how you will collect them. When sampling two different kinds of habitats, it is important that you use the same sampling method and the same amount of sampling effort. It is also important that you sample in a random manner. You should not select specific organisms because they are large or pretty or because they are easy to catch.
   a. Soil and litter organisms can be collected by taking samples of soil and litter into the lab and sorting through the material in a white enamel pan. A white enamel pan makes the tiny animals show up better.
   b. Stream insects can be collected with dip nets and taken back to the lab for sorting in a white enamel pan.
   c. Pit traps consisting of tin cans buried in the soil so that their tops are even with the soil surface, can be used to capture small organisms. The animals collected can be taken back to the lab for sorting.
   d. Insect sweep nets can be used to collect flying insects in open areas.
   e. Birds can be identified by direct observation. Simply watch a given area and log the kinds of birds and the order in which you observed them.
   f. Plants can be collected from specific plots and returned to the lab for analysis.
   g. Your instructor may have other suggestions.
3. Return to the lab with your organisms. You should have two collections—one from each of the two sites you assessed. Animals may need to be chilled to slow them down to make them easier to manage.
4. To determine the sequential comparison index, place all the organisms from one of your habitats together. Randomly choose organisms from the collection and determine whether each is the same as or different from the one preceding. Give each kind of organism a symbol or name. You don't need to know what the organism is, so "green bug," "spiny plant," or "large black bird" will do. Record your data in the data table on page 267.

   Each time the succeeding organism is different from it predecessor, begin a new run. Record the number of runs and the number of individuals in your sample. For example, if there were five kinds of organisms (A, B, C, D, and E) and they were selected in the following order:

D D	E E E	A	D D	B B	C C C C C
run 1	run 2	run 3	run 4	run 5	run 6

   there are 6 runs and 15 individuals.

   The sequential comparison index is calculated as follows.

   $$\text{sequential comparison index} = \frac{\text{number of runs}}{\text{number of individuals}} = \frac{6}{15} = 0.4$$

5. Calculate the sequential comparison index for your two habitats. The closer the sequential comparison index is to 1, the more species diversity exists. If you had 15 organisms in your sample and each was a different species, you would have 15 runs and 15 individuals and the sequential comparison index would be 1.
6. Record your data in the sequential comparison index table on page 267.

27 Species Diversity

Name ____________________ Lab section __________

Your instructor may collect these end-of-exercise questions. If so, please fill in your name and lab section.

Data Sheet, Analysis, and End-of-Exercise Questions

Data Table

Organism Number	Description	Symbol
1		
2		
3		
4		
5		
6		
7		
8		
9		
10		
11		
12		
13		
14		
15		

Sequential Comparison Index Table

Kind of Habitat	Number of Runs	Number of Individuals	Sequential Comparison Index Number

1. Which of your two habitats had the greatest species diversity?

2. Did other students in the class get similar results?

3. List three problems with your method that might have led to biased results.

4. How would you change your methods to make your results less biased?

5. What effect did disturbance have on species diversity?

LABORATORY

28

Frog Dissection

+ Safety Box

- Ether is highly explosive; keep it away from flames and do not agitate the container. Be sure to keep the container closed and refrigerated when not in use.
- Dissection tools are, of necessity, sharp. You must use them carefully to prevent cutting yourself and to maintain them in proper condition. When finished with a set of dissection tools, carefully clean, dry, and return them to their proper place.
- You will use a living organism for this exercise; therefore, treat it with the respect due any life-form.
- Make sure that the animal is anesthetized or pithed before cutting it. When you are finished with the dissection, make sure the organism does not come out of the anesthesia and feel pain. Therefore, you need to decapitate it and dispose of the remains as indicated by your instructor.

Objectives

Be able to do the following:

1. Define these terms in writing.

   isosmotic
   Ringer's solution
   dorsal
   ventral
   anterior
   posterior
   parasite

2. Locate these structures in a frog:
   a. External—anus, tympanum, nares, eyes, and pectoral and pelvic girdles
   b. Internal—liver, gallbladder, lungs, stomach, small intestine, large intestine, spleen, ovaries or testes, kidneys, adrenal glands, fat bodies, nerves, urinary bladder, heart, atrium and ventricle, and oviduct
3. Describe a function of each of the structures named in objective 2.

Introduction

A living animal is a much more interesting organism to study than a dead, preserved one. It is possible to show many types of phenomena with living animals that are just not possible in nonliving organisms. For this reason, we use an anesthetized frog for this exercise.

It is important in any study of living animals that the organism not be subjected to unusual pain or cruel treatment. Anesthetizing the frog effectively prohibits any feeling on the part of the frog while still allowing the internal organs to function normally. Alternatively, pithing is a technique that destroys the nervous tissue of the brain, so the frog feels no pain. Your instructor will provide pithed frogs if this technique is used.

When you are working with living tissue, it is important that the cells be bathed in a solution that is **isosmotic** to the cells. Why?

The solution we use is called **Ringer's solution,** which contains sodium and potassium salts in a 0.7% solution. During this exercise, moisten the tissues frequently with Ringer's solution.

Preview

During this exercise, you will work with a live frog. Because it is a living organism, treat it with respect.

During this lab exercise, work in groups of 2–4 people. You will

1. obtain a frog.
2. examine the frog's external structures.
3. anesthetize or pith your frog by closely following the instructor's directions.
4. study circulation in tongue or toe.
5. dissect your frog.
6. manipulate the frog's body temperature and count its heartbeats.
7. examine the frog's internal structures.
8. examine organs for parasites.
9. complete a quiz.
10. clean up your materials.

Procedure

Anesthetizing

Place some crumpled paper towels or cloth in a jar (with a tight-fitting lid) large enough to hold this material and a frog. Pour a few milliliters of ether into the jar, place the live frog into the jar, and close the lid. When the frog becomes very limp, it can be removed and your observations can begin. Be aware that occasionally a frog may show signs of reawakening during the observation. At the first sign of movement, return the frog to the jar until it is once more quite limp; then the observation can proceed.

Pithing

If the frog is to be pithed rather than anesthetized, watch your instructor carefully as this process is demonstrated. Remember, the object of pithing, as with anesthetizing, is to prevent unnecessary pain.

External Anatomy

The frog's basic body consists of the (1) head, (2) trunk, (3) pectoral girdle (the front legs and associated structures), and (4) pelvic girdle (hind legs and associated structures). Identify these structures on your frog. Also identify the **dorsal** (back), **ventral** (belly), **anterior** (head end), and **posterior** (tail end) surfaces of your frog. Locate these structures: the *anus* on the dorsoposterior surface between the hind legs; the *tympanum,* a circular area found posterior and lateral to the *eyes;* and the *external nares,* small, pitlike openings on the anterior of the head.

Studies of Circulation

You can observe the capillaries either in the frog's tongue or in the webbing between the toes. Place the anesthetized frog on its belly on a frog board and stretch the tongue or toes over the opening in the edge of the frog board. Secure the tongue or toes in place with pins and observe with low power through the compound microscope. You will be able to see three different kinds of blood vessels with blood flowing through them: *arteries, capillaries,* and *veins.* The arteries branch into smaller and smaller blood vessels until eventually the blood vessels are about the same diameter as the red blood cells. These smallest blood vessels are capillaries. These capillaries combine into larger and larger blood vessels called veins. The red blood cells are oval in outline and flow from the arteries through the capillaries to the veins. Usually, the blood flowing in the arteries flows in pulses, and the blood in the veins flows at a continuous pace. Compare the rate of blood flow in these three blood vessels.

After you have studied the circulation in the tongue or toes of the frog, open the body cavity of the frog and study the action of the heart. Place the frog on its back in a dissecting tray and use a pair of scissors to cut through the skin from a point where the hind legs join to the point of the chin. Make cuts to the side at the level of the shoulder and hip so that the skin can be peeled back on either side, as indicated in figure 28.1.

Figure 28.1 Frog dissection: cut on dotted lines.

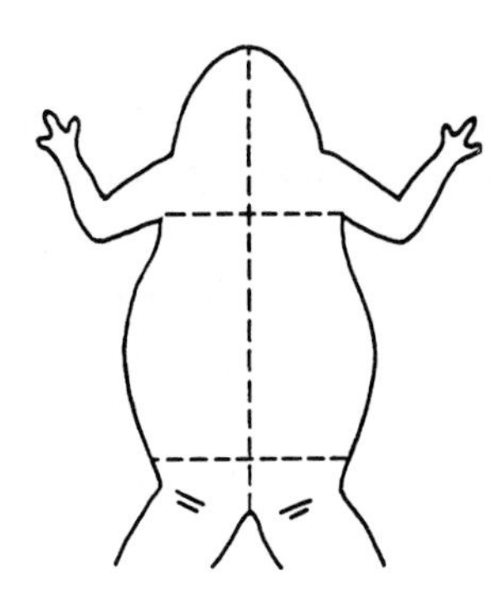

Notice the large number of blood vessels inside the skin. The skin is the major organ of gas exchange (oxygen and carbon dioxide). The whitish-colored, thin layer remaining over the belly of the animal is the belly musculature. Make a similar pattern of cuts through this muscle, but cut to either the right or left of the midline to prevent cutting the *ventral abdominal blood vessel.* When you get to the region of the arms, it is necessary to cut through the bones of the *shoulder girdle.* Immediately under these bones is the heart. It still should be beating. You will be able to see two major portions of the heart. The *ventricle* is the triangular-shaped structure beating. There are two smaller chambers at the anterior end of the heart called the *atria.* The atria are anterior to the ventricle and beat slightly before the ventricle. Locate and distinguish the atria and ventricle of the heart. Determine the heart rate in beats per minute and record. After a few minutes, place some cool Ringer's solution (10°C) on the heart and record the heart rate. Use several milliliters of Ringer's solution, not just a few drops. Repeat a few minutes later with warm Ringer's solution (30°C).

1. Warm: ______________________ beats per minute
2. Room-temperature: ______________________ beats per minute
3. Cold: ______________________ beats per minute

What effect does temperature have on heart rate? Why?

Internal Anatomy

After your experiments on the heart, locate the following structures and compare them with figure 28.2:

1. The *liver,* a large, reddish-brown organ located posterior to the heart, stores food and releases it as needed. The liver produces compounds that aid in the digestion of fats and assist in the coagulation of blood. It also interconverts carbohydrates, fats, and proteins and destroys toxins.
2. The *gallbladder:* Lift the lobes of the liver until you locate a bluish-green sac. This sac stores bile, which contains compounds that aid in fat digestion. If you have ever removed the internal organs of a game animal, you may have been warned not to break the gallbladder to avoid spilling this bitter fluid on the meat.
3. The two *lungs* are located on each side of the heart and should be seen protruding from under the liver. These are thin sacs, usually speckled with black. Some exchange of gases between the atmosphere and the blood takes place in the lungs. Remember that the skin is the primary organ of gas exchange.

4. The *stomach, small intestine,* and *large intestine* take up most of the abdominal cavity. The huge, J-shaped structure is the stomach. This can be seen at the frog's left side and just below the liver. Digestion starts in the stomach of the frog, rather than in the mouth, as in humans. The long, slender middle portion, extending from the stomach, is the small intestine. This is called the small intestine because of its small diameter, even though it is several inches long. Digestion is completed here. The small intestine empties into an abruptly expanded portion at the base of the abdominal cavity, which is the large intestine. Undigested food is held in the large intestine until the animal defecates.
5. The *spleen* is a small, reddish structure found near the small intestine. It is sometimes difficult to locate. The spleen stores and cleans the blood.
6. The two *kidneys,* located on opposite sides of the backbone, are about 2 centimeters long and are reddish in color. There may be a thin membrane separating the kidneys from the rest of the abdominal cavity; if present, remove it. The kidneys serve the important job of removing cell wastes from the blood.
7. The yellowish stripe on the surface of the kidney is the *adrenal gland.* This complicated organ furnishes the hormone epinephrine, which helps the organism respond to emergencies.
8. a. The presence of *testes* is the easiest way to verify that your frog is a male. These are the two white, bean-shaped structures located at the anterior end of the kidneys. As in humans, they produce the male sex cells.
   b. If you have a female frog with mature eggs, you see the enlarged, saclike *ovaries,* which contain the eggs. You may want to remove these to see the remaining parts.
9. On the right and left sides, you should see two coiled tubes. They will be very large in females with eggs. Females without eggs will have smaller oviducts. These *oviducts* conduct eggs from the ovary to the outside. In males, they are present but are much smaller and nonfunctional.
10. *Fat bodies* are irregularly shaped, yellowish-white masses of stored food. They may be found attached at the anterior end of the kidney. Their size varies greatly, depending on how well the frog has eaten.
11. The white threads that run from the backbone are *nerves.* Pinch some of them and see what happens. Nerves are most easily seen on each side of the backbone, running to the legs.
12. The *urinary bladder* is attached to the large intestine very near the point where the legs join the body. It is very thin and is easily damaged during dissection. It is usually empty and will look like a small, irregular mass of membrane. Its function is to hold the "urine."

Parasites

Most living organisms have **parasites** that live inside some of the organs of the body, and the frog is no exception. Survey the following organs for parasites. Be sure to wash your hands thoroughly when finished.

1. Remove a lung and place it in a dish of Ringer's solution. Using two needles, pull the lung apart and examine under the dissecting microscope for parasites. You might find a brownish-black lung fluke with whitish blotches and long, thin roundworms flexing about.
2. Open the large intestine and examine some of the contents for parasites. Some very small ones are common here, so examine with both the dissecting and compound microscopes.
3. Similarly, examine the
   a. small intestine.
   b. urinary bladder.
   c. gallbladder.
4. How many different kinds of parasites did you find?

Clean up all utensils and your work area. Your instructor will tell you how to dispose of your frog. Wash your hands before you leave the laboratory.

Figure 28.2 Internal organs of the frog.

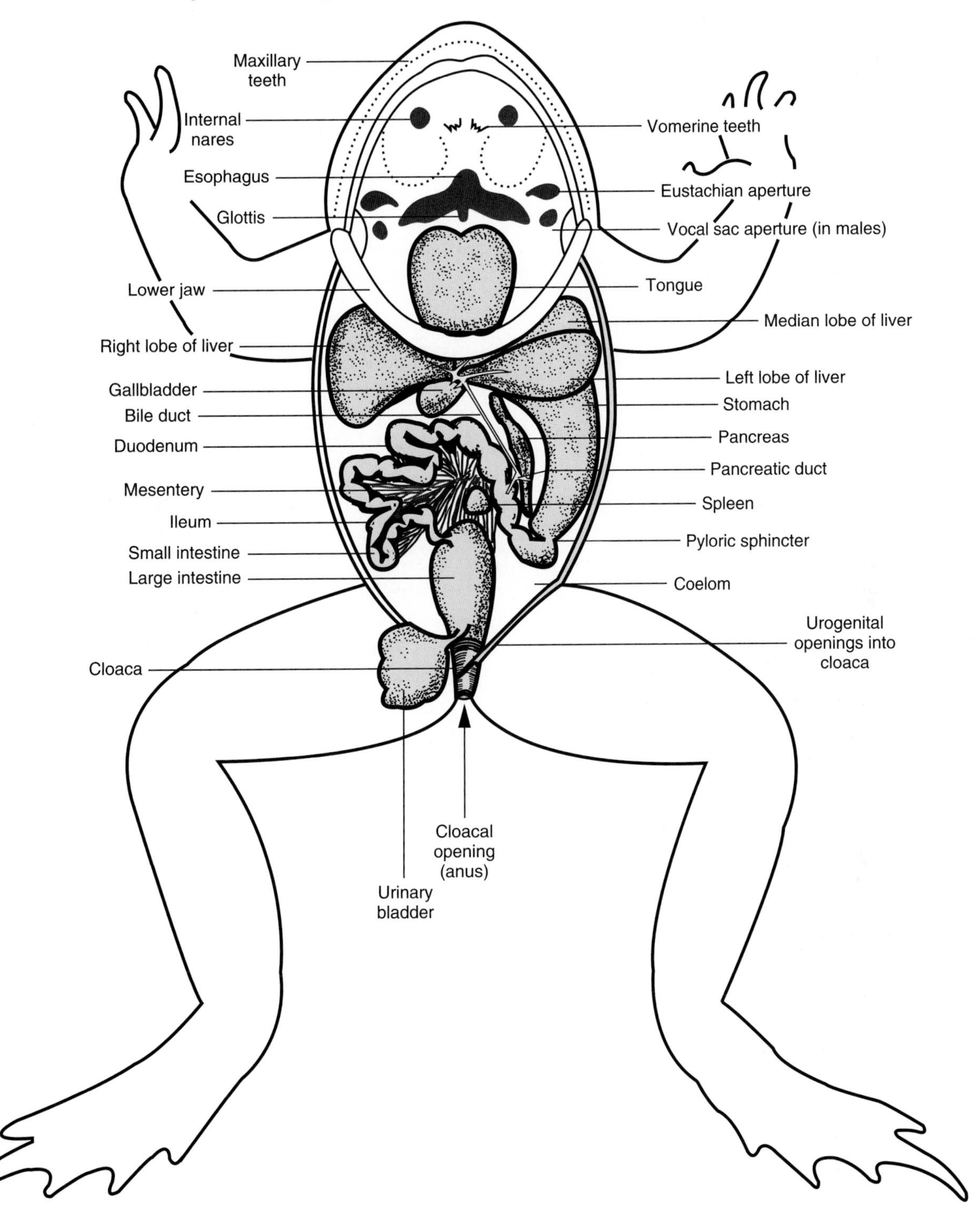

LABORATORY

29 Roll Call of the Animals

+ Safety Box

- No unusual hazards are associated with this laboratory experience. Please follow standard laboratory safety procedures.

Objectives

Be able to do the following:

1. Define these terms in writing.

   dichotomous key | segmented body | taxon
   bilateral symmetry | taxonomy | scientific name
   radial symmetry

2. Use a dichotomous key to identify specimens correctly.
3. Recognize major distinguishing characteristics used to classify animals.
4. List the classification groups from the most inclusive to the least inclusive.
5. Recognize the correct form of a typed scientific name.

Introduction

There is great diversity within the animal kingdom. No one can be expected to recognize all of the different organisms. You already know the differences between cows and dogs, birds and people, and snakes and frogs. You place these animals in categories based on certain differences in the characteristics that each possesses. If you know what critical traits to look for, it is possible to separate any animal into its proper taxonomic category. A taxonomic category is a group of closely related organisms that have evolved along similar lines. Recognizing the important characteristics that differentiate organisms into natural categories is the basis of the science of **taxonomy,** the classifying and naming of organisms.

The ranking order of classification groups **(taxons)** from the most inclusive through the least inclusive is as follows:

Domain
Kingdom
Phylum
Class
Order
Family
Genus
Species

There are three domains: Eubacteria, Archaea, and Eucarya. The Eubacteria and Archaea are prokaryotic organisms. All eukaryotic organisms are in the domain Eucarya. Therefore, all animals are in the domain Eucarya. Each domain includes several kingdoms. Animals are in the kingdom Animalia.

Organisms in the same kingdom are very broadly similar; those in the same phylum are more similar to each other than those in other phyla. Organisms in the same genus are quite similar, more closely so than those in another genus of the same family.

Each species of organism has a **scientific name,** which is often descriptive and uses two terms: the genus name followed by a specific epithet. Because two names are used to identify a species, the descriptive terminology is called a binomial nomenclature. The genus name is always capitalized and the specific epithet is never capitalized, and both the genus and the species names are underlined or set in italics. For example, the leopard frog is scientifically identified as *Rana pipiens.* You belong to the species *Homo sapiens.*

Preview

You can become acquainted with most of the important groups of the animal kingdom through this laboratory experience. By using a dichotomous key, you will become familiar with some of the important characteristics used to classify animals. Although it is impossible to demonstrate all the kinds of animals, you will have an opportunity to see examples of the major phyla and classes of animals.

During this lab exercise, you will

1. use the dichotomous key provided to identify the various specimens.
2. record the phylum or class to which each animal belongs.

Procedure

A **dichotomous key** is a tool used to help determine the taxonomic category to which a specific organism belongs. It consists of a series of pairs of statements that require you to place the organism into one of two categories. Each choice will lead you to another pair of statements until you have identified the animal's taxonomic category. For each of the organisms on display in the lab, begin at step 1 of the key on page 277 and proceed through the key until you have identified the specimen. When you have determined the name of the organism, write the underlined name (from the key) opposite the appropriate number on the answer sheet provided on page 279.

You may begin your work at any station and proceed to any available station thereafter; just be careful to place your answer at the appropriate number on your answer sheet.

Dichotomous Key

1. a. Irregular-shaped body; structure with many pores—Phylum Porifera (e.g., sponge)
   b. Regular-shaped body (with right and left halves or a cylindrical shape) 2
2. a. **Radial symmetry** (disk-shaped or barrel-shaped) 4
   b. **Bilateral symmetry** (similar right and left body halves) 3
3. a. Animal has internal skeleton 19
   b. Animal has external skeleton or no apparent skeleton 6
4. a. Body hard, arms extend from a central disk, or spines present—Phylum Echinodermata 18
   b. Soft body; little or no color—Phylum Cnidaria 5
5. a. Saucer-shaped transparent body with small tentacles—Class Scyphozoa (e.g., jellyfish)
   b. Barrel-shaped body, tentacles at one end—Class Anthozoa (e.g., sea anemone)
6. a. Hard outer covering 10
   b. No hard outer covering 7
7. a. Body flattened—Phylum Platyhelminthes 8
   b. Body not flattened. 9
8. a. Smooth, nonsegmented body—Class Trematoda (e.g., liver fluke)
   b. Apparently segmented, flattened body—Class Cestoda (e.g., tapeworm)
9. a. Nonsegmented 11
   b. **Segmented body**—Phylum Annelida (e.g., earthworm)
10. a. Body has jointed legs—Phylum Arthropoda 14
    b. Body inside of shell is soft, has no jointed legs—Phylum Mollusca 13
11. a. Tentacles or other appendages present 12
    b. Body long and tubular, no appendages—Phylum Nematoda
12. a. Appears as snail without shell—Class Gastropoda (e.g., slug)
    b. Tentacles and eyes present—Class Cephalopoda (e.g., squid, octopus)
13. a. Bivalved shell (two halves)—Class Bivalvia (e.g., clam)
    b. Univalved shell (single unit)—Class Gastropoda (e.g., snails)
14. a. Jointed appendages on most body sections 15
    b. Jointed appendages on certain body segments; not all appendages are legs 16
15. a. One pair of legs per body segment—Class Chilopoda (e.g., centipede)
    b. Two pairs of legs per body segment—Class Diplopoda (e.g., millipede)
16. a. Two pairs of antennae, large claws often present—Class Crustacea (e.g., crab)
    b. One pair of antennae or none, no large claws. 17
17. a. Four pairs of legs, no antennae or wings—Class Arachnida (e.g., spider)
    b. Three pairs of legs, wings present—Class Insecta (e.g., insects)
18. a. Arms present, body surface knobby—Class Asteroidea (e.g., sea stars)
    b. Many-spined animal, resembles a pincushion—Class Echinoidea (e.g., sea urchin)
19. a. Fishlike, flattened body, appendages finlike not jointed 20
    b. Not fishlike, body not flattened, appendages jointed or absent 21
20. a. Scales on body do not overlap, skeleton of cartilage—Class Chondrichthyes (e.g., sharks, stingray)
    b. Scales on body overlap, skeleton bony—Class Osteichthyes (e.g., bony fishes)
21. a. Body covered by scales, zero or four legs—Class Reptilia (e.g., snake, lizard, turtle)
    b. Body not covered by scales 22
22. a. Claws absent—Class Amphibia (e.g., frogs, toads, and salamanders)
    b. Claws or nails present on toes, skin covered with feathers or hair. 23
23. a. Feathered, claws present—Class Aves (e.g., birds)
    b. Hair present—Class Mammalia (e.g., mammals)

29 Roll Call of the Animals

Name ______________________________ Lab section ____________

Your instructor may collect these end-of-exercise questions. If so, please fill in your name and lab section.

Answer Sheet

1. ______________________________
2. ______________________________
3. ______________________________
4. ______________________________
5. ______________________________
6. ______________________________
7. ______________________________
8. ______________________________
9. ______________________________
10. ______________________________
11. ______________________________
12. ______________________________
13. ______________________________
14. ______________________________
15. ______________________________
16. ______________________________
17. ______________________________
18. ______________________________
19. ______________________________
20. ______________________________
21. ______________________________
22. ______________________________
23. ______________________________

LABORATORY

30 Intraspecific and Interspecific Competition

+ Safety Box

- There are no specific safety hazards associated with this exercise.

Objectives

Be able to do the following:

1. Define these terms in writing.

   competition

   intraspecific competition

   interspecific competition

2. Assess the effect of competition among plants grown at different population densities.
3. Compare the success of two plant species that are in competition with one another.

Introduction

Competition is an interaction between organisms in which both organisms are harmed by the interaction. **Intraspecific competition** is competition that occurs between members of the same species. **Interspecific competition** is competition that occurs between members of different species. Competition among animals can involve fights over food, water, nesting sites, or mates or may involve less overt forms, such as animals that have greater height being able to obtain food more easily than shorter animals.

Among plants, competition is more difficult to visualize. However, certain resources, such as water, soil nutrition, and sunlight, are in limited supply. Plants that are particularly efficient at obtaining these resources should be more successful in competition.

In this exercise, we will examine the impact of planting seeds of the same species at different densities. We will determine the effect of intraspecific competition on the average growth of individuals in the population. We will also look at interspecific competition by determining how two species of plants influence each other when they are planted together.

Preview

During this lab exercise, you will

1. plant radish seeds in pots at different population densities to assess the effects of intraspecific competition.
2. plant wheat seeds in pots at different population densities to assess the effects of intraspecific competition.
3. plant both radish and wheat seeds in the same pots at different population densities to assess the effects of interspecific competition.

Procedure

Initial Set-Up

Intraspecific Competition—Radish

1. Obtain radish seeds that have been soaked in water so that they are beginning to sprout.
2. Obtain three 8-cm (3-inch) pots and fill with sand to within 2 cm of the top of the pot.
3. Add water to the top of the pot until water runs out the bottom.
4. ***Gently*** use forceps to place germinating seeds on the surface of the sand.
   In pot 1, plant 1 seed.
   In pot 2, plant 10 seeds.
   In pot 3, plant 20 seeds.
5. ***Gently*** cover the seeds with a thin layer of sand.
6. Place in a warm, well-lighted place.
7. Examine the plants daily and water the plants when the surface of the sand is dry.

Intraspecific Competition—Wheat

1. Obtain wheat seeds that have been soaked in water so that they are beginning to sprout.
2. Obtain three 8-cm (3-inch) pots and fill with sand to within 2 cm of the top of the pot.
3. Add water to the top of the pot until water runs out the bottom.
4. ***Gently*** use forceps to place germinating seeds on the surface of the sand.
   In pot 1, plant 1 seed.
   In pot 2, plant 10 seeds.
   In pot 3, plant 20 seeds.
5. ***Gently*** cover the seeds with a thin layer of sand.
6. Place in a warm, well-lighted place.
7. Examine the plants daily and water the plants when the surface of the sand is dry.

Interspecific Competition—Radish and Wheat

1. Obtain radish and wheat seeds that have been soaked in water so that they are beginning to sprout.
2. Obtain three 8-cm (3-inch) pots and fill with sand to within 2 cm of the top of the pot.
3. Add water to the top of the pot until water runs out the bottom.
4. ***Gently*** use forceps to place germinating seeds on the surface of the sand.
   In pot 1, plant 1 radish and 1 wheat seed.
   In pot 2, plant 10 radish and 10 wheat seeds.
   In pot 3, plant 20 radish and 20 wheat seeds.
5. ***Gently*** cover the seeds with a thin layer of sand.
6. Place in a warm, well-lighted place.
7. Examine the plants daily and water the plants when the surface of the sand is dry.

Data Collection

After 3 weeks or a different period designated by your instructor, collect data on the biomass produced in each of the pots. Use the following technique for each of the pots.

1. Carefully invert the pot containing the plants over a newspaper and remove the soil and plants from the pot.
2. Place the soil and plants on the newspaper.
3. Gently pull the soil and plants apart, so that you have the roots as well as the stems and leaves of the plant.
4. Count the number of plants that survived.
5. Carefully weigh all the plants, including roots, stems, and leaves.
6. Calculate the average biomass.

$$\frac{\text{total biomass}}{\text{number of plants}} = \text{average biomass}$$

   Note: For the experiment dealing with interspecific competition (radish and wheat), you will need to separate the two kinds of plants and measure them separately.
7. Record your data on the charts at the end of the exercise.
8. Compare your data with those of other students in your class.

30 Intraspecific and Interspecific Competition

Name ______________________________ Lab section __________

Your instructor may collect these end-of-exercise questions. If so, please fill in your name and lab section.

End-of-Exercise Questions

Intraspecific Competition—Radish

Seeds per Pot	Total Biomass per Pot	Average Biomass = Total Biomass / Number of Seeds
1		
10		
20		

Graph your results as a bar graph.

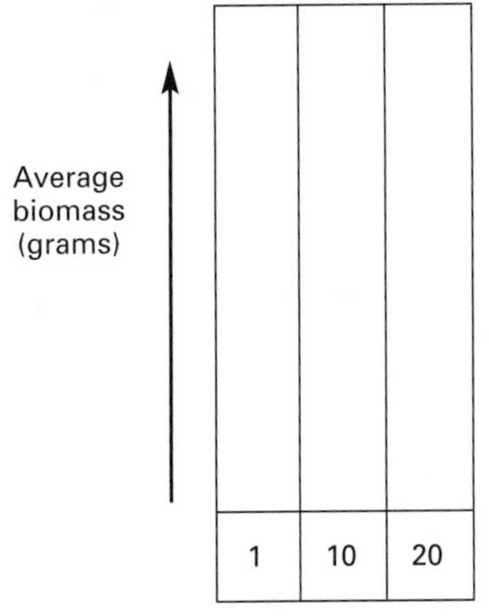

1. At what population density was the average biomass highest?

2. Was there evidence of intraspecific competition? Explain.

3. What resources might have been limiting?

Intraspecific Competition—Wheat

Seeds per Pot	Total Biomass per Pot	Average Biomass = $\frac{\text{Total Biomass}}{\text{Number of Seeds}}$
1		
10		
20		

Graph your results as a bar graph.

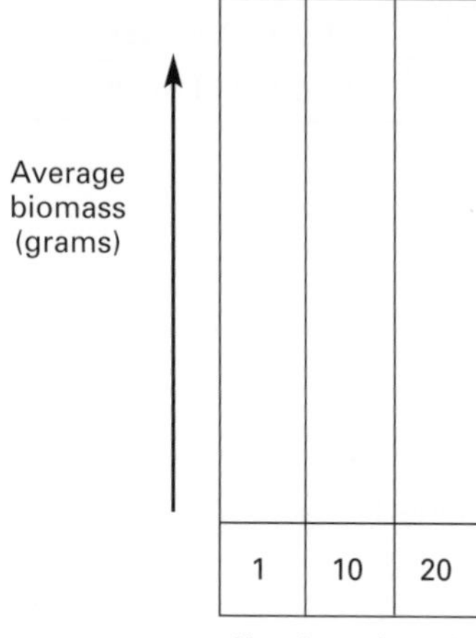

1. At what population density was the average biomass highest?

2. Was there evidence of intraspecific competition? Explain.

3. What resources might have been limiting?

Interspecific Competition—Radish and Wheat

Number of Plants per Pot	Radish		Wheat	
	Total Radish Biomass	Average Radish Biomass	Total Wheat Biomass	Average Wheat Biomass
1				
10				
20				

Graph your results as bar graphs.

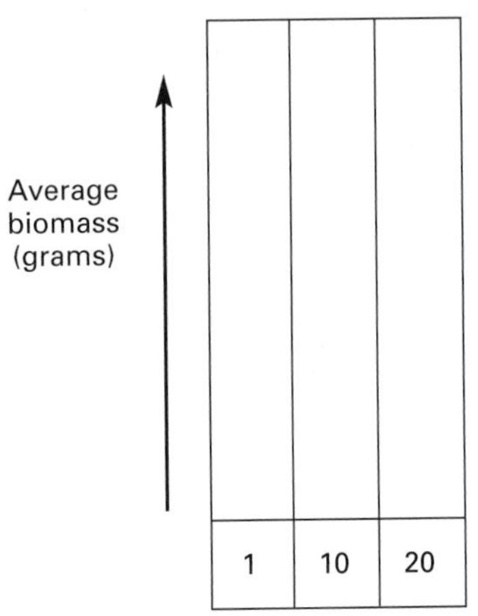

1 10 20

Number of wheat seeds

1. What kind of plant (radish or wheat) was most affected by interspecific competition?

2. List three factors that could have contributed to one plant being able to compete more effectively than the other.

Appendix

A

Chi-Square (Goodness of Fit) Test

The chi-square test is a statistical method of determining if the results of an experiment are close enough to what was expected to be considered valid. Basically, this test can be used to determine if deviations from the expected values are due to *chance* alone or to factors other than chance. Chi-Square is equal to the sum of the deviation squared divided by the expected:

$$\left[x^2 = \sum \frac{(O-E)^2}{E}\right]$$

Consider an experiment in which the observed values (O) were exactly equal to those expected (E) according to our hypothesis. If chi-square is equal to

$$\frac{(O-E)^2}{E}$$

and observed values are equal to expected, then O – E = zero. In this case, we have a perfect fit and a chi-square value of zero. Expanding on that idea, you could reason that small values for chi-square indicate that there is close agreement of the observed and expected values, whereas a large value for chi-square would indicate *significant* deviation from the expected values.

In experimental situations, small deviations from the expected rations are almost always bound to occur because of chance—rarely do we see a deviation of zero under natural conditions. Scientists use chi-square tests to answer the questions "Are the observed deviations within the limits expected due to chance?" "Is some other explanation necessary?" For example, if you were to flip a coin 100 times, you would *expect* 50 heads and 50 tails. If you actually do the experiment and *observe* 45 heads and 55 tails, should you assume that our original reasoning is faulty? Is the deviation from expected due to probability/chance, or is the coin weighted in favor of tails? The chi-square test helps answer these questions. Table A.1 shows how the chi-square value for this example is determined.

Table A.1 **Determining the Chi-Square Value**

Classes	Observed (O)	Expected (E)	O – E	$(O-E)^2$	$\frac{(O-E)^2}{E}$
Heads	45	50	–5	25	0.5
Tails	55	50	5	25	0.5

Chi-square value = 1.0 (0.5 + 0.5)

Degrees of freedom = 1

Probability = .30

After calculating a chi-square value of 1.0, we must ask the questions; 1. Is 1.0 a large or small value of chi-square? 2. Are the observed deviations within the limits expected by chance? To interpret what this chi-square value means, we must consult a chi-square table (table A.2). To use this table, we must first be able to determine the degrees of freedom in our experiment. *The degrees of freedom are equal to the number of classes minus one.* In this case, there are two classes (heads or tails). The degrees of freedom are equal to 2 – 1 = 1. Locate the degrees of freedom along the column marked *df* in the chi-square table. Look across the row for 1 degree of freedom until you find the chi-square value closest to 1.0 (in this case, it is 1.074). Now, look to find which probability column 1.074 falls under (in this case, it is .30). This number means that the deviation from expected that was obtained in this experiment can be expected to occur 30% of the time simply as a result of random variation (chance). The deviation from expected is within the acceptable range and our original prediction of 50 heads and 50 tails is still acceptable.

Table A.2 Table of Chi-Square Values

Distribution of X^2

							Probability (p)							
df	**.99**	**.98**	**.95**	**.90**	**.80**	**.70**	**.50**	**.30**	**.20**	**.10**	**.05**	**.02**	**.01**	**.001**
1	.00016	.00063	.00393	.0158	.0642	.148	.455	1.074	1.642	2.706	3.841	5.412	6.635	10.827
2	.0201	.0404	.103	.211	.446	.713	1.386	2.408	3.219	4.605	5.991	7.824	9.210	13.815
3	.115	.185	.352	.584	1.005	1.424	2.366	3.665	4.642	6.251	7.815	9.837	11.345	16.268
4	.297	.429	.711	1.064	1.649	2.195	3.357	4.878	5.989	7.779	9.488	11.668	13.277	18.465
5	.554	.752	1.145	1.610	2.343	3.000	4.351	6.064	7.289	9.236	11.070	13.388	15.086	20.517
6	.872	1.134	1.635	2.204	3.070	3.828	5.348	7.231	8.558	10.645	12.592	15.033	16.812	22.457
7	1.239	1.564	2.167	2.833	3.822	4.671	6.346	8.383	9.803	12.017	14.067	16.622	18.475	24.322
8	1.646	2.032	2.733	3.490	4.594	5.527	7.344	9.524	11.030	13.362	15.507	18.168	20.090	26.125
9	2.088	2.532	3.325	4.168	5.380	6.393	8.343	10.656	12.242	14.684	16.919	19.679	21.666	27.877
10	2.558	3.059	3.940	4.865	6.179	7.267	9.342	11.781	13.442	15.987	18.307	21.161	23.209	29.588

Source: Table IV of Fisher and Yates, *Statistical Tables for Biological, Agricultural and Medical Research,* published by Longman Group, Ltd., London (previously published by Oliver & Boyd, Edinburgh) and by permission of the authors and Pearson Education Limited.

Statisticians have set a limit of 1 chance in 20 (probability = .05) for drawing the line between acceptance or rejection of a hypothesis. Thus, probability values between .99 and .10 are conventionally accepted as a satisfactory fit and deviations can be attributed to chance. Probability values less than or equal to .05 indicate that deviations are significant enough to reject the hypothesis. Probability values less than or equal to .01 are considered highly significant.

What if we flipped the coin only 10 times and got 6 heads and 4 tails? (See table A.3.)

Table A.3

Classes	Observed (O)	Expected (E)	O – E	$(O - E)^2$	$\frac{(O - E)^2}{E}$
Heads	6	5	1	1	0.2
Tails	4	5	−1	1	0.2

Chi-square value = 0.4

Degrees of freedom = 1

Probability = .50

The chi-square value is 0.4. The number of classes is 2. There is 1 degree of freedom. Look along the row for 1 degree of freedom until you find the chi-square value that most closely approximates 0.4; then look above that column to determine the probability value. From the table, we can see that the deviation we obtained would be expected to occur 50% of the time as a result of random variation. Once again, our chi-square analysis suggests that there is no significant deviation from what was expected; thus, our original prediction is supported.

Application of Chi-Square to Genetic Crosses

Over the years, the Bugs Acres rabbit-breeding farm produced 922 black *(B)* long-haired rabbits and 282 white *(b)* long-haired rabbits from true-breeding stock. Is coat color inherited in a Mendelian fashion? ______________
Support your answer with a chi-square analysis by filling in table A.4.

Hypothesis: *A Mendelian ratio of 3/4 dominant to 1/4 recessive phenotypes is observed for the inheritance of coat color among Bugs Acres rabbits.*

Table A.4 **Determining the Chi-Square Value**

Classes	Observed (O)	Expected (E)	O – E	$(O - E)^2$	$\frac{(O-E)^2}{E}$
Black	922				
White	282				

Chi-square value = ______________

Degrees of freedom = ______________

Probability = ______________

Appendix

B

Math Review

Objectives

Be able to do the following:

1. Add and subtract whole numbers.
2. Multiply and divide whole numbers.
3. Reduce fractions.
4. Add, subtract, multiply, and divide fractions.
5. Convert decimals to fractions.
6. Convert fractions to decimals.
7. Add, subtract, multiply, and divide decimals.
8. Convert decimals to percentages.
9. Convert percentages to decimals.
10. Calculate percentages.
11. Calculate what percent one number is of another.
12. Calculate averages.
13. Use scientific notation.
14. Plot data on line, pie, and bar graphs.

If you are to succeed in science courses, you must have an understanding of fundamental math. The problems in this appendix are provided as a self-test. If you cannot correctly work these problems, you need to upgrade your math skills. Contact your instructor or counselor for assistance.

Practice Problems

Answers to practice problems are at the end of this appendix.

Addition and Subtraction

1. 17 + 327 + 0 + 6 =
2. 19 – 13 + 11 – 2 =
3. 493 + 32 – 9 – 6 =
4. 23 – 14 – 17 + 0 =
5. 47 – 20 + 387 – 43 =
6. 2 + 10 + 100 + 17 =
7. 462 – 279 – 9 + 98 =
8. 56 + 123 + 77 – 61 =
9. 91 + 17 + 3 + 12 =
10. 895 + 254 + 3,358 – 195 =

Multiplication and Division of Whole Numbers

11. 1,234 × 98 =
12. 752 × 42 =
13. 1,493 × 475 =
14. 372 × 372 =
15. 419 × 29 =
16. 529 ÷ 67 =
17. 92 ÷ 8 =
18. 6,391 ÷ 270 =
19. 2,855 ÷ 25 =
20. 472 ÷ 48 =

Reducing Fractions

21. 96/492 =
22. 125/875 =
23. 49/196 =
24. 90/150 =
25. 75/201 =

Addition, Subtraction, Multiplication, and Division of Fractions

26. 3/18 + 23/72 + 22/36 =
27. 45/48 + 3/12 + 21/24 =
28. 1/5 + 2/7 + 2/5 =
29. 24/65 + 1/65 + 22/65 =
30. 7/8 + 4/64 + 3/32 =
31. 1/3 – 1/15 =
32. 24/48 – 4/32 =
33. 4/16 – 8/64 =
34. 20/25 – 1/5 =
35. 18/21 – 1/7 =
36. 1/3 × 1/3 =
37. 1/2 × 9/12 =
38. 12/33 × 1/15 =
39. 13/16 × 3/4 =
40. 8/9 × 4/5 =
41. 1/7 ÷ 1/3 =
42. 1/18 ÷ 1/9 =
43. 45/56 ÷ 5/8 =
44. 8/9 ÷ 2/3 =
45. 14/34 ÷ 1/3 =

Converting Decimals to Fractions

46. 0.75 =
47. 0.3 =
48. 0.123 =
49. 0.13 =
50. 0.5 =

Converting Fractions to Decimals

51. 13/62 =
52. 15/75 =
53. 44/220 =
54. 7/8 =
55. 6/10 =

Addition, Subtraction, Multiplication, and Division of Decimals

56. 1.234 + 22.965 + 0.7213 =
57. 1874.2 + 81.742 + 1.8742 =
58. 46.72 – 2.68 + 13.44 =
59. 3.912 – 12.84 + 99.21 =
60. 65.370 + 54.6 + 11.1 =
61. 56.892 – 32.78 – 2.4892 =
62. 76.87 + 2.3 + 6.78 =
63. 98.47 – 38.6 + 24.9 =
64. 1,007.65 + 63.27 – 3.9999 =
65. 49.654 + 9.81 + 33.34 =
66. 32.1 × 75.43 =
67. 87.55 × 3.3 =
68. 24.0 × 35.1 =
69. 67.44 × 32.5 =
70. 67.8 × 2.2 =
71. 32.5 ÷ 0.5 =
72. 78.4 ÷ 1.444 =
73. 98.29 ÷ 23.4 =
74. 1,000.25 ÷ 0.25 =
75. 63.9 ÷ 0.33 =

Converting Decimals to Percentages

76. 13.0 =
77. 0.004 =
78. 9.23 =
79. 0.76 =
80. 0.05 =

Converting Percentages to Decimals

81. 89% =
82. 0.4% =
83. 15.1% =
84. 325% =
85. 0.008% =

Calculating the Percentage of a Number

86. 13% of 65 =
87. 0.8% of 120 =
88. 40% of 354 =
89. 210% of 600 =
90. 12.8% of 38 =

Calculating the Percentage One Number Is of Another

91. 18 is _____% of 72
92. 3 is _____% of 6
93. 15 is _____% of 75
94. 2 is _____% of 25
95. 624 is _____% of 445

Calculating Averages

96. The average of 78, 13, 35, 3 =
97. The average of 146, 78, 2, 111 =
98. The average of 12.3, 0.98, 78, 0.09 =
99. The average of 4, 678, 33, 45 =
100. The average of 2, 4, 6, 6 =

Scientific Notation

Express the following in scientific notation.

101. 0.78531
102. 675,500,000,000,000
103. 0.0097423
104. 34,500,000
105. 896,660,000,000,000,000

Express the following in numerical form.

106. 4.216×10^{6}
107. 6.53×10^{-4}
108. 9.75×10^{2}
109. 7.78×10^{-5}
110. 8.541×10^{9}

Plotting Graphs

111. Use the following data to plot a line graph:

Body Weight (kg)	20	30	40	50	60	70	80	90
Height (cm)	100	110	120	130	140	150	160	170

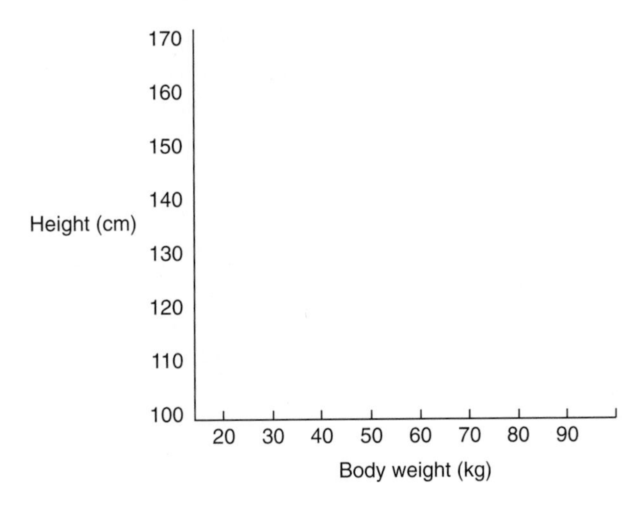

112. Use the following data to plot a bar graph:

The English population has 46% genes for blood type A, 10% for type B, and 44% for type O.

The Greek population has 45% genes for blood type A, 20% for type B, and 35% for type O.

The Russian population has 37% genes for blood type A, 28% for type B, and 35% for type O.

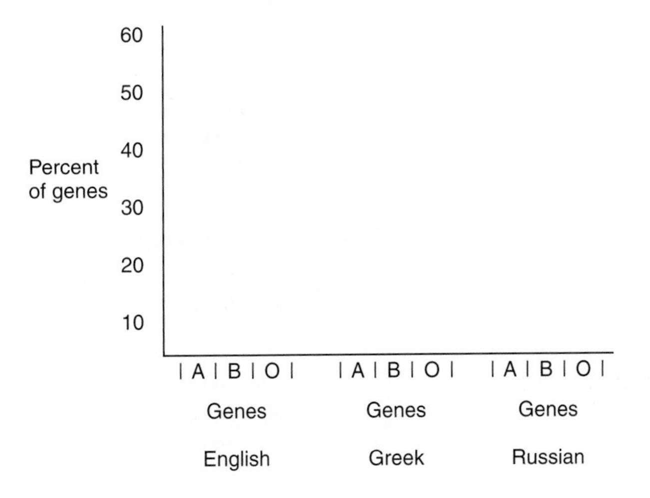

113. Use the following data to plot a pie chart:

Blood type	Percent in the population
A	41
B	10
AB	4
O	45

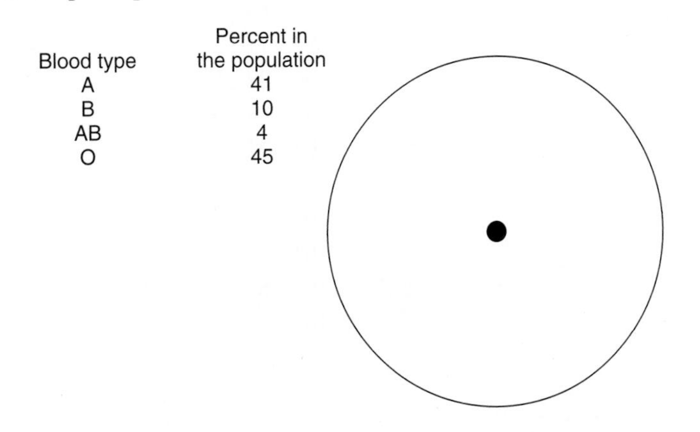

Answers to Practice Problems

1. 350
2. 15
3. 510
4. –8
5. 371
6. 129
7. 272
8. 195
9. 123
10. 4,312
11. 120,932
12. 31,584
13. 709,175
14. 138,384
15. 12,151
16. 7.9
17. 11.5
18. 23.67
19. 114.2
20. 9.83
21. 8/41
22. 1/7
23. 1/4
24. 3/5
25. 25/67
26. 1 7/72
27. 2 1/16
28. 31/35
29. 47/65
30. 1 1/32
31. 4/15
32. 3/8
33. 1/8
34. 3/5
35. 5/7
36. 1/9
37. 3/8
38. 4/165
39. 39/64
40. 32/45
41. 3/7
42. 1/2
43. 1 2/7
44. 1 1/3
45. 1 4/17
46. 3/4
47. 3/10
48. 123/1000
49. 13/100
50. 1/2
51. 0.21
52. 0.2
53. 0.2
54. 0.88
55. 0.6
56. 24.9203
57. 1,894.8162
58. 57.48
59. 90.282
60. 131.07
61. 21.6228
62. 85.95
63. 84.77
64. 1,066.9201
65. 92.804
66. 2,421.303
67. 288.915
68. 842.4
69. 2,191.8
70. 149.16
71. 65
72. 54.29
73. 4.2
74. 4,001
75. 193.64
76. 1,300%
77. 0.4%
78. 923%
79. 76%
80. 5%
81. 0.89
82. 0.004
83. 0.151
84. 3.25
85. 0.00008
86. 8.45
87. 0.96
88. 141.6
89. 1,260
90. 4.86
91. 25%
92. 50%
93. 20%
94. 8%
95. 140%
96. 32.25
97. 84.25
98. 22.84
99. 190
100. 4.5
101. 7.8531×10^{-1}
102. 6.755×10^{14}
103. 9.7423×10^{-3}

104. 3.45×10^7
105. 8.9666×10^{17}
106. 4,216,000
107. 0.000652
108. 975
109. 0.0000778
110. 8,541,000,000

111.

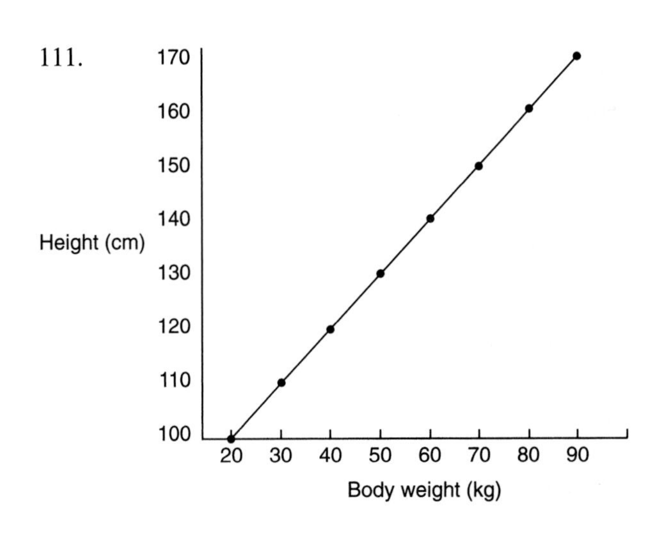

112.

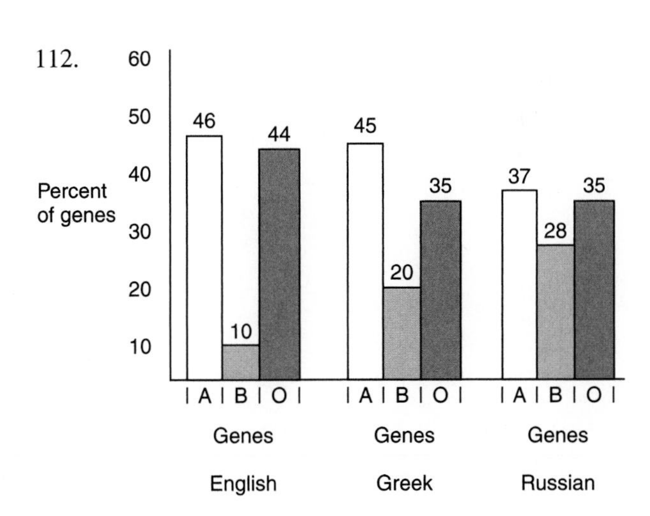

113.

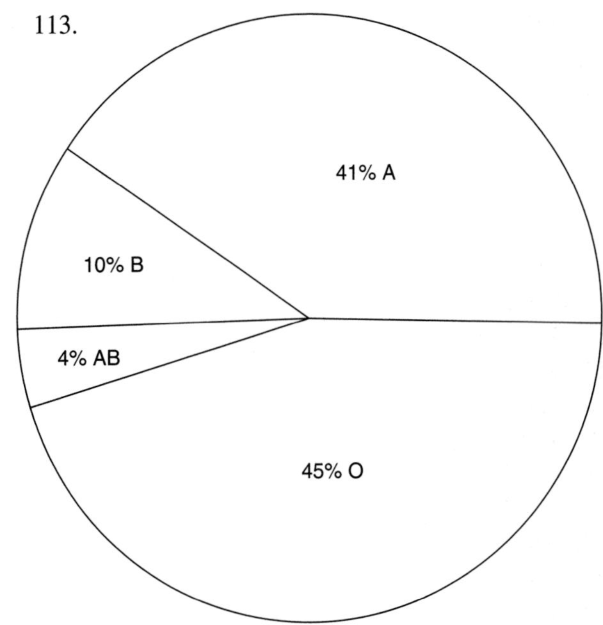

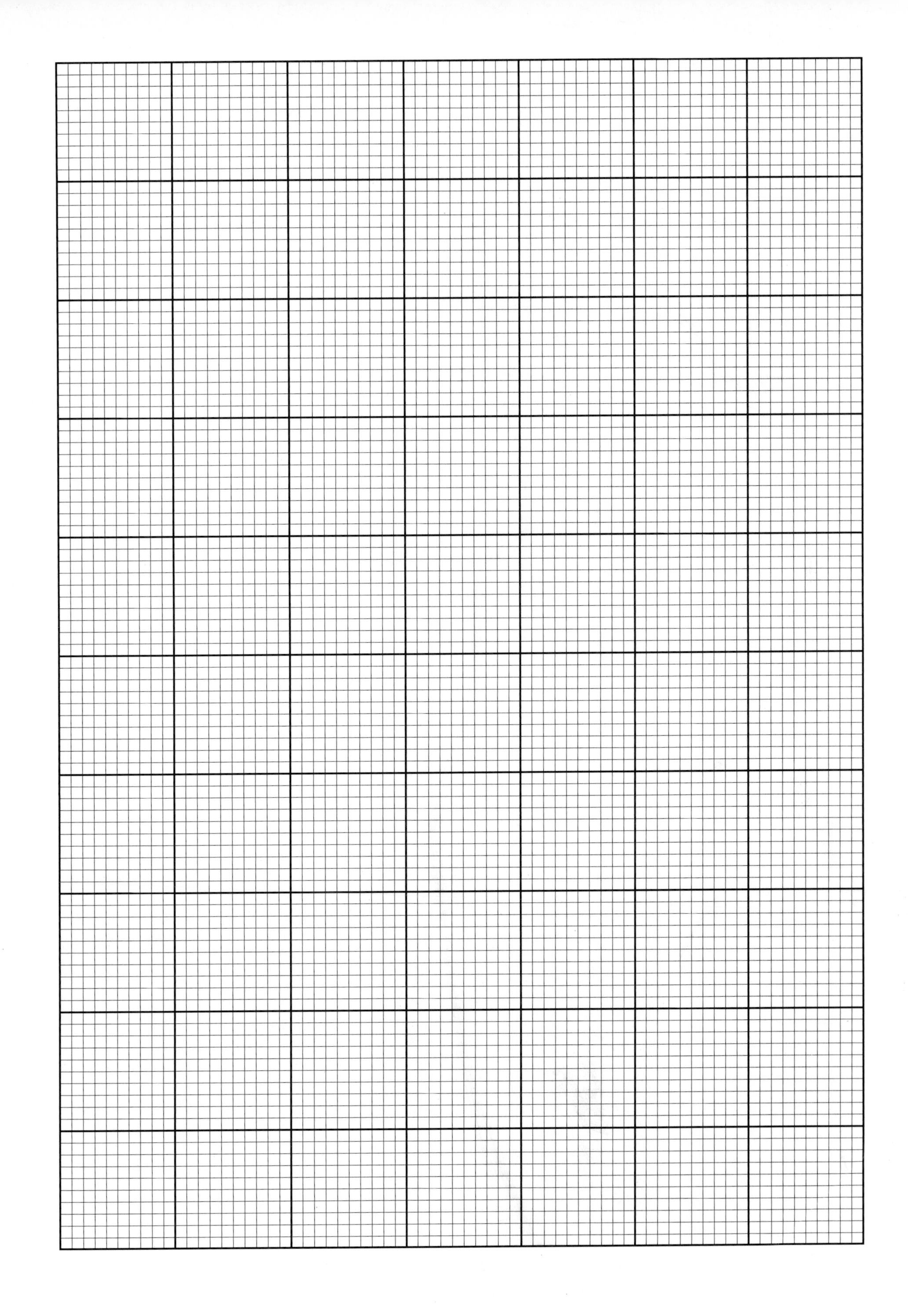

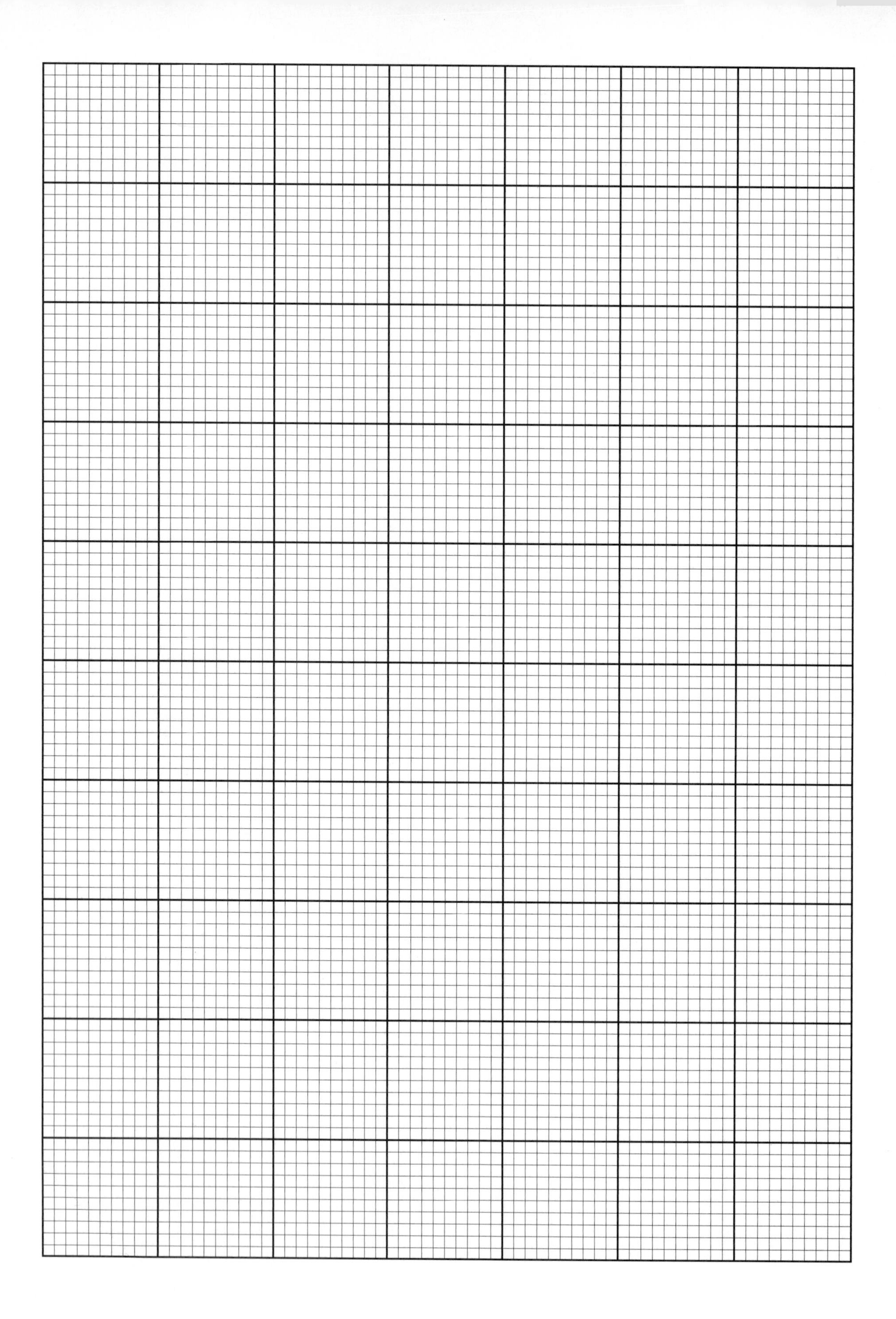

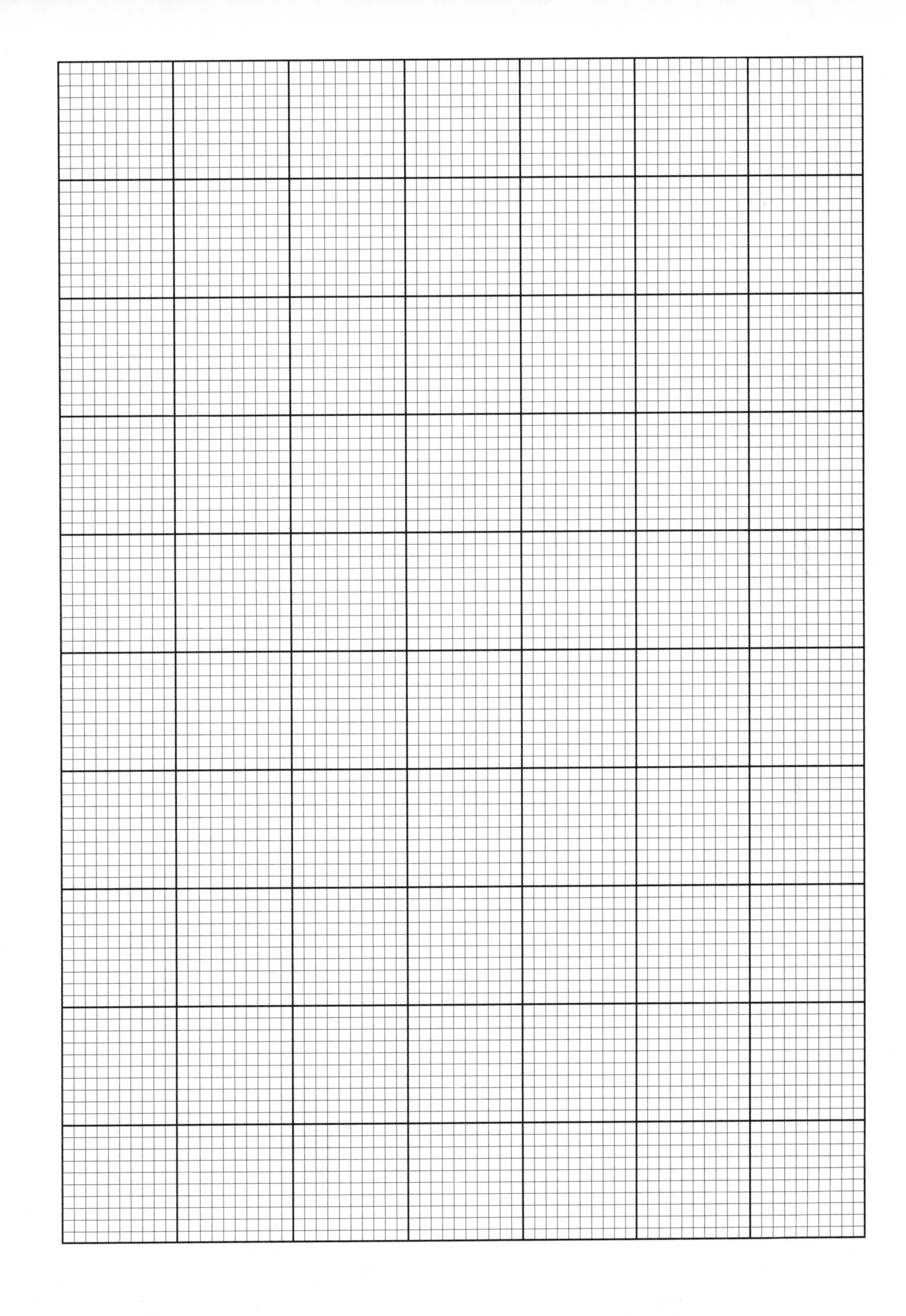

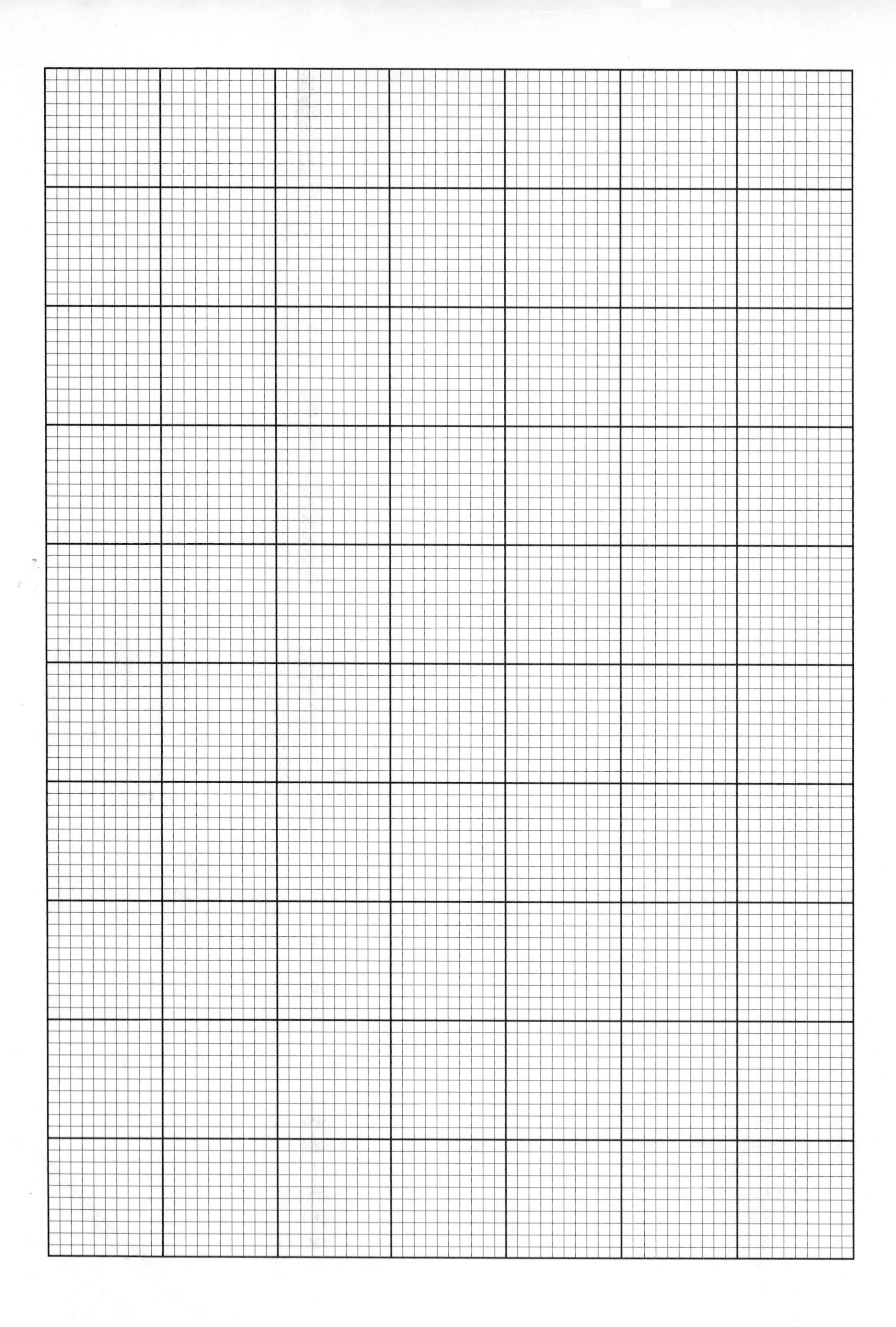